OSHA Inspections

Environmental Engineering Books

OSHA Inspections

Preparation and Response

Rick Kaletsky

McGraw-Hill

New York   San Francisco   Washington, D.C.   Auckland   Bogotá
Caracas   Lisbon   London   Madrid   Mexico City   Milan
Montreal   New Delhi   San Juan   Singapore
Sydney   Tokyo   Toronto

Library of Congress Cataloging-in-Publication Data

Kaletsky, Richard.
    OSHA inspections : preparation and response / Richard Kaletsky.
        p.   cm.
    Includes index.
    ISBN 0-07-033160-X
    1. Industrial safety—United States. 2. Industrial safety—
Standards—United States. I. Title.
    T55.K26   1996
    658.4'08—dc20   96-25460
                                                         CIP

McGraw-Hill

A Division of The **McGraw·Hill** Companies

   3 4 5 6 7 8 9 0 DOC/DOC 9 0 1 0 9 8

ISBN 0-07-033160-X

*The sponsoring editor for this book was Zoe Foundotos, the editing
supervisor was Penny Linskey, and the production supervisor was Don
Schmidt. It was set in Century Schoolbook by Dina John of McGraw-
Hill's Professional Book Group composition unit.*

Printed and bound by Donnelley / Crawfordsville.

McGraw-Hill books are available at special quantity discounts to use
as premiums and sales promotions, or for use in corporate training programs. For more information, please write to the Director of Special
Sales, McGraw-Hill, 11 West 19th Street, New York, NY 10011. Or
contact your local bookstore.

This book is printed on acid-free paper.

To Laurie, Amanda, and Darcy, whose love and patience helped see me through this project.

In loving memory of my mom and dad, whose constant encouragement remains as a great source of strength for me.

Contents

Foreword

In 1971, Congress created a government agency that has both improved safety and health in America and at the same time confounded many employers. Since it was created, volumes of material have been written about the Occupational Safety and Health Administration. It would be hard to find any other piece of work that captures the essence of OSHA and its interactions with American business as well as Mr. Kaletsky's book, *OSHA Inspections: Preparation and Response.*

Kaletsky's book is a true primer on OSHA. In an easy-to-read, common-sense manner, it gives an overview of how OSHA works, what employers need to do in order to have an effective safety and health program and how they should prepare for and respond to an OSHA work site inspection.

The information is presented without the government and legalistic jargon that in other writings on OSHA only further confuses issues surrounding this complex regulatory organization.

Mr. Kaletsky's breadth of experience in the field, both with OSHA and in the private sector, give the book a unique point-of-view. His background allows the text to achieve a rare feat for a book on OSHA, a combination of an insider's knowledge with an outsider's perspective. I worked for OSHA for over 20 years and after reading only a few pages, there was no doubt that the author had mastered the broad scope of this agency's responsibility. He knows exactly how OSHA operates. Mr. Kaletsky's experience in the private sector allows him to turn the tables on the book's perspective. At the same time, the internal workings of OSHA are outlined so that those who work outside of OSHA understand how the organization functions.

The first part of this book may be its most important, because here is where Mr. Kaletsky lays out the key components of an effective safety and health program. The important program elements of management commitment, employee involvement, hazard recognition, control and employee training are covered in easy to understand detail.

In the second part of this book, Kaletsky approaches OSHA with the intent of describing how a compliance officer will act when inspecting your workplace, how the agency is likely to interpret standards, and what the legal ramifications are to the business owner as a result of an inspection.

Safety and health professionals, as well as those who have collateral responsibilities for safety and health in an organization will benefit from this book. Those who have little experience with OSHA and would like to know how rules and standards are interpreted by the agency, how an inspector is likely to act during an inspection, or what happens when there are violations, will gain immeasurably by reading this book.

OSHA has now celebrated 25 years of existence and employers still do not have a clear understanding of how the agency operates. This timely text paints an excellent picture of what is likely to happen when the OSHA inspector knocks on your company's door and says "I'm from OSHA and I'm here to help you."

Alan McMillan
Vice President, National Safety Council
(formerly Regional Administrator
and Deputy Assistant Secretary for OSHA)

The Occupational Safety and Health Administration

The Occupational Safety and Health Administration (OSHA) was conceived on December 29, 1970, when the U.S. Congress passed the Williams-Steiger Occupational Safety and Health Act. The law became effective on April 28, 1971, when OSHA, an agency of the U.S. Department of Labor, was formally established. For more than a quarter of a century, it has required that every employer under its coverage furnish its employees with employment and a place of employment which are free from recognized hazards that are causing, or are likely to cause, death or serious physical harm to employees. Employers are obligated to comply with a wealth of occupational safety and health standards and regulations promulgated under the Occupational Safety and Health Act (the Act). These requirements are published in 29 CFR (Title 29, Code of Federal Regulations).

OSHA was created for the following main purposes:

- to encourage employers and employees to reduce workplace hazards and to implement new, or improve existing, safety and health programs;

- to provide for research in occupational safety and health and to develop innovative ways for dealing with occupational safety and health problems;

- to establish separate but dependent responsibilities and rights, for employers and employees, for the achievement of better safety and health conditions;

- to maintain a reporting and recordkeeping system to monitor job-related injuries and illnesses;

- to establish training programs to increase the number and competence of occupational safety and health personnel;
- to provide for the development, analysis, evaluation, and approval of state occupational safety and health programs;
- to develop and promulgate mandatory job safety and health standards and enforce them effectively through the conducting of workplace compliance inspections and the issuance of citations for alleged noncompliance.

Certainly, it is this last mandate with which OSHA is most frequently and closely identified. The chief objective of this book, then, is to aid employers in their efforts to comply with the Act and to be thoroughly versed in how to prepare for, and respond to, OSHA inspections. In so doing, the employer will also be in an excellent position to assure, insofar as is possible, that each and every employee is provided with safe and healthful working conditions. It is the responsibility of the employer, not of OSHA, to meet this important burden.

Part 1 of this book deals with preparation and is essentially comprised of numerous, specific elements of workplace safety and health programs. These elements, discussed in great detail, blend to form a reliable, cohesive foundation, firmly rooted in management leadership and employee involvement. They serve to significantly enhance employee protection and legal adherence to the Act.

Part 2 deals with the OSHA inspection process in depth. Emphasis is placed on how to respond (and to not respond) in words and actions during and after the inspection. A clear light is cast upon the rights afforded to employers so that they can, through keen knowledge and confident resolve, be ever vigilant to protect those rights. The goal is to minimize susceptibility to citations and fiscal penalties.

Many explanations of OSHA's operations and procedures are addressed throughout the book. In this section, for the most part, I merely highlight those that are later placed in context. There is no intent to glorify or denigrate the Agency.

Scope of Coverage

OSHA's coverage encompasses the fifty states, the District of Columbia, Puerto Rico, and all other territories under federal government jurisdiction. Those employed directly by any state or political subdivision thereof are not covered. Federal employees are not covered in the traditional sense, but federal agency heads are responsible for establishing effective and comprehensive occupational safety and health programs and providing safe and healthful working conditions for their employees. Inspections of federal workplaces are conducted because an executive order requires agencies to comply with standards consistent with those that OSHA issues for private sector employees. OSHA cannot propose monetary penalties against another federal agency for failure to comply with the standards, but unresolved compliance issues at a local level are raised to higher organizational levels until they are resolved.

Simply put, the Act applies to the great majority of private sector employees. Most OSHA inspections are conducted in the manufacturing and construction industries. However, virtually every type of business including stores, offices, longshoring, agriculture, hospitals, food preparation, miscellaneous distribution, publishing, restaurants, private education, and numerous types of services fall under the Agency's jurisdiction. The following are not covered under the Act: farms at which only immediate members of the farm employer's family are employed, working conditions regulated by other federal agencies under federal statutes, and self-employed persons.

As for other federal agencies, sometimes the lines are difficult to draw, but examples of non-OSHA coverage relate to the Federal Aviation Administration, the Coast Guard, the Mine Safety and Health Administration, the Federal Railroad Administration, the Federal Highway Administration, and the Nuclear Regulatory Commission. Thus, for the most part, OSHA does not regulate occupational safety and health with regard to trains, airplanes, over-the-road vehicles when actually on the "open road," mines (with some nuances), or vessels several miles out at sea. In some cases, there are elements of the business that are covered by OSHA and elements that are covered by another agency. For example, OSHA has jurisdiction over certain matters related to the safety of a dump truck when it is used on a construction site and to certain portions of an airport as opposed to the airplane itself.

Remember that OSHA deals with protection of employees. Therefore, it is not the business of the Agency to regulate the cleanliness of the product in a meat-packing plant unless the employees must taste the meat as part of their job. It is not the business of OSHA to sanction an employer who is polluting the air or water, where such pollution does not endanger the employer's employees as related to their work. One irony is that a company might do an excellent job of protecting its employees against respirable hazards by the use of respirators and strong exhaust ventilation but pollute the environment because the stacks are not functioning properly to catch, scrub, neutralize, or filter the toxics before they reach the atmosphere outside of the plant.

Of course, if OSHA comes upon indications of hazards to the consumer or to the general public, a referral to the appropriate agency can be filed. A reverse sort of circumstance could arise if the stacks are properly designed, maintained, and working adequately to protect the public from the airborne dangers, but work station exhaust ventilation units are not functioning properly or are not positioned as they should be. It could even be that the lack of exhaust ventilation results in only very small amounts of the toxics actually escaping the confines of the facility. In either case, it is conceivable that those in the plant are endangered, while the public is not.

Further, it is not within the mandate of OSHA to protect customers, visitors, and so on. Nevertheless, such persons may well reap the benefits of an OSHA inspection. For example, if an OSHA enforcement action causes a large restaurant, department store, or grocery store to maintain readily accessible exits, the shoppers will gain protection. As a sidelight in such situations, the OSHA factors considered in the assessment of the penalty to be proposed against the host employer should include the fact that panic- and confusion-related hazards can definitely be exacerbated by the presence of the public.

Regarding self-employed persons, if a person is the owner of a corporation (rather than being a legal sole owner), then he or she is an employee of the corporation and not exempt from the Act. This issue becomes a bit tricky when it involves contractors who claim that those doing work for them are actually independent contractors themselves. Although this book more broadly considers general industry issues, in the overwhelming majority of scenarios, construction concerns do apply. In fact, the issue of the relationship between a so-called "boss" and those who are performing work in some form of consort with him or her is usually a construction/contractor matter. Keep in mind that a general contractor is in jeopardy of being cited and penalized for OSHA violations even if it is only his or her subcontractor's employees who are exposed. Note, too, that a subcontractor whose employees are exposed to violative conditions can very seldom escape citation by blaming the general contractor.

Enforcement

Actual OSHA enforcement is provided either directly by federal OSHA or through an OSHA-approved state program that is substantially funded by federal OSHA. No state is forced to develop and implement an OSHA enforcement program. It is their choice to submit one, and they must provide standards and enforcement programs, as well as voluntary compliance activities, which are at least as effective as the federal program. So, in several states, OSHA enforcement is run by the federal government, and in the others, it is run by the particular state with federal OSHA acting as a monitor.

A few states have plans that cover the public sector (e.g., state, local, and similar government employees), but the private sector is under federal OSHA. In this book, I generally use terms that refer to federal OSHA positions or documents, while occasionally reminding the reader that there is a state equivalent. There should not be major differences, although state-plan states have the right to enforce standards that are more stringent than those on the federal books, as long as they are reasonable.

Sources of Standards

OSHA standards were originally derived from three sources: consensus standards, proprietary standards, and federal laws in effect when the Occupational Safety and Health Act became law. Consensus standards are developed by industrywide standard-developing organizations and are discussed and substantially agreed upon through consensus by industry. OSHA has incorporated the standards of the two primary standards groups—the American National Standards Institute (ANSI) and the National Fire Protection Association (NFPA)—into its set of standards. OSHA had essentially incorporated by reference the National Electrical Code, which is a part of the NFPA standards. Later, OSHA reformulated their electrical standards to stand alone.

The proprietary standards were prepared by professional experts within specific industries, professional societies, and associations. Those standards are determined by a straight membership vote, not by consensus. The best known such source used in the OSHA standards is the "Compressed Gas Association, Pamphlet P-1, Safe Handling of Compressed Gases."

Some preexisting federal laws are enforced by OSHA. These include the Federal Supply Contracts Act (Walsh-Healey), the Contract Work Hours and Safety Standards Act (Construction Safety Act), and others.

Standard-Setting Procedures

OSHA can begin standards-setting procedures on its own initiative or in response to petitions from other parties, including (but not limited to) nationally recognized standards-producing organizations, employer representatives, or labor representatives. One motivation for the Agency to consider the need for a new standard concerns Section 5(a)(1) of the Occupational Safety and Health Act. This is the section commonly known as the "general duty clause." I address this in detail later.

Basically, if an OSHA compliance officer observes and can properly document a serious hazard of which the employer was aware but for which there is no standard, the officer can simply write the functional equivalent of a standard. The wording begins with the general duty clause and continues "in that," followed by carefully elucidated particulars of the hazard. There is a heavy burden on the Agency regarding the legal sufficiency of its related documentation. OSHA's ground becomes firmer when it can be shown that the employer did not follow the manufacturer's recommendations or did not abide by the specific ANSI (or similar) standard that was known to the employer. The compliance officer may study operating manuals and other literature provided by the manufacturer or distributor of the subject equipment. Copies of alleged violations of the general duty clause are regularly sent to the OSHA national office. If that office receives several on the same hazard, it is time to evaluate the need for a new standard.

If for any reason OSHA determines that a specific standard is needed, any of several special advisory committees may be called upon to develop specific recommendations. The standing committees are the Advisory Committee on Construction Safety and Health and the National Advisory Committee on Occupational Safety and Health. At times, there are ad hoc committees appointed to examine special areas of concern to the Agency. All of the advisory committees must have members representing management, labor, and state agencies, as well as one or more designees of the Secretary of Health and Human Services. The occupational safety and health professions and the general public may also be represented.

OSHA may promulgate standards as recommended by the National Institute for Occupational Safety and Health (NIOSH), which is an agency of the U.S. Department of Health and Human Services. NIOSH conducts research on various safety and health problems, provides technical assistance to OSHA, and recommends standards for OSHA's adoption. While conducting its research, NIOSH may make workplace investigations (although it is not an enforcement agency), gather testimony from employers and employees, and require that employers measure and report employee exposure to potentially hazardous materials. NIOSH also may require employers to provide medical examinations and tests to determine the incidence of occupational illness among employees. When such examinations and tests are required by NIOSH for research purposes, they may be paid for by NIOSH rather than the employer. NIOSH generally deals with industrial hygiene concerns, emphasizing such subject matter as particularly worrisome levels of toxics and ergonomic/lifting considerations. NIOSH publishes criteria documents that address the results of studies regarding a wide body of specific substances.

Whenever OSHA develops plans to propose, amend, or revoke a standard, the information is published in the *Federal Register*. This is called a Notice of Proposed Rulemaking, and sometimes an early version is published as an Advance Notice of Proposed Rulemaking. The idea behind this second type of notice is to draw comments that can be considered for the drafting of the proposal. When the advance notice is issued, the public is given at least 30 days (but usually much more time) to respond. Frequently, public hearings are scheduled for proposal discussion. If no such hearing has been announced in the notice, anyone who submits a written argument and germane evidence may request the hearing. The *Federal Register* publishes the time and place of these hearings.

At the end of the period allowed for comment, the Agency will finalize the standard. The *Federal Register* will carry the complete final standard that was amended or adopted and the date it becomes effective. There is usually a notable gap in time between publication and effective date. Along with the standard, there is a lengthy explanation of it and the reasons for its implementation in a preamble. It can prove well worth the time to read the preamble not so much to gain a knowledge of the history (including incidents) relating to the hazard, as to learn the intent and interpretations of the standard. OSHA may publish its decision that no standard or an amended standard is needed. Through the rulemaking process, the employer, employees, trade associations, and others have an opportunity to take part in the process. Everyone has a chance to participate.

Although quite uncommon, OSHA is authorized to set emergency temporary standards that become effective immediately. They stay in effect until they are superseded by a permanent standard. These standards can only be considered when there is substantial evidence that employees are in grave danger of exposure to toxic substances or agents determined to be toxic or physically harmful, including newly introduced hazards. OSHA must demonstrate that the extraordinary steps are necessary to protect employees. The process is generally the jump-start result of the release of a heavy, disturbing body of data pointing to a major health risk, when exposures exceed certain limits. OSHA publishes the standard in the *Federal Register*, where it is presented as a proposed permanent standard. Then, the regular procedures for adopting a permanent standard apply, except that a final ruling is made in 6 months. The validity of an emergency standard may be challenged in an appropriate U.S. Court of Appeals.

If a person (often on behalf of a company) may be adversely affected by a final or emergency standard, he or she may file a petition within 60 days of the promulgation for judicial review of the standard, with the U.S. Court of Appeals, for the circuit in which the objector lives or has his or her principal place of business. Filing an appeals petition does not delay the standard enforcement unless specifically ordered by the court.

Categories of Standards

There are four major categories of OSHA standards: general industry, maritime, construction, and agriculture. The two most common sets of standards are broken down into construction and general industry. General industry standards cover factories, stores, hospitals, and basically everything not covered by the other standards. There are some general industry standards that apply in construction. In the great majority of cases within the general industry standards, the specific standards apply "across the board" regardless of the type of general industry involved. Thus, employers would not look up "foundry," "automotive dealership," "nursing home," or "candy manufacturing." Rather they would search for the type of hazards that could be addressed. Common examples include (but are not limited to) means of egress, guarding floor openings and holes, flammable and combustible liquids, powered industrial trucks, abrasive wheel machinery, arc welding and cutting, wiring design and protection, the control of hazardous energy, eye protection, hazard communication, and so on. Some standards, however, are only pertinent to a particular industry. These "special industries" include the following: pulp, paper, and paperboard mills; textiles; bakery equipment; laundry machinery and operations; sawmills; logging operations; agricultural operations; telecommunications; electric power generation, transmission, and distribution; and grain handling facilities.

Whenever researching standards, it is very important not to merely flip through pages of the standards, until you find what you believe are key words. It is imperative that you assure that you are in the correct section and the correct subparagraph. The general wording may seem to apply to your needs, but you may have come to the area after bypassing a heading that clearly limits the application of the standards before you. As one example, you may search for a standard that is relevant to a particular hazard under the broad topic of "flammable and combustible liquids." If you do not reach that area by way of an orderly outline breakdown, you may gather the wrong information. For instance, you may be in the middle of an area that only applies to processing plants, but your facility has no processing plant. You may find elsewhere in the section on flammable and combustible liquids that your situation is viewed differently by OSHA (and with different requirements) if it is not in a processing plant. It is also wise to tab and highlight the standards, as well as the definitions. It will be well worth it to become familiar with the format of the standards before you have a crisis brewing.

Obtaining OSHA Standards

No matter what literature you have concerning OSHA standards, there is absolutely no substitute for the actual complete set of standards. You may find these in a library, but you have a genuine need to have a copy readily available at all times. They can be purchased from the Superintendent of Documents, U.S. Government Printing Office. There are also private companies that sell the standards at a discount. Be sure that you are ordering exactly what you need. I strongly suggest that you also purchase OSHA's CD-ROM, which includes all of the standards and a great many other documents, such as letters of interpretation; however, it is not a substitute for the hard copy. Further, there are various similar private CD-ROM sources that are available. OSHA standards, interpretations, directives, fact sheets, news releases, and other pertinent information are available via modem on the Labor News Bulletin Board. The Agency makes a great deal of this type of information available through the Internet, on the World-Wide Web. Short documents can be obtained through fax, as well. My only reluctance to providing postal addresses, telephone and facsimile numbers, and Internet addresses is that the information may be in a state of flux.

The standards are theoretically issued each summer, but their actual availability may not be until near the end of the calendar year. To determine whether there have been any amendments since the revision date of the Code of Federal Regulations volume in which you are interested, you may wish to consult the "Cumulative List of CFR Sections Affected" (issued monthly) and the "Cumulative List of Parts Affected" (appears daily in the *Federal Register*). These two lists refer you to the *Federal Register* page containing the latest amendment of any given rule. The pages in the *Federal Register* are numbered sequentially from January 1 through the remainder of the calendar year.

There are also private update services that charge a fee. You can always check with the OSHA area office to find out about new or revised standards. In all cases, if the OSHA enforcement in your geographical location is through the state itself, you should ascertain the state's standards and other pertinent information. Do not forget that, although the state plan must be at least as effective as that of federal OSHA, it may be more stringent and have slight differences.

As for letters of interpretation, you may want to clear away the fog on an issue by writing to the OSHA national office. If there is a question as to what would constitute compliance, you can talk with some-one in the area office. You are not likely to receive a written response from an OSHA area office, and such an answer is doubtful even from the regional office. The regional office is administrative and oversees several area offices that handle the direct enforcement. Thus, any doubt about the response (a compliance officer's simple opinion will not suffice) and/or how your company reacts to it can have a major impact regarding monetary allocation for abatement or possible cita-tion jeopardy, and a letter to the national office would be prudent. Try to compose a short letter that includes all of the salient information. If possible, limit the letter to one page. Cut to the chase. Send it by registered mail, ask for a return receipt, and unambiguously request the courtesy of a prompt reply. Refer to the apparently relevant stan-dard(s) at the top of the letter. It is best if you can formulate specific questions and options.

There are few negatives to this type of letter. The national office is not in the habit of directing an area office to track down any such let-ter writers, even when their current compliance status may appear to be questionable. If you are nervous about this sort of concern, perhaps you can motivate a trade association, business organization, your insurance carrier, consultant, or similar group to write the letter. If so, the name of your company need not be included at all. For instance, it is conceivable, though not likely, that your company may be inspected several months later (unrelated to your letter), that the letter and OSHA's response may surface, and that you have earned yourself a one-way ticket to a "willful" violation if you just ignored the interpretation. Yes, this is possible. Yet, if you plan to act profession-ally and responsibly when receiving OSHA's interpretation, there should not be any problem.

In fact, in one important scenario, an OSHA compliance officer can visit your establishment and note an apparent violation about which you had written to the national office (not that you had necessarily viewed it in those terms). You carefully reread the interpretation and then, provided that its language verifies your defense, give a copy to the compliance officer. The compliance officer, armed with the "offi-cial" interpretation, is now in a position to defer to the national office.

High-Priority Inspections

Inspections take place without advanced notice, except when there are highly unusual circumstances. Aside from reports of imminent danger and when inspections must take place outside of regular working hours, the circumstances would have to be extraordinary for advanced notice to be given. The employer should plan on receiving an OSHA visit with no forewarning. OSHA may not conduct warrantless inspections without an employer's consent. Both of these issues are examined in detail later.

OSHA has established a system of inspection priorities. The first priority is when the Agency has reason to suspect that an imminent danger exists. An imminent danger, according to the Agency, is any condition where there is reasonable certainty that a danger exists that can be expected to cause death or serious physical harm immediately or before the danger can be eliminated through normal enforcement procedures. The most common examples of imminent danger occur in the construction industry and are highlighted by situations in which an excavation cave-in can occur or where there is a lack of fall protection from a roof, skeleton steel, or similar working station. Another type of imminent danger can involve exposed, live electrical parts, where employees are working with or near such hazards while wet. For health hazards to be considered imminent dangers, there must be a reasonable expectation that toxic substances such as dangerous vapors, fumes, dusts, or gases are present and that exposure to them will cause immediate and irreversible harm to such a degree as to shorten life or cause reduction in physical or mental efficiency, even though the resulting harm is not immediately apparent.

If the compliance officer, upon inspection, verifies that an imminent danger exists, he or she will ask the employer to voluntarily and immediately abate the hazard and remove endangered employees from exposure. If the employer does not do this, the compliance officer will advise all potentially exposed employees of the hazard and its gravity, and then post an imminent danger notice. If the employer still refuses to comply at once, OSHA (by way of the U.S. Department of Labor's Office of the Solicitor) may apply to a federal district court for appropriate legal action to correct the situation. This action is generally a temporary restraining order (TRO) which will cause the immediate shutdown of the operation.

Although leaving a workplace because of potentially unsafe or unhealthful conditions is not ordinarily an employee right, imminent danger situations fall in another category. An employee has the right to refuse (in good faith) to be exposed to an imminent danger. The employee can be protected under Section 11(c) of the Act, which is OSHA's antidiscrimination section. Employees removing themselves from an imminent danger is also a protected activity, although there can be an argument about what constitutes the imminent danger. Surely, employees cannot just stop working because there is a technical violation such as a machine danger zone that can only be accessed if the employees make a substantial effort to get around the inadequate guard. The chief question is whether or not the danger facing the employee is so grave that a reasonable person would conclude that there is a real danger of death or serious physical harm and that the danger is so imminent that there is not sufficient time to have the danger eliminated through normal enforcement procedures.

It is preferable for the employee to ask the employer to eliminate the danger. There are some situations in which the employee can request and be granted reassignment to another area or another job. At any rate, if there is even a reasonable consideration of an imminent danger, the employer should not worry about how the case would turn out, but rather take prompt action to assure the protection of employees. I discuss Section 11(c) a little bit later in this introduction.

So, imminent danger is the first inspection priority. Next is the fatality or catastrophe. Briefly, if OSHA becomes aware of an occupational fatality, an inspection is scheduled, unless it is indelibly clear that the matter is not in OSHA's jurisdiction. The Agency begins with the supposition that a fatality occurring at a workplace or in the pursuit of employment is an occupational fatality. If there is any doubt (which could relate to dual agency enforcement aspects), OSHA will begin an investigation. The same situation exists when OSHA gains knowledge of a catastrophe (e.g., three or more employees in-patient hospitalized as the result of one incident). The employer must report fatalities and catastrophes to OSHA within 8 h of the occurrence.

Complaints and Referrals

The third priority is given to complaints and referrals. In most cases, inspections are conducted when what employers have come to know as formal complaints are received. Although the term "formal complaint" may not be currently used by OSHA in the new system of classifying complaints, the old concept of a formal complaint remains. As a reminder, such a complaint must be reasonably specific, and it must be signed by a person with proper standing, with the most obvious and most common example being a current employee. Other responsible individuals include an authorized union representative where the union is certified in/for the company and facility in question on behalf of employee(s); an attorney for, or immediate family member of, an employee on behalf of (and with the knowledge of) that employee; or an employee of another company who is exposed to the hazards of the subject company. When OSHA receives this type of complaint, an inspection is scheduled, unless the allegations relate to very minor potential hazards.

If the complaint does not meet all of the foregoing criteria but alleges circumstances that include at least one of the following components, an inspection is usually scheduled: imminent danger situation, hazards or an establishment covered by a local or national emphasis program, permanently disabling injuries or illnesses, or discrimination. Further, whether pointed out by the complainant or otherwise discerned by OSHA, if the workplace has a significant history of noncompliance with OSHA, an inspection is likely to be scheduled. As explained later, a sanitized copy of the complaint (with personal identifiers, including special references and pet expressions, removed) is given to the employer upon inspection. The employee's name is withheld from the employer, unless the employee permits it to be released.

When OSHA receives a complaint that does not fit any of the aforementioned descriptions, an inspection is not scheduled. Instead, OSHA contacts the employer by letter, telephone, and/or fax. For the time being, this procedure is called an "investigation." The employer is informed of the complainant's allegations (with no suggestions as to their validity) and the most likely relevant standards. The employer is told that he or she must provide a detailed written response to the OSHA area office. Additionally, the employer must post a copy of OSHA's letter, as well as a copy of the response, which should include supporting documentation.

The Agency has started a policy whereby employers are notified of nonformal complaints very quickly and must provide at least an initial response within 5 days. This is not to say that, even if the employer agrees that there are hazards and/or violations, they must be abated within that short period. The point is that OSHA wants to keep close track of the file. Depending on the employer's response, several weeks may be granted for a complete evaluation and subsequent abatement by the employer. In virtually all cases, that sort of time period would only be granted in phases, with the employer regularly apprising OSHA of its progress.

The complainant, if not anonymous to OSHA, is updated on the case, including copies of the employer's response(s) or synopses of them. If the complainant offers a very specific argument to challenge the employer's contentions that no violation existed or that abatement has been attained, OSHA can schedule an inspection. Of course, if the employer fails to provide explicit responses to OSHA, or appears to drag its feet, this is also grounds for an inspection. Occasionally, despite the employer's good faith efforts to prove to OSHA (e.g., by letter, copies of paperwork, photographs) that a violation did not exist or that it has been abated, the Agency may have to perform an inspection as the only way to verify if the company is in compliance. In any case, it behooves the employer to encourage honest, two-way communication with employees. It is far better for the employer to hear a complaint from an employee before that employee even considers contacting OSHA.

A referral may come from another governmental agency, from a physician, or from some other medical professional who has become aware of a trend of injuries or illnesses, which could include cumulative trauma disorders, at a workplace. OSHA may also schedule a referral inspection if a report in the media appears to indicate the existence of serious violation(s) at a workplace. Sometimes, a photograph in a newspaper shows employees who are apparently in danger. Typically, this involves employees lacking fall protection on a construction site or lacking shoring (or equivalent protection) to prevent excavation cave-in. There have been referrals spun off of television footage as well. For instance, a human interest story about an emerging industry or some other story shot inside a plant can show unguarded machinery, a forklift truck without an overhead guard, the lack of eye protection where it is needed, or an employee with extremely long, unrestrained hair exposed to rapidly revolving machine parts. The lesson is that you should be very careful to avoid such publicity.

If OSHA compliance officers, just driving by or in a workplace for non-OSHA purposes, notice what appears to be a violation, they can submit a referral to their OSHA office or to the OSHA area office that has geographical jurisdiction. If there is a concern of imminent danger, the inspection should be conducted immediately without trying to contact the office. If the matter seems to be quite serious, although not of the imminent danger variety, the compliance officer generally calls the pertinent office, reports the situation, and is assigned right then to conduct the referral inspection.

Enforcement Personnel

Before explaining the next type of referral, this is a logical place to address the hierarchy of enforcement personnel. There are two basic disciplines (i.e., general areas of expertise) of compliance officers. In all cases, compliance officers are formally known as compliance safety and health officers (CSHOs). The two job classifications are broken down to safety and industrial hygiene. Generally and at least in a federal office, the safety CSHOs work for an assistant area director who is also known as a safety supervisor. The industrial hygiene CSHOs work for an assistant area director who is also known as an industrial hygiene supervisor. The area office is headed by an area director. All area offices (several at once) come under the umbrella of a regional office. The next step up is the national office, headed by the Assistant Secretary of Labor for Occupational Safety and Health. Subgroups within the safety heading are safety specialist and safety engineer.

Other Types of Referrals

If a safety CSHO, while conducting an inspection, feels that an industrial hygienist is needed, he or she can in effect put it in an intraoffice referral. Likewise for industrial hygienists. There are, theoretically, situations in which a safety CSHO might note circumstances that another safety CSHO would be better suited to inspect. Perhaps, a CSHO with heavy ergonomic hazard training is needed. In such a situation, a referral could be issued.

There is still another scenario whereby a referral can result. If an employer calls OSHA to report a serious injury or illness, even if not required to do so by the Act or the standards, an inspection may be scheduled. When such reports come into OSHA, or in some cases when an inspecting CSHO writes an intraoffice referral, a referral letter (similar to a nonformal complaint letter) may be sent. An inspection may still be conducted subsequently, depending upon the response from the employer.

Targeting Programs

Next in priority are programmed, or planned, inspections aimed at specific high-hazard industries, occupations, or health substances. OSHA's desire is to concentrate its enforcement efforts where they are needed most. The Agency is continually looking for ways to refine its targeting. At times, a High-Hazard Planning Guide has been based on standard industrial classifications (SIC codes), in which high frequencies of injuries were occurring. This is generally derived from data by the U. S. Bureau of Labor Statistics, but can also result from state-formulated statistics. Another method is to focus on SIC codes that have shown a high frequency of citations. That type of concept can relate to industrial hygiene concerns (e.g., illnesses) as well, whereas the prior method could not be readily applied because the illnesses being diagnosed may have developed from hazards that were abated many years ago.

OSHA experiments with pilot targeting programs, recently including the type where specific companies (not just industry classifications) in a state are identified through workers' compensation (or similar) records as having the highest rates of injuries and illnesses. Ultimately, OSHA desires to develop a high number of employer-specific profiles. This type of program is expanding, as it appears to be quite successful. It is generally referred to as a "CCP," or Cooperative Compliance Program. The government-employer-employee partnership concept is emphasized. After their listing, the companies are contacted in writing by OSHA. They are notified of their unfortunate standing and are given the option of voluntary abatement of hazards (and the building of a strong, effective safety and health program) or a probable OSHA inspection.

A new national CCP began via a survey of employers with more than 60 employees, in manufacturing and 14 other SIC codes. Worksites revealing lost workday injury and illness (LWDII) rates of about twice the national average or greater were put on a primary comprehensive inspection list. *Barring poor OSHA history and/or very high LWDII rates*, they were notified and offered an alternative. If they join CCP, they move to a secondary list. If they join and have 100 or fewer workers (in firms with 500 or fewer), *and* agree to work with an OSHA-funded State Consultation Program, they move to a tertiary list. No more than 30 percent of the secondary list and 10 percent of the tertiary list will be inspected *under this program*. Participants must send OSHA, once per year under CCP, OSHA injury and illness summary data. They must also commit to an effective safety and health program (emphatically punctuated by active worker involvement) of the type detailed in this book. *CCP is under legal challenge.*

Other types of targeting systems involve special emphasis programs. These programs may be on a local, regional, or national level. In this system, a small number of particular industries are targeted, or the focus may be on specific companies that, although involved in diverse businesses, share exposures to particular targeted chemicals. There may be several inspections in those industries in a relatively short period of time. Then, the emphasis can shift to other types of business. Examples that have fit into this tighter, often-changing category, as opposed to the wider, general category of programmed SIC codes, include excavations, roofing, foundries, meat products, longshoring, lumber and wood products, sawmills, paper mills, rubber mills, manufacture of car parts, manufacture of car bodies, food processing, petrochemicals, process safety management, hazardous waste, tunneling, logging, grain elevators, lead, mercury, silica, asbestos, and cotton dust.

Periodically, OSHA experiments with a small number of random selections of low-hazard, and nonmanufacturing types of businesses. This kind of programmed inspection at least assures an SIC code mix. It can serve to remind employers in those covered lines of work that they are not exempt from OSHA inspections. It is also a way for OSHA to keep an eye out for newly arising hazards in businesses that were not traditionally viewed as being particularly hazardous. This does not imply that nonmanufacturing fixed sites are necessarily among the safest. For example, auto body shops and similar workplaces frequently have significant or potential exposures to many serious hazards.

Focused Inspections

With programmed inspections and at times with others, OSHA may conduct a focused inspection. In these cases, only certain specific types of operations or types of anticipated hazards at the workplace are targeted. This is an experimental system. Instead of detailing the criteria for these pilot programs, it is more sensible to remind you that your entire workplace—from top to bottom and inside and out— may be inspected. Why not be prepared for a complete inspection? In fact, regardless of the scheduling mechanism used by OSHA to bring a CSHO to the establishment, a comprehensive inspection may be conducted. OSHA often uses the term *wall-to-wall* as a synonym for "comprehensive."

Follow-up Inspections

The last inspection priority is generally considered to be the follow-up. The principal purpose of this type of inspection is to determine if previously cited violations have been corrected. The CSHO can also determine the employer's progress toward meeting long-term abatement dates. One special type of follow-up, with the main purpose being to track long-term abatement visually, is actually a monitoring inspection. A follow-up inspection may "jump" the priority of a complaint, referral, or programmed inspection, especially when violations concerning grave hazards were cited, a very high number violations were cited, or the employer has been lax in providing documented evidence of abatement.

The entire inspection process is expounded upon in Part 2. There are detailed explanations starting with the arrival of the CSHO and including the opening conference, the walkaround inspection, and the closing conference. The subjects of warrants and subpoenas are studied, as are specific elements of care that should be used when conversing with the CSHO. Rights, responsibilities, and procedures— including postinspection options, forms of appeal, and what the employer interprets as redress—receive an in-depth examination. This last grouping includes amendments, petitions to modify abatement dates, informal conferences and informal settlement agreements, and contests and formal settlement agreements.

Citations and Penalties

Citations and penalties are also discussed in Part 2, with a special dissection included in the chapter covering defenses. The fine definitions of each citation classification are broken down in that part of the book. As a simple introduction, "serious" citations can carry penalties of up to $7000 per alleged violation. The $7000 figure is the highest gravity-based penalty that can be proposed. Gravity is determined by an evaluation of the anticipated injury/illness probability and severity, which often causes the base to be less than the maximum. Then reductions (i.e., adjustment factors) can be applied. The three factors, which are figured separately, are "good faith," history, and size of the company. The applicability and graduated percentages of reductions aligned with these factors change periodically as OSHA tweaks its system. With certain citation classifications (most notably, "willful," "repeated," and "failure-to-abate"), particular factors do not apply.

Although "other than serious" citations can legally carry penalties as high, there is seldom any proposed penalty attached to an "other" citation, unless the alleged violation is regulatory in nature. Regulatory subjects include the posting of the OSHA notice, the posting of citations, the posting of the annual injury/illness log summary, and access to these records. Other alleged violations that fall under the regulatory heading include errors or omissions on (or failure to maintain) the injury/illness logs and supplementary reports of those injuries/illnesses and the failure to report a fatality or catastrophe to OSHA in a timely fashion. One more regulatory category deals with the employer's obligation to tell employees about advance notice of an inspection. Since advance notice is extremely rare in any case, there are very few alleged violations in this category. There are specific assigned proposed penalties that apply to each type of alleged regulatory violation. The penalties can range from $1000 to $7000.

For alleged violations on a *"repeated"* citation, the proposed penalty is determined not only by the gravity of the circumstances, but by the size of the company (relates to coefficient of base penalty, not an adjustment factor) and the number of times that the alleged violation has been repeated. Proposed penalties can be as high as $70,000 and are usually assessed at several thousand dollars, with "double figures" being common.

Each alleged violation of a *"willful"* can earn the employer a proposed penalty of up to $70,000. The minimum that will be proposed after any consideration of gravity and adjustment factors is $5000. Most of the time, the actual number is much higher. If an employer is convicted of a "willful" violation of a standard that has resulted in the death of an employee, the offense is punishable by a court-imposed fine, by imprisonment for up to 6 months, or both. A fine of up to $250,000 for an individual, or $500,000 for a corporation, may be imposed for a criminal conviction (legally speaking, all of the other alleged violations are in the civil arena). OSHA is pushing for Congress to agree to a broader use of criminal sanctions that, for instance, would not necessarily have to involve a death.

"Failure-to-abate" (also called "failure-to-correct") alleged violations can result in proposed penalties of up to $7000 for each day that the noncompliance continues beyond the prescribed abatement date. As with a "repeated," a "failure-to-abate" involving several days of clear

noncompliance beyond the abatement date can be cited as "willful," where the alleged violative situation was blatant. The line may be quite thin, as interpreted by OSHA, when all of the details are analyzed. Note, too, that even an "other" (including a regulatory matter) can be cited as a "willful," "repeated," or "failure-to-abate."

When extremely large penalties are proposed (e.g., several hundred thousand dollars), OSHA has usually employed a special *"egregious"* penalty calculation. The enhanced penalty is derived by way of a formula that considers each instance of alleged noncompliance with a particular standard (or an allegation of a specific breach of the general duty clause) to be a separate violation. Whether deemed to be "egregious" or not, a citation can include more than one allegation that the employer violated that Section 5(a)(1) clause. The allegations may have to do with totally different hazards.

A special note of caution is due for those who contemplate acquiring a company and/or a facility. I recommend that you or your attorney take all necessary steps to ascertain the history of the company as determined by OSHA or a similar agency. This can include whether there has been an OSHA inspection for which citation(s) may still be "in the works." Further, I urge you or your attorney to insist within a contractual framework that the owner, party, or parties in control of the company or facility, which includes land and structures, make known any and all details of any unresolved matters of OSHA citations. It is a fairly common practice for attorneys to request information to determine, for example, if a company has any liens, orders, actions, citations, and so forth, pending against it. You must know what you are walking into. In fact, OSHA has been known to cite "successors" based, at least in part, on the concept that the new owners should have known of the earlier violations. There may be some legal arguments to be made on behalf of the new employer, but it is preferable to avoid being caught in the trap.

Part 2 also explains penalties that may be proposed for reasons other than alleged violations of standards or the "general duty clause." The first category involves falsifying records, reports, or applications; the second relates to assaulting, opposing, intimidating, or interfering with a CSHO.

Employee Rights under Section 11(c)

There are many rights that the Occupational Safety and Health Act affords to employees. Of great importance, the employer should make certain to not abridge the rights that fall under Section 11(c) of the Act. Section 11(c) has nothing to do with race, religion, color, gender, age, national origin, sexual preference, or disability. It protects employees from their employers taking adverse action against them, if that action would not have been taken but for the employee's exercise of rights guaranteed by OSHA. Examples of these rights include participating in OSHA inspections, conferences, hearings, or other OSHA-related activities; participating on a workplace safety and health committee or in union activities concerning job safety and health; filing safety or health job grievances; complaining to an employer, a union, the media, OSHA or any other government agency about safety and health hazards; requesting to see material safety data sheets or other documents that the employer is required to share with employees upon request; and so on.

As for forms of retaliation against an employee, OSHA fully realizes that firing is not the only adverse action. The Agency is also concerned with demotion, holding up a promotion or raise, removal of seniority or other earned benefits, transferring the employee to an undesirable job or shift, threatening or harassing the employee, and forcing overtime when it is not desired or taking away overtime opportunities when they are desired. The true test is whether or not the action or purposeful inaction would have been leveled against the employee if he or she had not been involved in the protected activities.

OSHA investigates claims of 11(c) discrimination when employees who believe that they have been punished for exercising their safety and health rights have contacted the OSHA area office within 30 days of the time they learn of the alleged discrimination. Union representatives can file such complaints for employees. The OSHA investigator, who may be a regular compliance officer or may be a dedicated 11(c) investigator, looks for factors such as unequal treatment. The employer may claim that the adverse action was taken due to the employee's misconduct (e.g., violating safety rules!), poor work performance, tardiness, or other reason that is normally covered by a company's disciplinary policy. For instance, if OSHA finds that other employees were equally tardy or slow at production, but no action was taken against them, then the 11(c) protected activity stands out. Further, if the alleged victim of the discrimination had never been warned about the supposed reasons for the discipline, even though he or she had a history of violating a nonsafety and nonhealth company rule (for example, the employee was running a football pool on the shop floor), then why is the disciplinary action taking place now? Would it have taken place if that employee had not spoken freely with the OSHA compliance officer during an inspection, or asked the supervisor to add a railing to the steps, or requested a copy of the company's respiratory protection program?

OSHA's 11(c) investigations are meticulous. The investigators are adept at recognizing guises and ruses. They are also evenhanded and do not presuppose guilt on the part of the employer. If the Agency determines that there was 11(c) job discrimination, it asks the employer to restore the employee's job earnings and benefits. If necessary, OSHA will take the employer to court. The employer, upon settlement or unfavorable court decision, is generally obligated to post documents that elucidate the findings and the employer's commitment to no longer discriminate against employees who exercise the rights afforded to them by the Occupational Safety and Health Act.

By the way, there is no financial burden placed upon the employees by OSHA. The alleged victims of the discrimination do not pay any legal fees. In fact, they do not even have to fill out extensive, confusing government forms. For the most part, OSHA makes it easy for employees to offer the claim of discrimination. Whether the claim is, in the final analysis, considered legally valid is another matter.

The best and most obvious employer defense against such allegations is to assure that employees are permitted to exercise their OSHA-related rights. However, you may have good reason to take adverse action against employees for various reasons. You may even have employees who have exercised their OSHA-related rights, perhaps frequently, but otherwise have been very poor employees with regard to attendance, product sabotage, quality of work, sexual or racial harassment of other employees, and so on. You should not fear taking action. Your protection is to document, document, document. Be prepared to show OSHA (or other agencies that investigate other types of discrimination allegations) the detailed evolution of the employees' unsatisfactory behavior. Have the case built on paper. Just because employees threaten to call OSHA does not forever insulate them from being the recipients of fair, consistently enforced company disciplinary action.

Employee Rights under Other Acts

Truckers are also protected against safety and health discrimination in another comparable way. Section 405 of the Surface Transportation Assistance Act (STAA) provides protection from reprisal by employers of truckers and certain other employees in the trucking industry involved in activities related to commercial motor vehicle safety and health. These cases often include allegations that a driver was discriminated against because he or she complained about being coerced to drive for consecutive and/or cumulative hours that were in excess of those permitted by the federal laws applicable to commercial motor vehicle operation. Sometimes, the allegation also deals with the reporting of allegedly doctored time and mileage logs.

Another common subject of these allegations involves the employees' refusal to operate a vehicle when drivers have a reasonable apprehension of serious injury to themselves or the public due to the unsafe condition of the equipment. The alleged victim of the discrimination must file the 405 complaint with OSHA within 180 days of the claimed discriminatory incident. The rest of the 405 investigation and resolution system are akin, although not identical to, the 11(c) process.

OSHA also enforces employee whistleblower protections under the Asbestos Hazard Emergency Response Act (AHERA); International Safe Container Act (ISCA); Clean Air Act; Safe Drinking Water Act; Solid Waste Disposal Act; Toxic Substances Act; Federal Water Pollution Control Act; Comprehensive Environmental Response, Compensation, and Liability Act; and Energy Reorganization Act. Section 211 of AHERA protects primary and secondary school employees who complain of asbestos exposure in their work areas. Section 7 of ISCA protects employees who report unsafe intermodal cargo containers designed to be transported interchangeably by sea and land carriers and moving (or designed to) in international trade.

Compliance with OSHA Standards

Having knowledge of the basic operations and authority of the Occupational Safety and Health Administration is imperative. Yet, there is no reward for the attainment of that knowledge in itself. Instituting policies and practices that are designed to secure and maintain compliance with the standards (and the general duty clause) is more important. It takes work—a lot of work. There is much more involved than simply installing guards on machines, collecting material safety data sheets, insisting on the proper use of appropriate personal protective equipment, and having unobstructed exits. There is the whole matter of cultivating a pervasive climate of safety and health, and accepting nothing less than its fruition. Each and every employee must be infused with an understanding of, an appreciation for, and an allegiance to the principles and values of the resultant robust culture. In the real world of government enforcement, there is also the business of having a working comprehension of how to deal with OSHA and its personnel in a manner that is civil and cooperative, while ever skillful and self-protective.

If you are to prepare properly for an inspection (or lawful intrusion, if you prefer) from the Agency, now is the time for your wholehearted, unwavering acceptance of three unalterable facts. First is that feeling the full force of OSHA's legal power can be devastatingly costly and unpleasant. Second is that it is within your control to ensure that a visit from OSHA should be nothing more than a relatively inexpensive lesson in the fostering of occupational safety and health (a vital element of a successful business). Third is that employees must be recognized and treated as the most valuable resources of a company, and their exposure to risk of injury and illness must be minimized.

Preparing for OSHA

1

Top to Bottom

Occupational safety and health belongs to every employee of the company. It must be the concern of everyone, regardless of their job title, position in the hierarchy, or salary. The desire to work safely must be instilled in each person in the company. This desire should be fed and sustained by a well-established culture of safety that is evident throughout all company operations. It should be reinforced continuously, with all employees realistically being able to feel that they are working within a caring and supportive organization that places great worth in safety. The recognition of that worth must be reflected in all aspects of the business and at all levels. This includes the highest executive levels, top management, midmanagement, and line employees.

Each employee should be respected and valued as both an individual and as a member of a team. As with any sound and effective management style, strong and steady leadership must be in place. That leadership should be characterized by explicit direction and respectful inspiration, rather than by vague commands and oppressive control. The commitment to assure that a safe and healthful environment is provided and maintained must be constant and unwavering. For the worth of that commitment to be validated, unmistakable evidence of its depth and authenticity must be made clear to all employees.

The proper examples must be set at the top and reflected at each and every level. If the CEO walks through the plant, his or her safety-related compliments to employees can be instrumental in reinforcing their recognition of the sincerity of the company's commitment. By the same token, if the CEO walks through an area where industrial eye protection is required, he or she had better be wearing the protection. If the CEO moves a carton away from an extinguisher, removes a coiled (unused but available) electrical extension cord that is missing the grounding prong, or reminds employees to wear eye protection, the appreciation of that commitment is heightened. I do not suggest that CEOs perform the daily work of safety directors and supervisors, nor do I suggest that they bellow and scowl any time they notice a noncompliant or unsafe condition or behavior. I do suggest that their genuine concern be demonstrated. The tone can be serious, while hospitable, even if a foreperson must be told that a problem has been observed.

No one is too big or too small to be concerned about safety. It is certainly important that employees know that someone is looking out for them, for instance, that safety directors and/or supervisors are unceasingly fervent in their desire to reduce risks that can be encountered in the workplace. In a broader sense, it is important to make sure employees know that the company intends to do everything feasible to preclude injuries, illnesses, and accidents. However, be sure that employees (including line management) do not infer that they do not have to worry about safety issues because someone else is doing it for them. Too often, a safety director uncovers evident hazards in a department, brings it to the attention of a foreperson, and that individual responds by explaining that he or she figured that the safety director would eventually find it and deal with it. That may be a function of the safety director's job description, but the foreperson is a closer line of defense. That foreperson is unequivocally not supposed to wait for the inspection.

Give all employees the fundamental security and confidence in knowing that they are being looked after, but be absolutely certain that they know that they must take individual responsibility for themselves and, quite often, for actions that could affect others. Give employees a visible stake in the process. Give them an opportunity to exhibit pride in their workplace and give them an understanding that they are ultimately the ones who control their own minute-to-minute actions. Their direct involvement in the safety effort provides the means by which they develop and/or express their commitment to safety. Employee commitment is essential, just as management commitment (including strong financial backing) is essential. They are inextricably linked. The complementary association can only point to success.

2

Responsibility and Authority

Every employee has a responsibility to work safely and do what can reasonably be accomplished to assure that others work in the same manner. Yet, there are certain employees with particular responsibilities related to safety. Of course, safety directors or people with similar titles immediately come to mind. They should not simply be personnel specialists who were saddled with peripheral duties, or given another title, so that OSHA would be impressed that the company has a "safety director." Common sense and diligence have a strong place in the field of occupational safety and health, but a safety director must be equipped with much more than those admirable qualities.

A Safety Director's Qualifications

The field of occupational safety and health has evolved into a complex, multifaceted discipline. There are highly technical aspects and there are strong requirements for good management techniques. The safety director must be formally trained, which is not to say that an occupational safety and health (OSH) college degree is obligatory. It is not sufficient to claim that the safety director has been adequately educated by many years on the factory floor or in the carpeted offices. These individuals must be specifically trained in the discipline. They must possess a detailed and broad-scope comprehension of hazards and risks in the particular workplace. This involves behavioral considerations as well as mechanical and other technical factors. Safety directors must enjoy an intimate knowledge of regulatory standards, with an emphasis on OSHA. Further, they must have a well-balanced understanding of abatement approaches and methods and seek to keep abreast of new developments in the field.

The safety director does not have to be an exceptional engineer who is capable of designing, constructing, and installing all physical aspects of an abatement method. He or she is not the person to actually design a complete ventilation system or the layout for a series of open surface tanks, to build a complicated machine guard or a perimeter railing, or to install an overflow valve or a magnetic disconnect. However, the safety director must be prepared to converse with the engineering, maintenance, and electrical staff (or outside contractors) about the interpretation of the standards and what net abatement effect must be achieved.

At times, the safety director should dialogue with the purchasing staff and an on-board chemist. Preferably, the safety director can discuss alternative solutions and, depending on the feedback received from technical personnel, offer additional or fine-tuned ideas. The safety director should not simply tell someone that a machine guard is needed. He or she must be able to explain what the purpose of the guard is and what special considerations in its construction and function are required.

Safety directors should be in a position of practical knowledge to address state-of-the-art options and their basic underlying principles. They must know what types of means are available to get the job done. They should be able to bring to mind several possible avenues toward abatement. Then, if either the safety director or a member of the staff with which he or she is working finds lack of practicality, the choices will become more focused. As an example relating to the hazards of moving parts of machinery, the safety director can bring up "hard" guards, interlocked guards, barriers, presence-sensing devices of the radio frequency or light-beams variety (including proximity bars), sensing mats, restraints, pull-backs, and two-hand controls. Thus, the safety director is presenting a large portion of the available menu.

Responsibilities and Authority

For companies and their employees to benefit fully from the knowledge and skills of safety directors, they must be given complete freedom to access all operations and all personnel. Their presence and input should be welcomed and frequently sought. They are professionals and must be viewed as such by all. Others with assigned duties and responsibilities closely related to safety generally include supervisors, forepeople, and team leaders. It is also common that plant engineers and maintenance heads have specific, major roles in the day-to-day implementation of projects relating to safety. Further, do not neglect the role of purchasing agents.

In all cases where the responsibilities are specific and documentable, those responsibilities should be recorded. Examples of safety responsibilities to be delineated include but are not limited to the following: Who is responsible for the overall plant? Who is responsible for buildings and grounds? for particular departments? for particular operations, processes, and procedures? for procuring safety equipment? for assuring that only in-compliance (and adequately safe for the specific use and environment) machines, apparatus, containers, and so on are brought into the facility? for general orientation of new or transferred employees? for standard-specific training? for accident investigation? for compiling and maintaining injury and illness records? for compiling and maintaining material safety data sheets? for fire extinguisher inspections? for crane inspections? for discipline? for air sampling and testing? for noise sampling and testing? for in-house inspections? for heading the safety committee? The examples go on and on.

In many cases the responsibilities are often redundant and/or interrelated. Be sure to avoid ambiguity as to who has ultimate responsibility in each case. This does not necessarily mean that there *must* be only one person to "answer for" each concern or problem. It is conceivable, but ultimately difficult, to successfully assign and monitor team leadership responsibility. In those cases, despite the encouraging of brainstorming and the fostering of confidence, the reliance on the other team leader might cause a gap in "reins holding." It is preferable that there be a "go to" person to authoritatively and maturely report on progress, explain glitches, and take charge to see a project through to its fruition. This is so even though that person need not be the main contributor to or facilitator of a project. Delegation of temporary authority has a fair place in the system. This is often to a person or persons with particular expertise that the "authority" lacks. The important thing to remember is that, although authority can be delegated, responsibility cannot. In the final analysis, however, whoever bears the burden of the highest responsibility for a specific or overall issue must have authority to back up that responsibility.

For example, when a safety director feels that an operation must stop or a machine must be shut down, he or she should not have to go through several layers of management to effectuate that action. Strangely, this concept is controversial in some supposedly forward-thinking companies and corporations. Is there a danger that the safety director might needlessly cause the ceasing of an operation, greatly slowing production and costing the firm many dollars of downtime? Is there a danger that stopping a process at the wrong part of a cycle can, ironically, cause a grave safety hazard in itself? Sure. That is why safety directors must be well trained and, if there are reasonable concerns, such as those just mentioned, they can seek input from others. Those others might include a supervisor, a chemist, or even a machine operator. Still, there may be occasions when a safety director feels that there is no time to seek the preferred input. This could be in a perceived imminent danger case. In such cases, lost time and/or ruined product can be a small price to pay for eliminating the hazard of explosion, electrocution, or other grave potential consequences.

Safety directors must be accorded a significant degree of professional respect. This does not mean that they are to be worshipped or feared. It does mean that they are to be sought whenever their guidance, active listening, and specific knowledge can even remotely be of value in solving a safety problem. Employees must feel that they are strong partners in the safety and health concerns of their facilities and that they are there, in the final analysis, to protect employees from human suffering. In so doing, the safety director is also helping the company reduce risk, loss, financial burden, and legal jeopardy.

His or her position in the management hierarchy or corporate ladder must not be smothered by several levels of flow-charted titles, departments, divisions, and offices. The safety director must be afforded easy access to the main on-site authority—the person in charge of that plant, for instance. Generally, the safety director should not be more than two beats away from that main on-site authority. This often plays out so that the safety director reports to a vice president, who reports to the president. Clearly, the main authority need not have to be apprised of every action taken by the safety director. That authority does not have to approve every move. Still, the access must be there. The president or CEO (in large corporate situations) should be personally acquainted with the safety director.

3

Accountability and Evaluation of Management

All employees should be fully aware of their responsibilities. They should be held accountable for their actions, and in some cases inactions, that relate to those responsibilities. All employees should be evaluated. Relating to production, this traditionally involves such criteria as quality and quantity of work. In other types of work, these criteria for evaluation may be joined by elements relating to how well the worker gets along with co-workers, how well the worker performs under pressure, whether or not objectives are met, verbal and written communication skills, supervisory skills, attendance, knowledge of the process or product, demonstrated skill and tact in dealing with the public, and numerous other categories thought to directly affect the successful completion of the company's mission.

Certain types of behavior, although sometimes a part of those I have listed or added to evaluations, are not necessarily part of a formal job description. I refer to general courtesy, nonviolent behavior (unless your job happens to be with the World Wrestling Federation!), and working in a safe and healthful manner. However, it is a good idea for all formal evaluations to give some consideration to whether an employee is a safe worker.

For anyone with specific safety responsibilities, it is clearly advisable that they should be evaluated and held accountable for how well those responsibilities have been met. Where written evaluations are given, categories are often reviewed in short narrative form, in columns, in blocks, or in similarly focused, compartmentalized sections. Far too often, there is no reference to safety. Even though safe work is an across-the-board element that should be understood as a requirement for continued employment, and especially for professional advancement, it should be a specifically addressed part of any formal evaluation for those saddled with spelled-out safety responsibility. How well they perform those duties should be addressed in a dedicated block (or equivalent) of that evaluation, not merely as one small part of the overall review.

When evaluating a safety director, it is understood that safety will be the overwhelming theme of the whole evaluation; safety will be broken down into several distinct subelements. When evaluating positions such as plant manager, maintenance chief, operations manager, project manager, vice president of engineering, and division head, safety should certainly be a major element. Sometimes—but not often enough—it is given its due. A major lapse occurs far too often when forepersons and others with significant floor responsibilities are evaluated. Again, there should be a dedicated block relating to safety, which helps eliminate the foreperson waiting for the safety director to uncover hazards. It certainly sends a powerful message that upper management expects that person to run a safe department.

Upper management should be concerned with a foreperson running a department that has experienced a high frequency and/or severity of injuries, illnesses, accidents, unsafe conditions, or unsafe acts. Further, that foreperson will have to answer for violations of OSHA standards or violations of sound safety principles and practices in his or her department. Buck passing and shoulder shrugging must be eliminated. Who was responsible? Remember this during evaluation period. Besides regularly scheduled evaluations, interim or special evaluations may be necessary. Keep the forepersons continuously accountable.

The stated job may be to produce the most widgets with the fewest defects, but if that goal is met at the expense of pain, suffering, and unsafe, unlawful conditions, the price may well be far too high. In fact, putting aside the difficult-to-quantify aspects of those concerns, we can turn directly to the actual dollar costs that stem from an accident. The financial costs reach far beyond the initial medical bills. Those costs are examined in my chapter on accident investigation.

Do not neglect praise. This can be in the form of kind words, but it should not simply be an implication by the lack of unkind words. Remember, too, that a formal evaluation does not just dwell on the negative; it should praise where such accolades have been earned. The accolades may have been earned through a dearth of injuries, illnesses, accidents, unsafe conditions, unsafe acts, OSHA violations, and violations of sound safety principles and practices. Consider all of these points.

An individual's evaluation relating to safety in his or her jurisdiction and control should never be simply drawn from the number and nature of injuries and illnesses. I have made it clear that those statistics should be reviewed, but there are deeper questions relating to the spawning incidents. I am not suggesting that each case be documented on an evaluation. That would not make sense. However, when considering those injuries and illnesses, which may have been charted in many different ways, look at the entire situation. Again, see the accident investigation chapter.

For evaluation purposes, the question is not simply who was in charge of the department or cell where the injury or illness occurred. Other considerations focus on what steps had been taken to prevent the accident from occurring and what steps were or will be taken to avoid a recurrence. In plain English, do not simply count up the number of days lost as a result of workplace injury and illness and evaluate on the basis of that sum. For example, three near misses (with no lost time) or several minor injuries (with no or little lost time) at an unguarded point of operation on a mechanical power press should contribute to a negative or low rating on an evaluation. Such occurrences are much more significant than an employee who trips on his own shoelaces and sustains an injury resulting in 30 days lost time.

4

Employee Awareness, Acceptance, and Participation

Awareness

Many companies have excellent written safety and health programs. Well, at least they have committed to print all of the necessary rules, warnings, technical data, inspection records, and so on. The problem is often who knows about these loose-leaf books and computer programs, and how to access them. Not all employees need to know where all of the program documentation is, but employees must certainly be aware of the existence of programs, the details of those germane to their protection, and the philosophy, sincerity, and desires of the company. I am mainly thinking about nonmanagement employees. However, it is (sadly) not unusual for a foreperson being questioned by OSHA not even to know of the existence of a particular documented program element. Later, upper management needs to search for a dusty notebook containing that information.

Acceptance

Once employees are aware of the program, as well as the genuine concern of management, it is of paramount importance that those employees buy into the system—that they accept and appreciate the purpose and specifics of the program and that they do all they can to live within its rules and its spirit. Each member of management, starting with the highest levels, must vigorously demonstrate total, sincere commitment. Wherever and whenever any element of the work or workplace poses a threat to any employee(s), it is incumbent upon management to take an open and active interest without hesitation. That interest cannot be manifest in mere rhetoric. Do not just express concern or talk about the need for abatement. Take decisive action, make it known (not merely believed) that you plan to see it through, and then do so.

Find out what barriers stand in the way of your safety and health goals. Remove them or, if that is not feasible, work around them so that the same positive net effect is realized. These are examples to be set for employees. Dispel any notion held by employees that management is too busy to care and act. Leave no doubt that you categorically reject the cop-out adage that employees aren't really interested in on-the-job safety and health.

The program and its full support by management must be clearly communicated to each employee, including part-time, temporary, and contract workers. All those shelves of hazard communication information, lockout/tagout procedures, general operating practices, and company rules and procedures are worthless unless the employees are properly aware of them, accept their value, and recognize that they must comply. Simply put, compliance with safety and health rules, conditions, practices, and so on is a condition of employment.

To achieve the necessary level of acceptance by employees, they should participate in the program on a regular basis. Management should be coactive, eagerly and frequently drawing upon their considerable resources. Management is the ultimate decision maker. No, there need not (and should not) be a vote on every occupational safety and health policy. No, management has not handed over the keys to the factory. The employees must know that they are viewed as valuable human beings, not just as parts of machinery. More significant is the plain fact that line employees can quite often provide practical suggestions that even the engineer, Ph.D., and CEO cannot. They frequently possess a better understanding of the day-to-day operation of their machine. They may know how to make it better and safer. They may know why accidents can or do happen on it, even if management just cannot figure out how Marie could have gotten her hand caught in that section of the machine. They may know how to beat it, a bad habit often motivated by piecework.

The wisdom and experience of these employees should be sought when designing new protection systems or new procedures, when conducting accident investigations, and by way of safety committees and suggestion boxes. Their energies and insights can be of extreme positive value in the driving of a safety program. It must be made indelibly clear that employee input is greatly encouraged and appreciated and that management will be enduringly receptive; this is not just a fleeting phase. Then, and only then, can employees maintain their feeling of trust and pride in the organization.

Although managers exercise authority when necessary, it must be understood that they are not in that position to be authority figures. They should operate in the role of a coach. A coach teaches, guides, inspires a willingness to do the right thing, and fosters a genuine feeling of trust. When acting in that role and encouraging participation in the system, employees are not only safe to speak out; they feel obligated in a positive way to speak out. Similarly, such a coach can, in a nonauthoritarian manner, help employees to establish and secure a prosafety attitude and a consistent pattern of safe behavior.

The communications between coach and line employees should be characterized by openness, mutual respect, and a climate of trust. All communication with employees relating to safety, even if not exactly perceived as a trainer-trainee relationship, should be in the form of dialogue. It should not be a matter of the trainer or foreperson force feeding a homily to an audience of one that is forbidden to contribute verbally.

Remember to act as a coach. It is not just what you say. It is how you listen and how you use body language to reveal your interest, good motives, and respect versus impatience, patronization, and condescension. Be eager to show your concern for the feelings of the employee. Be fully attentive and do not interrupt unless absolutely necessary. As an active listener, verbalize empathy when you can genuinely do so, but do not say that you understand an explanation, an attitude, or a feeling if you truly do not. If you do not understand, say so and show your personal interest by asking for clarification. Paraphrase what you believe you are being told or asked, to demonstrate your concentration and interest and to confirm your comprehension. If it is obvious that you are listening to, and truly hearing and absorbing, the employee's comments (as opposed to fidgeting, trying to fill every sound gap with your profound wisdom, and straining to feign interest), the odds are in favor of the employee more readily receiving and accepting your words.

Avoid putting the employee in an unescapable corner, even if you are providing corrective comments. You can point out mistakes without robbing an employee of his or her dignity. The conversation does not require a winner and a loser. Each of you can gain from the interaction. Remember the partnership between management and employees—the coactive philosophy. You need not search too deeply to counter every misdeed that you point out with a reminder of some positive behavior on the part of the employee. That can become transparent, almost childish, and make it appear that you are neither comfortable nor competent in your position of authority.

You can be friendly and caring without the employee losing professional respect for you. Safety is serious business, but humor has a place. A light remark, now and then, can serve to break down barriers in communication. When both parties feel more at ease, concentration should improve. By all means, you can point out some positive actions of the employee, and you can make sure that his or her self-esteem is left intact. Yet, do not end the conversation until you have unambiguously explained what your concerns are, why they exist, and how the employee is expected to react.

Participation

If you communicate properly, employees will want to correct their negative behavior, and not just because a manager said so. They will want to do it because they feel that the conversation was a positive experience, they were not stifled in their efforts to tell their side of the situation, and they legitimately understand the reasons for the corrective feedback. When the participation of employees through suggestions or actions is openly appreciated, the employees gain self-respect and, generally, have more reason to be committed to the company because the company recognizes their high value to the system.

Safety is supposed to be a nonadversarial subject within a company. Real employee involvement—not just cosmetic involvement—must be promoted and fostered consistently. In my discussion on training, I address several issues that relate to this employee involvement, their awareness of safety rules (including why the particular rule is in effect), and their acceptance of their role as individuals who will benefit from compliance and who are in a position to act accordingly. Safety concerns should be kept out of the political arena and vice versa. This is particularly critical where there is a union. The concerns for a safe and healthful workplace must be shared by management and labor (organized or nonorganized).

Thus, the commitment to and participation in the safety system and the very essence of the safety culture must be shared. Workers must know the mechanism of providing that input. This is not to say that all reporting of potentially hazardous situations must be made within a formal structure. Rather, employees must know whom to notify (for one, a foreperson will always suffice). They should have absolutely no fear of reprisal, and they should expect timely and appropriate responses to their concerns. They should not be stifled by doubts that their explanation might not be perfectly worded or that it might reveal a lack of knowledge on their part.

Do not wait for employee input. Be proactive and seek it aggressively. There is every reason to believe that your active interest will be rewarded.

5

Attitude, Behavior, Motivation, and Philosophy

Attitude and Behavior

Employees must be instilled with the proper prosafety attitude. It is much easier to modify behavior than it is to modify attitude. Management can, with relative ease, cause an employee to behave in the dictated manner, whether that behavior is to directly follow orientation or whether it is to be changed due to inappropriate behavior that has been detected even after an employee has worked at the establishment for a long period of time. That behavior can be affected by the person in authority insisting on it, with the threat of discipline as a blunt instrument: "I'm the boss. That's why. Do it that way or work somewhere else." Or the authority can blame OSHA, but in so doing, these cop-outs do not get to the heart of the matter. Of course, some employees may operate out of fear alone. Yet, to make a particular behavior stick—to cause it to be self-regenerating—attitude must be modified.

A good general attitude is in itself not enough. It must be translated into specific behavior that can be readily observed and assessed. Each task must be performed in a safe manner. For one thing, a person may have the proverbial good attitude but not understand how to transfer attitude to actions. That must be taught. This does not mean simply explained in a philosophical framework; it means that very specific finite steps of clearly delineated tasks must be addressed, and the related safe behavior must be made perfectly clear. To complete the circle, once an employee behaves consistently in a safe fashion as ingrained habit, the right attitude is reinforced. The idea is, at some point following well-rounded training, to have employees act safely not simply because they are being ordered to, but because they have accepted an important responsibility and have been empowered to make critical performance decisions. The employees become self-initiators, and their safe behavior becomes self-perpetuating.

Figuratively speaking, the tools have been given to the employees, and they are in the very best position to use those tools properly. Recognizing the attendant, ongoing responsibilities and the position of empowerment, employees should become self-regulators. Thus, the head is turned in the right direction to want to see the light. Then the head is taught how to find the light, how to approach it, and how to function occupationally within the beam.

Some academicians regard the achievement of a good attitude as a habit that has been learned. Once again, that does not go far enough. Once the attitude is in place, detailed, safe work habits must then be inculcated. To begin with, the formation of a good general safety attitude requires a firm commitment to the construction and maintenance of five major building blocks:

1. focusing and concentrating on the tasks to be conducted;
2. taking the time to do the job correctly;
3. demonstrating the mental strength to do things in a safe way, despite pressures imposed by others;
4. taking responsibility—thinking of yourself as part of a team and acting to correct an unsafe situation—even when that situation is not a direct part of your job;
5. avoiding placing yourself and others at risk.

Motivation

Once these building blocks are on a stable foundation, then the specific safe behaviors must be learned. For an attitude to be effectively and lastingly cultivated, employees should really know why they should behave in a certain manner. This does not simply mean because the boss or OSHA says so. Employees must realize that they and their co-workers are potential victims of occupational injury and illness. Certain behavior can preclude or greatly reduce the chances of the employee becoming a victim. The adverse consequences of inappropriate, unsafe, and unhealthful behavior should be spelled out in simple language.

One example can relate to why the wearing of hearing protection is necessary in certain areas. An employee can be merely told, upon initial assignment, that hearing protection is worn because it is a company rule. The desired behavior—the wearing of that protection—will probably be achieved. Unfortunately, without constant monitoring, that behavior may change after a few days, and the muffs or plugs will come off. The employee might feel that the rule is simply a result of power-hungry management, bureaucratic red tape, and/or a far too cautious policy without any real purpose. The employee might feel that he or she is not in a position to question the order, but when management is out of sight, he or she will simply remove the hearing protection. The employee might feel that management does not realize how uncomfortable that protection is. There can be situations where an employee has not worn the protection for several months, although management should have noticed this and corrected it. The employee reasons that he or she can no longer hear the noise, not meaning this literally, but that the noise is such a constant part of the work environment that it can easily be blocked out. Further, the employee might reason that if he or she does ever sustain hearing loss, a doctor will reverse the damage.

The employee's attitude must be altered, with a large part of the convincing nature of management's argument lying in an explanation of what can happen, why it can happen, and why the hearing protection can make a critical difference. These components of the indoctrination are essential, as the employer should not assume that the employee automatically knows the answers. Briefly, in a case such as this, employees should be enlightened as to the type of damage that high decibel noise can cause; that if they already seem to not hear the noise, they may actually have lost some ability to perceive that noise (i.e., they already have hearing loss); that the damage is generally irreparable; and that the loss may take several years to develop. Employees must not feel that, just because they have no apparent hearing loss after 2 years of not wearing the protection, they will never be affected. They might well develop that false sense of security.

Following this example further, the employee's attitude might receive an even stronger positive jolt after being taught that noise can constrict blood vessels, cause heart problems, make victims irritable, and cause problems with digestion and sexual activity. After all this, there will probably be an attitude change, presenting the potential victim with strong motivation to wear the hearing protection.

Motivation is the bond between ability and attainment. Management has the ability (and, in effect, the burden) to assure safe behavior. To attain that lofty, critical objective, employees must be properly motivated, and in essence, the correct behavior should stick if proper motivation has been accomplished. This concept may seem too academic, but it is accurate.

Looking more briefly at a hazard that is simpler to explain, employees must be made to understand that if a machine is operated in a specific wrong way, they or other operators can sustain a very serious finger injury. Management should clearly explain, without a dangerous demonstration that becomes too graphic, how the finger can enter the danger zone. In such a situation, employees may wish to make a strong argument that they are intelligent, experienced, and trained. Therefore, employees should be given examples of how even intelligent, experienced, trained individuals have sustained amputations in mechanical power presses, power transmission chains, and similar apparatus. It is best to give real "horror stories." It is important to make it indelibly clear that even "tough guys" die on the job.

Philosophy

Employees and even applicants may indicate, by implication or direct language, that they are so strong and tough that they are willing to take risks. These employees or applicants may have thought that they were depicting an element of loyalty to an employer—that they were communicating a personal quality that would make them seem desirable for promoting or hiring. Management can explain that the company does not want that type of risk-taker on the payroll and that such attitudes do not fit within the company's philosophy of safety and health in the workplace. Further, the company should make every effort, including by review of employment history, to assure that hostile, impulsive, cavalier, or daredevil types are not hired. Screen them out. Potential employees should understand the company's safety and health philosophy from the day they are interviewed for initial employment. That philosophy is addressed in a later chapter.

6

Engineering, Education, and Enforcement

There are three Es that comprise the essential core elements of any full-blown occupational safety and health program. However, as a cautionary note, these three Es cannot assure a safe environment in themselves. The very real human beings who are involved must have achieved the right attitude. They must be consistently ready to practice the appropriate behaviors that go hand-in-hand with the purposes of solid engineering, education, and enforcement.

Engineering

Once attitude and behavior can be trusted, the three Es have a much greater chance for success. Number one is engineering, viewed from a slightly more liberal definition than one would find in a standard dictionary. I refer to this element because the employer should start with the right physical components. Plan a facility that is designed, engineered, and constructed properly with safety in mind from the start. Do not plan to make the general space and fit in safety considerations later.

During the incipient planning stages, decide on all physical considerations. In other words, build it right. Some of the major safety considerations that must be addressed in great detail up front include the well-planned location, construction, and the maintainability, serviceability, and adaptability of the following: exits, stairways, ventilation, electricity, plumbing, relative proximity of operations to each other (based on proper process flow), all types of ergonomic considerations, bathrooms, first-aid rooms, observation points, and storage areas (most notably for hazardous materials). Surely, there are other considerations, but the point should be clear.

For example, it would be a shame to construct a plant whose amperage potential is made obsolete a year after plant start-up due to the acquisition of new equipment. If a plethora of ground fault circuit interrupters would be advisable, do not plan to add them later; do it initially. It would be a major nuisance to find that plumbed eye fountains and deluge showers are needed but that there is no piping in the areas where the fountains and showers are to be located. Similarly, after completion of the facility, it would be costly to realize that storage areas for chemicals are so close to each other that they do not allow for proper separation and isolation of noncompatible chemicals. Also, it could take much work to reverse a situation in which it has become evident that the only reasonable places to store fuel gases or flammable liquids are near exits. The proper balancing of exhaust systems can be of utmost importance, so the location of welding operations, plating tanks, and similarly hazardous operations must be decided upon before duct work is built.

Get it right from the start. This may be a slight oversimplification, but don't just plan sufficient square footage and fill in later; engineer it correctly from the computer design or blueprints. A practical physical plant (including layout) that can fit into the fostering of an efficient, effective, and safe operation is essential.

My liberal reference to engineering also relates to purchasing the right equipment, chemicals, and supplies. When you bring in ladders, machines, tools, electrical apparatus, and so on, be sure they are proper for your very particular business and facility from the safety standpoint.

Education

The second essential core element is education. Never assume that employees or applicants have the suitable skill, knowledge, and attitude for the specific duties they have to perform. Be as certain as is reasonably possible that they know what you expect of them and that they have the ability to deliver. I expand on education in a separate chapter.

Enforcement

The third essential core element is enforcement. If you do not assure that employees have absorbed and heeded the education that you have provided, then the fact that you have provided them with the proper physical tools—fully equipped buildings and grounds, equipment, chemicals, and supplies—will be moot. They must use those tools properly and in a safe manner. Progressive discipline is discussed in a later chapter.

7

Resources, Library, Audiovisuals, and Instruments

Resources and the Library

A company should develop a system of easy access to all necessary resources. A safety and health library of sorts should stock OSHA, NIOSH, ANSI, NEC, NFPA, and similar literature. There are several magazines, available by subscription, that are dedicated to occupational safety and health. Although I am not directly lending my endorsement to any particular ones, they include *Occupational Hazards, Professional Safety, Occupational Health and Safety,* and *Workplace Ergonomics.* I suggest that you receive them all and create a file system so that you can quickly locate major articles on particular topics.

The library should include the catalogs of companies that deal in safety supplies. In some cases, the companies are purely in the business of occupational safety and health. In other cases, the companies and their catalogs are only partly devoted to occupational safety and health needs. The types of products that can be located in the catalogs include (but are not limited to) the following: personal protective equipment (including harnesses), fire protection apparatus, testing and sampling instruments, alarms, signs, tags, lockout/tagout devices, paints and tapes for demarcation, noise attenuating items, emergency eye fountains and deluge showers, medical supplies, emergency lighting, compressed air gun psi-restricting nozzles, spill containment and cleanup items, flammable liquids storage cabinets, safety cans, bonding and grounding equipment for flammable liquid transfer, and an ever-expanding number and variety of products designed to relieve ergonomic stressors. Other catalogs concentrate on computer software, CD-ROMs, literature, and audiovisuals.

There are sections of more general industrial supply type catalogs in which ladders, slings, scaffolds, numerous types of safety valves or similar devices, electrical extension cords, ground fault circuit interrupters, compressed gas transport cradles, trash receptacles, and similar products are available. There are also catalogs dedicated to machine safeguarding. They often include not just a variety of guards and their components, but safety devices such as two-hand controls, restraints, pullbacks, presence-sensing units, stop-actuating cables, and much more.

The library should also be stocked with the general company rules, specific programs such as lockout/tagout and hazard communication, technical manuals, and audiovisuals. You should have immediate access to names, addresses, and telephone (voice and facsimile) numbers of emergency medical services, emergency spill services, governmental agencies, insurance carriers, vendors, and consultants including safety specialists, ergonomists, industrial hygienists, and environmental engineers.

You should be able to access computer records promptly, which can greatly enhance the ease with which numerous types of safety-related information can be organized, made available, and presented. Computer programs are extremely useful for the logging and storing of documents concerning general orientation, training related to specific standards and operations, material safety data sheets, tracking of safety projects, internal audits and surveys, injury/illness records, inventory, medical clearances, disciplinary actions, and more.

By using a computer for these purposes, a special bonus is the ability to cross-reference the categories by dates, departments, and individuals. In that light, material safety data sheets can easily be cross-referenced by chemical name, brand name, synonym, area of use, compatibility, and other criteria. The computer can serve as a valuable aid for "tickling" when deadlines are about to approach for recertification, medical examinations, preventive maintenance, completion of projects, extinguisher inspections, crane inspections, finalization of reports, and so forth.

Audiovisuals

Although some general audiovisuals, usually in the form of videotapes with a steady and crystal clear pause capability, have value, you should have available videos that are closely related to the exact training that you are trying to pass on. I also suggest having a camcorder ready and loaded to record inspections, jobs with significant ergonomic implications, and for other germane uses. A still camera can also have value. There is more on the use of audiovisuals later.

Instruments and Other Resources

Testing and monitoring instruments (calibrated as needed) should be on hand, commensurate with your expertise in their maintenance and use. Such equipment can help detect potential dangers and graphically illustrate the existence of those dangers. Some instruments are inexpensive, easy to use, and relatively safe. This subject is also expanded upon later.

Safety directors or persons with related direct responsibilities should make an effort to become a part of the occupational safety and health community. They can apply to be a member of the American Society of Safety Engineers and similar groups or associations. They can then be very close to a wealth of knowledge that can be applied to bolster your program. The networking can be invaluable.

Appendices in this book should be consulted to help form your lists of resources. Included are selected safety and health organizations, associations, and federal agencies. Also referenced are selected safety and health Internet mailing addresses and World-Wide Web sites. Consider, too, such valuable resources as state Department of Labor and workers' compensation offices, trade associations, unions, and employee groups.

8

Everywhere, Everybody, Everything, and Every Minute

Everywhere

The occupational safety and health program must cover everywhere in and on the establishment. The program must surely cover all areas where employees may be exposed to any hazards. This is automatically expanded to mean that the program must cover all areas where the employees can be. It also means that the program must cover all areas at which there can be hazards that can affect the employees, even if the employees are not directly in that area. For example, if there is an explosion potential in an area accessible only to outside contractors, not employees, and if such an explosion can harm employees in adjacent areas, then the area of the potential explosion must be covered.

It is understood that the program must cover production areas and all other areas related to manufacturing and normal process flow. This generally brings to mind storage areas, laboratories (not always present), shipping, receiving, and so on. Office and administrative areas must not be overlooked. Common areas—including hallways, bathrooms, cafeterias, conference rooms, and first-aid rooms—are also included. Further, do not neglect "out" buildings and structures and those attached to the main facilities. This may include areas such as storage sheds and stationary trailers, compressor shacks, garages, silos, tanks, guardhouses, and treatment facilities. An often overlooked type of hazard that is especially relevant in these sorts of areas involves harborage and infestation of birds, rodents, and other pests.

The grounds themselves are important. Some concerns include impeded egress, parking lots, walkways, roads, stairs, hydrants, recreational sectors, transformers, drains, barbed wire, small bridges, picnic-type tables, and even the lawn. Do not forget the outside walls of the building, where hazards may relate to fan blades, electrical receptacles, sprinkler pipes, hose pipes, hazardous materials piping, balers, compactors, and dumpsters.

Do not forget the roof itself, where hazards frequently involve HVAC (heating, ventilating, and air-conditioning equipment), particularly unguarded or inadequately guarded power transmission equipment in the form of fan blades, belts, chains, and pulleys. Other common roof hazards involve the lack of perimeter protection in the vicinity of items, when those items are near the edge of the roof: HVAC, various pipes, access points from ladders or hatches, antennae, signs, lighting, and security cameras. This is not to imply that roofs in general require full railings; I am referring to where there are legitimate hazards, and this is not unusual! How about unguarded skylights (fall-through hazard)? Electrical hazards are often uncovered on roofs, as well. The reason that I have chosen to be specific with the types of hazards most commonly occurring on roofs is simply to dispel any thought that such surfaces are not truly worth the time to inspect.

The program must also cover janitorial supply rooms, closets, vending areas, basements, dead space behind stairways, areas that are considered obsolete, and surely small rooms labeled "For Authorized Personnel Only." Why shouldn't authorized personnel be protected? Yes, they may have a particular expertise that could allow them to enter at less risk than other persons, but that reason does not suffice to give them a free pass. (Note: I have conducted inspections while accompanied by the safety director and led that person into areas where he had never before set foot. That is a problem!)

Although more difficult to control, your program should cover your employees when on duty for you off-site. Three good examples of this are when your employee is working at another facility (possibly performing maintenance or repair), making deliveries or pickups, and driving or riding in an over-the-road vehicle.

In all cases, consider a plan of action for when your off-site employees could be confronted with potential risk from crime. How are they to respond? Do not leave them on their own to make a decision as to whether they are supposed to resist the threat of violence, if they face an imminent theft of their vehicle, its contents, or items or cash in their personal possession. Make your wishes—your instructions—crystal clear. In the vast majority of cases and with very few exceptions (e.g., for armed security staff), resistance should not be attempted. You should not couch your words in any way that can be remotely construed as even passively condoning resistance.

When your employee is working at another facility, some of the most common considerations involve the wearing of personal protective equipment, the use of the proper ladders in a correct fashion, assurance that electrical extension cords provided by the host company are grounded (better yet, bring your own), and seeking, absorbing, and heeding necessary input on the hazard communications program of the host company. When making deliveries and pickups, common considerations involve proper manual lifting techniques, the use of wheel chocks, and the correct use of pallet jacks and similar devices.

For the operation of over-the-road vehicles, common considerations involve various aspects of safe driving skills and assuring, as far as is practicable, that the vehicle itself is safe, but most important is the absolute requirement to wear available lap and shoulder belts. Although OSHA has little jurisdiction over safety factors concerning over-the-road vehicles and their operations, do not fail to comply with your obligations under standards regulated by the United States Department of Transportation and similar state agencies. Whether relating to federal motor carrier safety regulations or others, there are many areas of concern. In many cases, these include the following: licensing; driver qualifications; parts and accessories necessary for safe operation; hours of service for drivers; transportation of hazardous materials (includes driving and parking restrictions, as well as marking/placarding); capacity and securing of loads; accident reporting; drug abuse; vehicle inspection, repair, and maintenance.

Everybody

What if the roof, utility rooms, certain fill points of outside tanks, or other areas are supposedly accessed only by outside contractors? They must be covered also. In fact, your susceptibility to legal action and financial burden, other than through OSHA, may well be greater with nonemployee victims. This can also include customers and any other visitors. In many states, if your employees are hurt, they are limited as to what they can collect from you, their direct employer, by the workers' compensation charts. If outside contractors are hurt on your premises, they are limited as to what they can collect from their direct employer by those same charts, but may seek uncapped amounts from you. Customers and other visitors may find ways to seek even greater fiscal damages from you. So, everyone who might be on your property should be protected (i.e., should be covered by your program).

When thinking of your employees, particularly with regard to training, be sure to include temporary help, part-time help, and those hired through an agency. Forget about any idea to excuse yourself from training an individual because he or she is mentally retarded, does not speak English, or is illiterate, hearing impaired, vision impaired, dyslexic, and so on. In a related vein, disabled employees and visitors may require special considerations. Often, existing hazards put them in exceptional jeopardy. While not all disabilities are of a physical nature, I am mainly addressing that type of condition now. If an employee's hearing is impaired significantly, supplement audible alarms with visual alarms (e.g., strobes); if vision is impaired significantly, assure the audible alarms. I stress the need for one-on-one buddies, available at all times, for each employee who might need personal attention to exit an area quickly and safely.

There are also tasks for which accommodation (e.g., by ergonomic redesign) can preclude or greatly reduce the inordinate risk factors. The Americans with Disabilities Act, while not primarily a worker safety law, serves as an additional motivation for noting and addressing such special concerns. One special warning related to accommodation is to be very careful not to create a hazard for workers who are not disabled in a conventional physical sense, while seeking to protect those who are so disabled. How can this happen? Funny you should ask. If an employee has only one arm, the employer might retrofit the machine so that it can be operated by only one hand. However, if a two-handed person operates that machine, he or she can have one hand available to be in harm's way. This can result in a serious injury and an OSHA violation. In such cases, you can devise supervisory controlled activators (usually by a key switch) that set the machine to be operated by one button or by two. This calls for exceptional monitoring, but it can be accomplished.

Everything

The program must cover everything in the establishment. OSHA can cite the employer for any hazard to which his or her employees are exposed, even if that exposure is only temporary or sporadic. This is the case, for instance, whether or not equipment was rented, leased, borrowed, or the personal property of an employee or contractor. Even if the hazard is not solely under the employer's control, he or she can be cited. For example, there may be dangerous situations for which the achievement of abatement involves the addition or significant modification of ventilation systems (including duct work), exits, structural supports, stairways, or other major improvements to the physical plant. At times, this can only be accomplished with the aid and/or approval of a landlord. In such cases, the employer whose employees are exposed cannot circumvent the legal obligation to comply merely by pointing to the landlord's power of control. One way or another, the employer must comply.

He or she may have to begin this process by documenting communication (e.g., requesting assistance or approval for construction or major facility modification) with the landlord. Do not simply write one letter requesting that another side-hinged exit door be installed. Be persistent and follow through. It would certainly be helpful to reference particular OSHA standards. Once the company is cited in circumstances such as these, the documentation can serve to show good faith, and to eradicate the possibility of a "willful" citation. When discussing possible abatement dates with the compliance officer, be sure to explain why you feel that an extended period of time is needed due to the necessity of working with and through the landlord.

It does not matter if the equipment was recently obtained or obtained before OSHA existed. Employers are often angry and frustrated when OSHA cites because of an inadequately guarded machine that the company had recently purchased. The employer often wants to know how that can happen.

One good example is a compressor (as referenced above) that, in this case, was recently purchased and lacks adequate guarding for the v-belt and pulley drive. In almost all cases in general industry, OSHA must deal with an employer-employee relationship; the Agency cites the company that employs the exposed person, not a third party. Therefore, barring novel, creative scenarios (and possibly legal stretches), OSHA cannot cite the compressor manufacturer or distributor unless employee(s) of that manufacturer or distributor are exposed to the hazards created by the inadequately guarded power transmission section. If, by some chance, OSHA discerns with legally sufficient documentation (e.g., as compiled during an on-site inspection of that manufacturer) that an employee of the compressor manufacturer winds the v-belt onto the inadequately guarded pulley, stands near it while it is tested (with not even a temporary barricade or similarly reasonable control), and is therefore exposed to the danger, OSHA can cite that manufacturer. Perhaps some day, the Consumer Product Safety Commission, OSHA, and other germane agencies will work together to preclude or severely limit the manufacture or distribution of products that on face value would present OSHA violations if used as designed and delivered.

Meanwhile, the employer can take steps to avoid the delivery of products that could easily lead to a violative condition. One notable step is to include special wording on purchase orders and work orders. You might wish to check with an attorney on the exact wording if you have a lot of available space on those forms. The wording would indicate that receipt of the item and completion of the sale will only be finalized if the item is in compliance with OSHA standards. You might refer to very specific OSHA standards by number and to ANSI, NEC, or other standards and codes.

If you have taken these steps but still put into use a machine that presents hazards and documentable violations of OSHA standards to employees, your efforts will probably not stop the issuance of a citation. Yet, you may very well have eliminated the citing of a "willful" violation. You may also be in a very good position (again, check with an attorney) to return the item to the manufacturer or distributor for a full refund or a fully OSHA-safe replacement at no additional cost, or you may be able to force that manufacturer or distributor to bring your machine up to speed (e.g., by retrofitting with a compliant guard). Furthermore, you might be in a position to bring suit against that manufacturer or distributor if your employee was injured as a result of contact with the noncompliant section of the machine.

It is important to note that there are further glitches in your purchase order protective system. These depend on exactly where you intend to use a product and for what exact purpose. You must make certain that the manufacturer or distributor is fully aware of your particulars. For example, you might purchase a droplight that has UL (Underwriters' Laboratory) approval and meets the general requirements of the NEC (National Electrical Code), has a well-insulated heavy-duty round cord, is grounded or double-insulated, and is equipped with a cage. This sounds pretty good, and for the vast majority of cases, it seems like a product that will be OSHA-safe. However, if that droplight, as with most such products, has a conventional metal cage, electrical receptacle in the base of the lamp, and a regular on/off switch, *and it is used in the pit of a four-bay automotive repair shop,* its use would be in violation of OSHA standards. Generally, droplights used near the floor of these types of facilities, where vapor ignition could occur, must have the cage insulated, no receptacle as described, and no switch as described. Further, a droplight used in a pit must be generally of a very special enclosed type.

Consider where you will use other purchased electrical apparatus. A metal ladder might be in compliance with OSHA standards, as it meets the general specs for that type of ladder (A-frame and extension are the best examples). However, if you use that metal ladder in close proximity to exposed live electrical parts, there will probably be a major hazard and an OSHA violation.

Other examples abound. For now, I add just two categories. You may purchase a chemical that can be used in an OSHA-safe manner. However, its use in contact or in near proximity to another particular chemical may result in a catastrophic explosion. Compatibility is always a concern. A particular respirator may be OSHA-safe for use (with all the normal qualifiers) to protect against the respirable hazards presented by a particular airborne toxin. However, it might be totally ineffective and even present a false sense of security when used to protect against the respirable hazards presented by a different airborne toxin, which happens to be the one with which you are dealing.

I have used the fabricated term *OSHA-safe,* and its meaning should be evident. I caution that products are not *OSHA-approved,* however. They may *meet OSHA requirements* if used and maintained as designed and intended, which includes use in the proper location. You should be dubious when hearing salespersons state or when reading in sales literature that an item is OSHA-approved. When considering that your program should cover every*thing* in your facility, individual items such as machines, tools, electrical extension cords, fire extinguishers, forklift trucks, and ladders quickly come to mind. It is important to remember that other *things,* as used herein, include floors, columns, doors, and other elements that generally define the space and construction of the facility.

Every Minute

It does not matter how often or for what duration an unsafe area is entered, an unsafe practice is performed, or an unsafe condition is allowed to exist. There is no acceptable time period for allowing a hazard and violation to exist. An exit that is blocked for 5 min can spell tragedy; fire lacks the ability to tell time. Persons have suffered blindness when operating an off-hand bench grinder without wearing proper eye protection for only a minute or so.

At times, an employer is annoyed when an OSHA compliance officer explains that a citation may be forthcoming related to an inadequately guarded power transmission belt and pulley drive in a small room. The employer argues that there is very little exposure, as the employee only enters the room for a few minutes per day. The problem is that the room is very often remote, congested, dark, noisy, and has a slippery floor due to poor drainage. Other exacerbating circumstances usually include that the compressor starts automatically and that the employee does not enter the room at a regularly scheduled, easily predictable time. So, when the person is in the room, many factors can increase the probability of exposure to the moving parts, and the severity of the injury can be multiplied due to lack of discovery of the victim for a long period of time. It matters not if only a few very limited employees enter the area. The boss should be protected, even if he or she is the only one who enters an area or operates a particular machine and only for a few minutes. Cover all possibilities!

9

Inventory and Control

Inventory

It is extremely important that you keep accurate records of what safety items you have on hand. Know your inventory and control it. Computer records should be backed-up for this and any other safety-related information purpose by disks or equivalent means. This can relate to articles of personal protective equipment (including attachments and components), fire extinguishers, testing and sampling instruments, spill-neutralizing agents, locks and keys, tags, labels, batteries, signs, flashlights, first-aid supplies, and similar things.

There are also items that may not, at first, be thought of as safety related in themselves but that should be stocked so replacements are immediately available. Examples in this category include hoisting/rigging slings and electrical extension cords. Situations could arise, and certainly have arisen, in which a hoisting/rigging sling requires replacement because it is badly frayed. Even if new ones are ordered immediately, poor judgment fueled by a tight and heavy production schedule, which is no excuse at all, results in further risk in the use of the damaged sling. Similar situations arise with electrical extension cords, where the insulation has been severely stripped and split and the grounding prong has been broken off. It is essential to have a system whereby procurement steps are initiated before the items are depleted. As obvious as this may seem, this principle is violated far too often. It is not just a matter of having the supplies available, but a matter of knowing that they are available, where they are, and with what are they compatible.

Control

Labeling and logging are very important not just for the reasons just mentioned but also to help assure that items are not missed during inspections. There are other safety-related reasons as well. If a maintenance person is sent to repair or replace an item, he or she had better deal with the relevant item. Otherwise, the damaged (more important, the unsafe) item is not made safe, and yet word may spread that it has been corrected, creating a false sense of security. This can be deadly when it relates to lockout/tagout, ventilation (particularly for permit-required confined spaces), chemical tank valves, machine actuators, emergency stops, and interlocks.

Individuals might be sent to repair an exit door that sticks badly, the brakes on a forklift truck, a malfunctioning emergency eye fountain, or the hoist-limit switch on a crane. If they address the wrong area or piece of equipment due to a lack of labeling or ambiguous labeling, the consequences can be dire. Further, absolutely clear designations just make life easier. Rather than refer to "the electrical breaker panel a little bit to the left of where Juan usually works," which could be ambiguous, it makes a lot more sense to refer to the "electrical breaker panel P2." Too often, machines have a serial number, model number, and company number. Within your facility, there must be a universally understood way to refer to a specific machine. There might be six machines with the same model number, and I have witnessed other forms of ambiguity that troubled company personnel when attempting to discuss a particular machine. It is even important to know exactly what part of the plant or structure is the subject of conversation and concern.

I recommend a labeling and logging system that uses a nonredundant number and/or letter for each tool, machine, ladder, fire extinguisher, fire hose, alarm, electrical disconnect, chemical tank, sprinkler control, forklift truck, hoist, crane, sling, deluge shower, emergency eye fountain, first-aid kit, material safety data sheet (MSDS) station, and hazardous waste cleanup station. I suggest the same system be applied to each exit, stairway, spray booth, loading dock, mezzanine, and column. Yes, the simple labeling of columns can often help pinpoint a problem area in a hurry. Rooms, including offices and closets, should be unambiguously labeled, but it might be more practical to use words than numbers or letters. The same applies to departments, sheds, shacks, and buildings. Pipes should be labeled as to their contents, indicating whether the pipe transports hot water, cold water, steam, natural gas, cryogenic material, or some other specific chemical. It is preferable to do this for hoses as well, unless the identity of the contents is totally obvious. If relevant, include an appropriate hazard warning.

The hazard communication standard, which is examined briefly later, addresses the labeling of virtually all substances, that is, the containers that store them. That standard generally mandates the labeling of compressed gas cylinders, abrasive (grinding) wheels, and even such things as metals that are to be exposed to welding heat that would liberate a fume. In the case of those pieces of metal, which could be regarded as "articles" when their composition was undisturbed and no realistic potential industrial hygiene hazard had existed, the label or sign can be located nearby. Under the hazard communications standard, the key elements to the label are "identity" and "appropriate hazard warning."

A special problem arises when the container label does not match the material safety data sheet (MSDS). I have been in situations where the label was clear and there was a corresponding material safety data sheet, if only it could be found. The problem of locating it was caused by the fact that the MSDS was filed under a chemical synonym, or it was filed under a chemical name, whereas the container displayed a brand name.

Assure that any and all kinds of labels can be understood by all relevant employees. I have seen labels written in languages other than English. I do not mean in addition to English; I mean that there was no English on the label. I have seen weight capacities labeled in kilograms, linear dimensions in meters, and volumes in liters. If your employees do not understand metric, what is the point? Guesswork on the part of those employees could result in serious injuries.

A special note is due for "flammables" (i.e., "flammable" liquids). "Nonflammable" is generally understood. However, "inflammable" is ambiguous. Many persons think that "*inflammable*" means "nonflammable," but it does not. It means "flammable." The misunderstanding can easily stem from the strange language that we call English. If insane is the opposite of sane, then "inflammable" must be the opposite of "flammable." That is incorrect. I strongly advise changing any labels indicating "inflammable" to read "flammable."

Of course, there are times that labels printed in language(s) in addition to English are a very good idea. In many cases, symbols are valuable. In the case of labels required by the hazard communication standard, diamonds or triangles with numbers do not, in themselves, suffice. They must have an easy-to-understand key adjacent, or the employees must be so well trained that they can explain what the codes mean without going to a chart or asking someone else.

Chapter

10

General Program and Policies

The General Program

The general occupational safety and health program must be concisely documented, readily available, well organized, and precisely worded. This is fundamental. Too often, programs cover entire shelves, and although all of the necessary information is available, it is in a verbose form. Details can be very important, but extraneous language can frustrate and confuse readers rather than enlighten them. Get to the point.

Be certain that all sections of the written program are on hand as needed. They should not be locked away during third shift, and they should not be at the maintenance chief's home. They should not be on loan to your insurance carrier or another branch plant. When the program (revisions included) is printed, make multiple copies. If copies of those masters (in this case, I refer to each original set as a master) are made for any reason, other masters should be available on hard copy and, ideally, on computer. Yet never rely on a computer alone. There must be an adequate number of full-set hard copies.

Contractor programs address lockout/tagout, personal protective equipment, hazard communication, electrical cords, fire protection, ladders, fall protection, smoking, etc. I also endorse visitor rules.

If all of the information exists somewhere in the volumes of material comprising the program but cannot be easily located, a serious problem exists. It is not a matter of information-seekers being lazy. They should not have to spend a great deal of time looking up a particular policy or rule. The program (including the contents of individual books and sections within books) must be organized in a simple, logical fashion. No one should be saddled with the tedious obligation to run through a million cross-references and codes. Where is the section that addresses sampling and testing procedures? Where is the section that addresses accident investigation procedures? Where is the section that addresses standards and rules concerning the use of flexible electrical cords? The answers to these questions should be evident with minimum research.

Precise wording is absolutely necessary. It is not just a matter of being concise. It is more than a matter of the number of words used. It is choosing those words wisely. There must be no misunderstanding. Get to the point following a logical sequence and in terms that can be—will be—comprehended without a lot of head scratching. Spell it out, so there is no doubt. Even in safety-related memos or simple correspondence, the form and contents are critical. Spelling and grammar do matter. When a document appears to have been written and printed in a haphazard way, it sends a signal that the author lacks interest and that, maybe, the contents are not too important. I am not suggesting that an English teacher be consulted regularly but merely that adequate time and effort be utilized to formulate the correspondence.

For letters that do not involve complex issues or several different issues, it is preferable to strive for a one-page limit. Naturally, this cannot always be accomplished. There are no extra points for extra words, though. Open with the purpose of the correspondence. The first paragraph should clearly and simply state what it's all about. Concisely explain what you want the readers to know and what you want them to do as a result of having that knowledge. Avoid unnecessary jargon. Shoot for simple words and short sentences.

If there is any doubt as to the clarity of the message to be conveyed regarding a technical situation, consider asking a person who is not technically oriented to read it. That individual will not necessarily understand the whole process that is the subject of the message but should be able to state what you expect of those who read it. He or she should be able to explain, for example, where the reader is supposed to store a widget and what button on the widget machine should be pushed if there is an emergency, even if your special reviewer does not know exactly what a widget is.

No matter how attractive your program is, do not lose track of its purpose, which is to identify potential hazards, make thorough assessments, and act to eliminate, mitigate, or control those hazards. The next step is to gather feedback and determine the degree of success that has been achieved and is likely to be maintained. It may then be necessary to react by implementing another plan or by simply improving the original. Effectiveness audits are indispensable.

The central underlying objective of an occupational safety and health overall program is to minimize risk. In turn, all elements of the program should be directed toward one unqualified goal: accident prevention. The following are key elements: instilling the proper atti-tude and assuring the proper behavior, a high degree of active employee participation engendered by the management-labor part-nership approach, delineating responsibility and authority, unam-biguous written policies and rules consistently applied, training and education, personal protective equipment, accident investigation, pro-gressive discipline consistently enforced, preventive maintenance, inspections to recognize and abate hazards, setting priorities, accountability and evaluation of management, and effective labor-management safety and health committees.

Formal job hazard analysis can also be a very good tool. Incentive plans can yield excellent results, but only if the reporting of all injuries and illnesses is strictly required. It is preferable to require the reporting of all accidents, even if no injury or illness occurred. Of course, if a total integrated safety program is properly functioning, the number of injuries, illnesses, and even close calls will decrease. This may not always be the case in the short term, but it most cer-tainly should be in the long run.

Other important elements include medical case management (with an emphasis on ergonomic-related cases), employee assistance pro-grams, and wellness programs. This chapter addresses written poli-cies and rules. I have devoted separate chapters to each of the other elements just mentioned.

Policies

There may be some quick fixes that are called for, but the goal is to view potential safety problems and methods to deal with them from a perspective that looks far ahead. Strategies should ultimately focus on achieving sustained improvements that will continue to narrow the gaps and traps that frustrate the efforts to go accident free. The long-term performance of the safety program must always be kept in mind. Think ahead. Predict risks and, as much as possible, eliminate them. Be proactive and coactive, not just reactive. Do not be lulled into thinking that you have a great overall program because you have a great accident investigation system. Yes, you need a great accident investigation system. However, the higher priority is to have a great accident prevention program.

Too many companies permit their safety program to be driven pure-ly by mod rates and pie charts. This type of program tends to be dys-functional. These companies suffer the disability of tunnel vision and only see neatly diagrammed statistical indicators. All too often, they overlook accidents waiting to happen because they have concentrated too keenly on the types of accidents that have already occurred.

The correct approach is not to lean too heavily on the blinding numbers gleaned from accident frequency and severity data. When tracking trends, the analysis should span considerable time periods (i.e., years) so that patterns can be discerned and common causes uncovered. That does have value, but the main thrust should be to observe and measure adherence to proper safety conditions and practices.

Look upstream to evaluate the frequency and degree of conformity and nonconformity. Many psychologists speak of the superiority of influencing employees to go toward positive reinforcement as opposed to away from negative reinforcement. Thus, it would appear that it is better to discuss or publish information on the degree of conformity rather than the degree of nonconformity. In general, I agree. Nevertheless, everyone in the company must know that 80 percent or even 95 percent conformity is still not totally satisfactory. This is most important when considering an employee who sustained a severe injury. What relief does he or she find in the fact that there was even 99 percent conformity?

Regarding OSHA inspections, employers often hastily point out to OSHA how many things the company has correct (i.e., in compliance). That only means that the company will be cited for fewer "things"; the 1 percent that is not in compliance will not be erased by the 99 percent that is in compliance. So praise the conformity, the safe conditions, and the safe practices, but aim to identify what still has to be abated, and then get it done.

How much risk remains? In what specific areas is there nonconformity that allows unacceptable risks to exist? Seek long-term solutions to hazards and even potential hazards that have been identified through diligent scrutiny.

Maintaining a safe and healthful workplace and workforce is a matter of good business. Just as with production and quality, safety should be managed very carefully. Essentially, the safety program and system should be handled as an extension of the company's TQM (total quality management) process. With TQM, systems and process are put into motion to foster excellence. That is far more complex and lasting than simply plugging leaks. Long-term goals are emphasized. Enthusiastic leadership and participative management should be continually in view, and continuous improvement should be sought. When and if it appears that compliance with OSHA standards and good practice have been achieved, the next step is to look deeper.

Seek the help of employees in solving problems, improving processes, and so on. As with all TQM applications, drive out fear so that employees report hazards and potential hazards, and seek solutions, without even being asked to do so. Teach them the value of brainstorming, divergent thinking, being creative, and being innovative—team-based problem-solving skills fostered by the Odyssey of the Mind program. It would serve an employer well to learn more about the development of these stimulating and sometimes spontaneously exhibited proficiencies embraced by this international program.

Odyssey of the Mind is concerned with how atypical or surprising ideas are received by others. Do not allow mocking, sarcastic, or insulting criticisms of ideas. That reaction can be very disconcerting to contributors, who may effectively be stifled, view themselves as unworthy of bringing up another suggestion or observation, feel humiliated and angry, and turn away from the whole process. It is quite different, and acceptable, for a listener to introduce opinions or information as to why an idea might need tweaking or why it may not be practical in its current form. Offered with the proper tone, everyone can feel comfortable to at least investigate the possibility of building on the idea or spinning off from it. Some wonderful solutions to problems of all kinds have started with some seemingly wild or strange notions. Explain the concept of synergistics.

Traditional, time-honored, successful methods of hazard prevention and hazard abatement must be considered, but do not be afraid to be progressive. That is not to say that all movement or change yields positive results. It does not mean that all experiments, even those that have been well-designed and gauged against real-world situations, will succeed quickly if at all. It does mean that skepticism has a rightful place in the formation or evaluation of safety programs and concepts. It does mean that open minds can flourish on safety teams. Forget the expression "it has always been done that way." Ask why it was always done that way and how it should be done now. Maybe the new idea was tried before and did not prove beneficial. Well, with a few adjustments and a review of the lessons of past failures, it may be time for another try. All right, so the risks have been lowered. Why can't they be lowered further? The number of injuries, illnesses, accidents, and near misses has been lowered. Why can't it be lowered still more? No accidents are tolerable. No number other than zero is acceptable. This is not to suggest that a company is doing a good job of creating a total safety culture only if there are no accidents. It means that zero accidents are the goal.

The International Organization for Standardization

Companies are starting to incorporate their safety and health program (in fact, the whole integrated safety and health system) into their ISO program. ISO is a series of generic quality management and quality assurance standards issued by the International Organization for Standardization, which is centered in Geneva, Switzerland. Companies that meet ISO standards—that fulfill all of the requirements for certification—are deemed to have proven that they can be relied upon to consistently ensure the production of very high-quality products. ISO is most often associated with its 9000 series quality standards, which are designed to aid manufacturers in building and maintaining superior quality systems throughout the company. The ISO 9000 system follows a product through every single phase, from design and development inception, through production, inspection, and servicing. In a simple sense, if a company is ISO registered, customers can trust that they will receive no less than what was ordered. The part or total product should be defect free and ready to go; it should be in perfect working order and on the money as to its being built to specifications. The registration only comes after a third-party approval following the in-depth ISO audit. Registrants are then audited regularly to assure that they uphold their high standard of quality assurance processes.

To tie in our subject with ISO, the elements of close operations reviews are implemented to absorb safety and health considerations at, for instance, every microstep of a product's life cycle at which an employee can be involved. The chief ways that safety and health are inserted into the certification process are by way of employee orientation, training, and documentation of that training. Safety and health becomes an integral part of the quality objective. As a main avenue for this inclusion, ISO 9000 includes an employer obligation to thoroughly document the correct, detailed methods of machine operation. Thus, there is a natural relationship between ISO 9000 and safety and health.

Safety Throughout the Company

Safety should be designed into products, machines, processes, systems, and so on. Potential risks in all these regards should be anticipated at design conception. In fact, safety in general should be built into the whole demeanor of the company. It should be totally integrated into the mainstream of the organization. The safety program should not be superimposed, like a template, over parts of the company's structure. It must be inextricably interwoven into the lifeblood of the company's way of doing business. Ironically, some companies purposely avoid the labeling of distinct safety rules or references because they absorb safety directly into standard operating procedures. The idea is that safety cannot be separated from any aspect of the work. It cannot stand alone. That type of sophisticated philosophical manifestation can be a bit confusing, but the foundation for the integration is solid.

I certainly believe in a detailed, written safety program, but it should not be thought of as emanating from an isolated segment of the company or applicable to only certain aspects of the business at hand. The desire to maintain a safe and healthful workplace must be ingrained in all personnel, in all facets of the business. That mind-set must pervade every decision-making process that can be remotely related to a need for risk assessment. At every juncture of that process, thought patterns should be permeated with the sincere desire to consider safety ramifications automatically. That desire should be instinct induced. To think of safety considerations must be second nature. No one should have to be reminded by a safety director, for instance. That is the essence of the argument.

"Work safely" should not just be a cheer during special campaigns, drives, or seasons, nor the fallout of a tragic accident. Safe conditions and safe practices must be in effect on a daily basis. Create a climate in which the employees want to contribute and in which they eagerly anticipate every opportunity to do so. Then, management should maintain credibility with employees, causing an upward spiral of good faith and definable, concrete prosafety actions by those employees. Promote ongoing teamwork among all persons and hierarchal levels. Strive to make it a matter of routine to turn to line employees when even considering significant changes or the initiation of new processes. Instill in them the secure feeling of empowerment with which they will perform at a higher level as individuals and as team members.

Management is not surrendering its authority by fostering such dynamics. It is simply using the experience, interest, and energy of employees as a living tool to clear the path and pave the way for risk reduction. This is neither abdication nor exploitation. So there is no misunderstanding, solicit the input of employees in the planning stage; do not wait to herald their spirit of cooperation after a management decision has been finalized and changes are about to be implemented. Management must continue to maintain a posture whereby it will plan, monitor, and lead.

Policy or Mission Statements

At the beginning of any worthwhile written safety program, there should be a policy or mission statement. The entire program may be too lengthy to include in the company's general handbook. If considering such specialized and detailed step-by-step sections and regulations as those that cover lockout/tagout and permit-required confined spaces, the program would certainly not be included, in its entirety, in the general company handbook. I am referring to a document of several pages that is given to all employees upon hiring. That sort of document typically addresses work hours, rates of pay, attendance, union matters if relevant, benefits, vacations, parking facilities, occupational safety and health, and similar across-the-board issues. That document should include at least the safety-related policy or mission statement. Then, general safety rules would be addressed with a clear reference to the overall program. If such a statement does not appear very close to the beginning of the handbook, then the wrong message has been sent from day one.

Model policy or mission statements are not difficult to come by. You may prefer your special twist on the statement. To start you out, the Occupational Safety and Health Administration has printed a model policy statement. It is most interesting that so few employers have actually seen this statement. There is nothing wrong with any company copying it, adapting it, or customizing it. I have, in effect, already implied some modifications, but the OSHA sample serves as a fine example. It follows:

> The Occupational Safety and Health Act of 1970 clearly states our common goal of safe and healthful working conditions. The safety and health of our employees continues to be the first consideration in the operation of this business.
>
> Safety and health in our business must be a part of every operation. Without question it is every employee's responsibility at all levels.
>
> It is the intent of this company to comply with all laws. To do this we must constantly be aware of conditions in all work areas that can produce injuries. No employee is required to work at a job he or she knows is not safe or healthful. Your cooperation in detecting hazards and, in turn, controlling them is a condition of your employment. Inform your supervisor immediately of any situation beyond your ability or authority to correct.
>
> The personal safety and health of each employee in this company is of primary importance. The prevention of occupationally-induced injuries and illnesses is of such consequence that it will be given precedence over operating productivity whenever necessary. To the greatest degree possible, management will provide all mechanical and physical facilities required for personal safety and health in keeping with the highest standards.
>
> We will maintain a safety and health program conforming to the best practices of organizations of this type. To be successful, such a program must embody the proper attitudes toward injury and illness prevention on the part of supervisors and employees. It also requires cooperation in all safety and health matters, not only between supervisor and employee, but also between each employee and his or her co-workers. Only through such a cooperative effort can a safety program in the best interest of all be established and preserved.
>
> Our objective is a safety and health program that will reduce the number of injuries and illnesses to an absolute minimum, not merely in keeping with, but surpassing, the best experience of operations similar to ours. Our goal is zero accidents and injuries.

Our safety and health program will include:

—Providing mechanical and physical safeguards to the maximum extent possible.

—Conducting a program of safety and health inspections to find and eliminate unsafe working conditions or practices, to control health hazards, and to comply fully with the safety and health standards for every job.

—Training all employees in good safety and health practices.

—Providing necessary personal protective equipment and instructions for its use and care.

—Developing and enforcing safety and health rules and requiring that employees cooperate with these rules as a condition of employment.

—Investigating, promptly and thoroughly, every accident to find out what caused it and to correct the problem so that it won't happen again.

—Setting up a system of recognition and awards for outstanding safety service or performance.

We recognize that the responsibilities for safety and health are shared:

—The employer accepts the responsibility for leadership of the safety and health program, for its effectiveness and improvement, and for providing the safeguards required to ensure safe conditions.

—Supervisors are responsible for developing the proper attitudes toward safety and health in themselves and in those they supervise, and for ensuring that all operations are performed with the utmost regard for the safety and health of all personnel involved, including themselves.

—Employees are responsible for wholehearted, genuine cooperation with all aspects of the safety and health program including compliance with all rules and regulation—and for continually practicing safety while performing their duties.

11

Specific Programs Required by OSHA

Several OSHA standards obligate the employer to develop and maintain written programs on specific subjects. These programs are generally tied into training standards. Not all of these standards and programs are relevant to each employer. In other words, they are required only if you are involved in the pertinent activities. The most common of these programs are the following: hazard communication, lockout/tagout, electrical safety-related work practices, permit-required confined spaces, hazardous waste operations and emergency response, bloodborne pathogens, emergency action plans, and fire prevention plans.

There is a curious nuance to some standards that are thought of as performance standards. The best example involves the requirement to assess the need for personal protective equipment and to train the employees who wear the protection. The employer must have a certification of the assessment and training, but that does not mean that you are required to show OSHA the actual assessment or training programs. This is a fine point, but it can have significant legal implications. It would be wise to have the programs available, as long as they were developed and implemented properly, and your company has complied with the standard. Regarding personal protective equipment, in general, one would be hard-pressed to find an employer (certainly any type of factory, in loosely defined terms) that is not covered by the hazard communication standard and the personal protective equipment standard.

Additional required programs involve respiratory protection, hearing conservation, medical surveillance (which can include substances such as lead, asbestos, cotton dust, acrylonitrile, ethylene oxide, arsenic, cadmium, coke oven emissions, and 1,2-Dibromo-3-chloro-propane), ergonomics, laboratory chemical hygiene, and process safety management. In the overwhelming majority of situations, medical surveillance is not to be carried out as a single reaction to an incident or as a one-time spot-check for overexposure. A goal is to discover any health problem as close to onset as possible.

Medical surveillance programs, then, are to be an ongoing, regular process. The process does not only include direct health monitoring. It encompasses a balanced approach to avoiding overexposure to particular substance(s) or groups of substances. This combines personal monitoring with ambient level testing, training the employees in safe behavior, the consistent and effective use of personal protective equipment (if applicable), and the design, installation, and use of engineering controls.

There are numerous standards that relate to programs, but more closely in the form of inspections instead of narratives. The most notable regard mechanical power presses, overhead cranes, and slings. As with general programs, it is paramount that your specific programs are concisely documented, readily available, well organized, and precisely worded. At times, the employer can "max out" of a standard by changing operations, procedures, equipment, chemicals, or volumes of particular chemicals. In many cases, hiring outside contractors to perform certain work can remove your employees from exposure and remove your company from coverage by a particular standard. I urge caution. Whether or not you are covered by OSHA for a certain standard and whether or not your employees are in jeopardy, you should still want the contractors to be protected. In addition, you should still want to assure that those contractors function in a manner that would not allow your employees to be harmed as a result of a contractor's deeds or omissions.

The Need to Review Programs

Programs must be reviewed regularly, sometimes as dictated by the standard but always as a matter of good practice. Have new processes or equipment been added? Have procedures changed? Have new chemicals been introduced into the workplace, are old ones combined with each other for the first time or in a different way, or are old ones exposed to greater heat? Have you ceased contractor assignments, for instance, in favor of your personnel doing emergency cleanup of hazardous spills, electrical work, or work in permit-required confined spaces? Have you increased the volume of germane in-house chemicals? Surely, if accidents occur and are in any way tied into these programs, a thorough review of your specific program is called for.

12

Personal Protective Equipment

The absence of wearing personal protective equipment is often cited by OSHA. Citations frequently involve the eyes and face (impact and chemical concerns), the ears, the respiratory system (there is an in-depth, dedicated standard for this), the feet (usually relating to toe impact, occasionally for metatarsal impact, and sometimes for chemical concerns), the hands (cuts or chemical concerns), and the head. Other citations have addressed legitimate hazards of hand knives or chemicals, where aprons (metal mesh or leather in the first case; material properly impervious in the second case) are required.

Occasionally, there are other types of personal protective equipment required. I have addressed the importance of assessing the need for personal protective equipment and training employees in its proper use. Keep in mind that the training includes not only general wearing instructions, but also how to don, doff, and adjust it, as well as the particulars of the equipment's limitations, storage, handling, sanitation, care and maintenance, useful life, and disposal. Then, upon evidence of lack of compliance, retraining should result.

Who Pays for Protective Equipment?

Where the wearing of the equipment is required by OSHA, employers have the obligation to assure that it is worn whenever and wherever OSHA dictates. You are not in compliance simply because you make the equipment available. OSHA has interpreted its general personal protective equipment standard, as well as specific standards, to require employers to provide and to pay for the equipment needed by the company and for the employees to do their jobs safely and in compliance with OSHA standards. However, if equipment is very personal in nature, and is usable by employees off the job, the matter of payment may be left to labor-management negotiations.

Examples of personal protective equipment that would not normally be used away from the workplace include welding gloves, wire mesh gloves, respirators, hard hats, specialty glasses and goggles (e.g., those designed to protect against the hazards of lasers or ultraviolet radiation), specialty foot protection (e.g., metatarsal shoes or lineperson's shoes with built-in gaffs), and face shields. Also in this category are rubber gloves, blankets, cover-ups, hot sticks, and other live-line tools used by power generation employees. Examples of personal protective equipment that OSHA considers to be personal in nature, and often used away from the workplace, include nonspecialty safety glasses and safety shoes (generally with ANSI-approved toe protection).

If shoes or outerwear are subject to contamination by carcinogens or other toxic or hazardous substances and cannot be safely worn off-site, the employer must pay for them. In the case of safety shoes, the common practice is for the employer to pay for up to two pair per year or to subsidize (usually contributing $50) the cost of the shoes. This can be done by reimbursement against a receipt or, for instance, by arrangement with a company that brings a shoemobile to the workplace. Regarding nonspecialty safety glasses, the employer almost always pays. I recommend that employees requesting replacements for equipment that has been damaged in the line of duty be obligated to turn in that equipment. There is no OSHA requirement that the employer pay for industrially safe (ANSI-approved for impact) prescription glasses. However, the employer generally subsidizes that equipment.

Monitoring Compliance

It is often easier to monitor compliance if you require particular personal protective equipment for broad areas, although OSHA might not require it of all employees in all portions of those areas. That does not mean that an employer should rush to put everyone in the facility in a hard hat. If there is some concern about a hazard, you should opt for the protection. Yet, with the hard hat as an example, there can be problems if you are blindly requiring the protection. If your facility is extremely hot by the nature of its normal operations, and there are no realistically discernible hazards that can be mitigated by the wearing of that head protection, then you should think twice about requiring it. You may be adding a heat stress hazard. Similarly (and this type of concern must be covered during training), an employee may wear one type of heavy gloves when handling sharp sheet metal and a different type when handling chemicals that can be corrosive to the skin. Fine.

Just make sure that those gloves are not worn, for instance, during the operation of a drill press or off-hand grinding machine.

Eye Protection

In many cases, the wearing of personal protective equipment is a bit uncomfortable, but the pros outweigh the cons. Lack of comfort is another matter, and it can be dealt with in various ways. Give employees a choice of industrial eye protection styles and safety shoe (ANSI-approved) styles. I do not suggest that you necessarily give totally free reign. You need to exercise some control, but choices can make a difference in actual comfort and even in perceived comfort.

If an employee wears prescription glasses and works in an area where ANSI-approved industrial safety glasses (for impact) with side shields are required, there are two choices. One is that the employee wear prescription glasses that are ANSI-approved for that purpose. The other is that the employee wear the relevant ANSI-approved goggles over the regular prescription glasses. If eye protection is only needed for short periods (e.g., when making a few cuts on a radial saw), the goggles could work out okay. If eye protection is needed continually, it is unreasonable to expect an employee to wear goggles for long periods. The general discomfort, heat and humidity buildup, and fogging (even if treated) can lead a person to remove the goggles when they should be worn or can even cause a hazard related to lack of vision and/or attention. In any case, install lens-cleaning stations.

Another problem arises when goggles are used to protect against chemicals entering the eyes. Sometimes, employees are given goggles that have ANSI approval, but that approval is for impact and not for chemical splash. In such cases, the vents on the goggles are often on the top, and configured and positioned so that a splash onto or near the forehead can easily enter the vents and then the eyes.

Hearing Protection

When considering different types of hearing protection, consider the general cleanliness of the areas in which your relevant personnel work. If the hearing protection will be going in/on and off frequently, you do not want to have employees with dirty hands frequently pushing plugs into their ears. Muffs or other options may make more sense. The muffs can be undesirable when it is very hot and humid in the plant. Of course, all types of hearing protection can be somewhat uncomfortable, but again, the pros can certainly outweigh the cons. Still, the choice of which exact equipment to use should be considered. In fact, if plugs are used and inserted correctly, the dirty hands can be more of a concern.

I have often seen employees simply taking disposable plugs out of the original small bag and sticking the plugs into their ears. There was a minimum of handling. In most cases, instructions on the bag or on accompanying literature indicate how to take the plug and roll it between the thumb and index finger, then stretch the ear lobe gently up and out by reaching with the opposite arm over the head and pulling the lobe, and then inserting but not letting go of the lobe for a few seconds so that the plug could expand in the ear canal as designed.

Other Types of Protection

Too often, boots used to protect against chemicals, even molten metal, are worn with the pants tucked in. Wear pants over boots, or use spats, to avoid the boots' becoming receptacles. For hard hats, those with rear-mounted ratchets for suspension adjustment can be a lot easier to use and fit than the type without such adjustments. Naturally, the ratchet adjustment must be checked regularly to assure that it will not slip. Although hard hats should not be generally shared unless they are sanitized before passing on, whenever they are shared, the suspension will most likely require adjustment for the proper fit.

"Engineering Out" Problems

The professional approach is to try to avoid or reduce the need for personal protective equipment. This would be done, for the most part, by work practice and by "engineering out" problems. For example, seek to lower noise levels at the source so that hearing protection is not needed or so that source hazards are mitigated. Another example of engineering out is to provide properly positioned ventilation systems (often localized) that exhaust vapors or fumes so that respirators are not needed. A warning concerns that positioning. Effective exhaust ventilation may adequately protect employees spraying in a booth, welding, or working at a plating tank. However, if that employee shifts position so that he or she is now between the danger source and the exhaust, the hazard can be grave. This is not a matter of merely neutralizing or negating the value of the exhaust system, which can be a major problem in itself. It is a matter of completely turning the value of the exhaust against the employee. The employee is in a position where the exhaust is actually contributing to his or her ill health by drawing the hazardous vapors or fumes into the breathing zone.

Respiratory Protection

It is not my intent in this book to go into great detail concerning specific hazards or standards. Nevertheless, some special comments about respiratory protection are called for due to major misunderstandings by many employers. OSHA can obligate a written respiratory protection plan, even if one employee wears a cartridge respirator. The full plan applies if the permissible exposure level (PEL) is exceeded, or you *require* employee(s) to wear respirators even if below the PEL. If you *allow* (not *require*) an employee to wear a respirator, but the PEL is not exceeded, medical evaluation, storage, and cleaning elements of a written program must be in place. Tight-fitting face pieces must be fit tested at least annually, if the PEL is exceeded or you require employees to wear respirators in any case. Note that facial hair can impede seal integrity.

Respirators are a complex matter, almost like a machine. Some of the concerns involve sanitation, storage, proper use, compatible parts, medical suitability, proper types of filters, and of course training. Use of the wrong respirator component(s) or the improper use of any respirator can result in very serious hazards, often exacerbated by a false sense of security on the part of the wearer. Such false security often causes wearers to bring themselves closer to the danger source than they would normally be. The wearing of any respirator can place an added burden on the lungs. It also poses the threat of dermatitis or, if a respirator is shared, which should definitely not be the case, the transmission of disease.

Appropriate Clothing

What clothing is and what clothing is not generally appropriate are usually evident. Three-inch platform heels would be a bad idea when working at the plating tank! There is a particular group of concerns that OSHA has addressed on citations, and well they should. There are many situations where loose, droopy, billowy clothing, jewelry, and unrestrained long hair must be avoided. The concern is usually that a person can be pulled into a machine or tool with catastrophic results. Do not place hair in a conventional ponytail. The chances of the hair being grabbed may be reduced, but if the hair is grabbed all of it can be pulled in. Good examples of machines and tools where these concerns are most relevant are lathes, drill presses, grinders (except meatball), portable drills, rotating saw blades, and wherever in-running rolls (or materials running onto rolls) are present.

Be cautious about claims of "isolated incident" or "employee misconduct" in cases where OSHA documented that the required personal protective equipment was not worn. Such defenses are very seldom successful. I address the negative implications of this type of defense later.

Preventive Maintenance

The Maintenance Staff

A fully staffed and totally competent maintenance staff is a necessity. The size of that staff can vary greatly depending on operations, square footage, general occupancy hazards, and specific equipment to be serviced. Some state-of-the-art machines, ironically, can be finicky enough to need a better trained maintenance staff than would be needed for very old machines. On the other hand, although the types of repairs and the attention needed for older machines may be less sophisticated, the frequency of monitoring and hands-on work can be incredibly frustrating. Either way, the maintenance staff has an important role in assuring the safe condition and operation of machines, miscellaneous equipment, and building structure.

If your staff can fix it right, that's a good thing. Better yet would be to take every feasible step to assure that fixing is not required. Of course, some of the factors that can lead to breakdowns and damage are out of the control of the maintenance staff. Yet, for maintenance purposes, the most important strategy is to devise and diligently follow a strict regimen of preventive maintenance. This cannot be exaggerated.

Preventive maintenance traditionally involves such mundane tasks as lubrication, tightening (occasionally loosening), and tweak-type adjustments. Regular inspections of equipment should not be overlooked. Where testing is relevant and can be accomplished in a practical manner, it should be performed. Some inspections are so informal and frequent in nature (a careful tug here and there when passing by, perhaps) that logs are not needed or expected. In many cases, logs are required. These logs should always be dated and signed. In some instances, even the time of day is important. If it was important enough to record, then responsibility should be assigned. The date, possibly the time, and signature help emphasize that responsibility; checkmarks are not really adequate.

Periodic Testing

OSHA requires inspection records in many cases and testing in others. Testing can be a form of preventive maintenance. If you do not believe that the definition suits, then testing is at the very least a key related mode of attack to uncover the problem before it is too late (e.g., after the amputation, electrocution, blindness, explosion, or fall). So semantics aside, whether such testing is conducted by the maintenance staff or others, it is essential.

Required by OSHA or not, some of the more important items to test are the following: interlocks, presence-sensing devices, pullbacks, restraints, emergency shut-off buttons, stop-actuating cables, ground fault circuit interrupters, quality of electrical grounds (quantifiable), emergency lighting, vehicle horns and backup alarms, emergency eye fountains, deluge showers, manometers, drains, fire alarms, and evacuation alarms. The testing must be performed under monitored and controlled conditions to avoid creating a false sense of imminent danger and to avoid causing an accident.

Some other tests require extra diligence and planning to avoid an ironic tragedy. It should be evident that some of these tests would be best suited for the industrial hygiene staff or the company's equivalent, as opposed to being a regular duty of the maintenance staff. In many cases, it would be prudent for maintenance personnel to accompany the industrial hygienist so that, in the event of an apparent emergency discerned by the testing, an emergency shut-down or adjustment can be accomplished promptly. Examples include the following: high temperature, overflow, leaks (liquids, vapors, fumes, gases), oxygen deficiency, excessive oxygen, concentrations above the lower explosive limit (sometimes referred to as the lower flammable limit), and dangerous levels and/or concentrations of various airborne toxics. Among other tests that require extra care are braking systems for vehicles (including cranes), various material handling equipment concerns (capacity being a major factor), and the adequacy of devices to reduce or preclude fall hazards (e.g., nets, safety belts or harnesses, lanyards, life lines).

An additional type of preventive maintenance—if you really break down the words—can be accomplished by any employee. This does not refer to mechanical or technical concerns. It involves unimpeded access to fire extinguishers, fire hoses, exits, main aisles, alarms of any type, electrical disconnects (including breaker panels), first-aid supplies, fire blankets, emergency respirators, hazardous waste cleanup stations, emergency eye fountains, and deluge showers.

Eye Fountains and Deluge Showers

Regarding emergency eye fountains and deluge showers, I recommend that each unit (at least each plumbed unit) be equipped with an alarm on a spring-loaded bypass. The purpose of the alarm is simply to alert others of the emergency. It is important that employees promptly respond to assist the employee who has been sprayed, splashed, or otherwise contacted by the dangerous substances. The bypass allows employees to easily test the units without setting off the alarm. If the bypass is not there, employees might be reluctant to conduct the test, feeling that it takes too much effort to alert all of the relevant persons to the test. As a result, the inadequacy of the flushing system would go undetected. By having that bypass on a spring-loaded system, the person who conducts the test cannot fail to reset the alarm; it is reset automatically.

When inspecting the eye fountain (i.e., eye-wash station), note that there should be two streams meeting at a vertex for a constant flow of at least 15 min, with controlled temperature and pressure. Upon initial activation, caps or similar devices, which had been in place to help avoid a blast of dust or dirt, will be displaced. Immediately following that initial activation, the flow will continue without the need for constant foot or hand pressure. There must be a clear sign to indicate the location of the unit. Note that, if OSHA would cite relating to the need for an emergency eye fountain, a squeeze bottle would not suffice. If you are using a nonplumbed device, your inspections will have to assure that the unit is adequately filled. Usually, there is a line to indicate this. In these cases, if you do test the unit, you will probably have to replace some liquid. With these types of reservoirs, biocides may be needed.

For deluge shower requirements, a regular bathroom or locker room type of shower will not suffice. Such units require the adjusting of water temperature and pressure. Further, the head must be pointing in the correct direction.

Material Safety Data Sheet Stations and Chemical Storage

Material safety data sheet stations must be readily accessible. This is partially to avoid a frustrated employee who cannot easily access the material safety data sheet from taking a guess and using a chemical in a dangerous fashion. For instance, an employee may rely upon his or her faulty memory and create a dangerous compound of noncompatible chemicals. Another reason to keep the material safety data sheet station clear is so that, in an emergency, the sheet can be tossed into the ambulance and provide valuable information to the medical staff. Relating to this issue, various employees can perform another type of nonconventional preventive maintenance. If an employee notices that noncompatible chemicals are stored together, he or she should bring that fact to the proper authority at once. It is understood that not all employees will be able to recognize that situation, but proper training should put many employees in a position to do so. For example, do not store flammable liquids with chlorine, bromine, nitric acid, sulfuric acid, or oxygen. Additionally, you might consider substituting safer chemicals to eliminate various hazards associated with noncompatibility, or concerns regarding vapor ignition, or very high or very low pH. Similarly, you should consider substituting or modifying procedures and using different processes, tools, or machines where those in use could not be reasonably made safe.

Gauges

Another tidbit regarding preventive maintenance concerns gauges. It is quite common, in my experience during inspections, to find gauges that are missing bezels. The reason that this can be a hazard is that, if the stylus or general mechanism is damaged, a false reading may be given. Such a reading may result, for instance, in an employee boosting air pressure or the level of liquid in a tank far beyond the safe limit. When conducting preventive maintenance tasks, check those gauges and replace missing or damaged bezels. It is a trite but accurate adage that "an ounce of prevention can be worth a pound of cure." The impact is trebled where safety is concerned.

Safety and Health Committees

Safety and health committees can be extremely valuable. Sure, that is a fair premise, but far too often, such committees are basically cosmetic. To be of value, they must be run effectively and efficiently, and they must be taken very seriously. They cannot be viewed by management as a necessary evil and a tool of the bleeding hearts liberal political establishment. They cannot be viewed by line employees as a management charade to make it appear to the government and to naive employees that the company really cares. These employees must also not view the committee as a simple opportunity to get some time off from the rigorous duties of the regular job tasks of the day.

Of utmost importance, if a safety and health committee has been operating for several months and the number of occupational injuries and illnesses has not dropped, there is a problem. The one exception would be if the injury and illness rate was extremely low when the committee went into effect. Does that mean that the committee was not needed? No. The committee may well be doing a wonderful job. There may be fewer accidents and near misses occurring, and the committee may be engaged in dialogue or effectuation of action that is far-reaching and serving to assure that the good record is not reversed.

If the committee is not respected, there is less likelihood that employees or management will go to that committee to submit complaints, point out hazards, or try to work with the committee in any regard. Anyone who submits an idea or gripe that has any relevance to the committee should get an initial written response, and if action or even consideration goes beyond the initial review, additional written responses should be received until the matter is brought to an apparent, successful conclusion.

The workings of a safety and health committee can help bring about improvements in the safety and health conditions and culture of almost any facility. Committees can be assertive promoters of safety ideas that go beyond the legal standards. They can play a pivotal role in the establishment and proven value of special programs, such as those involving contests, incentives, suggestions, slogans, bulletin boards, posters, handouts, and so on. Although you may have an executive safety and health committee, it is imperative to have a labor-management safety and health committee, which should be your main committee. Essentially, that is the one that I am addressing.

Committee Membership

There should be at least as many line or nonmanagement employees as management representatives. Virtually all laws requiring committees call for this composition. The nonmanagement members should be selected by the nonmanagement employees of the company. The management members should be selected by the employer. All of the members should be given training specifically related to how to work on such a committee in a cooperative, productive fashion. They must receive some training in occupational safety and health as well. What are they to look for in conducting a fruitful accident investigation? What are alternative approaches to different types of workplace safety and health issues? How do they conduct a walk-around inspection? Where do they look and what are they looking for?

It is solid logic to stagger terms of office so that not all experienced committee members are replaced at the same time. Besides the standing members, you may wish to enlist a volunteer from a different department or discipline at each meeting and then keep the rotation flowing. The enlistee brings a fresh set of ideas and finds out what is really going on behind those doors where colleagues are supposedly formulating plans to help protect so many co-workers.

The committee must have at least one member who has significant authority in the company. I do not mean just a supervisor. I do mean a safety director, a vice president, or person on a similar level. Thus, there is someone present who does not have to go through three levels of higher management to discuss each potential action to be taken as a result of the meeting. There must also be at least one person who, simply put, "knows how to fix things." This individual does not literally have to know how to fix everything, but he or she must be able to address the feasibility of suggestions that involve physical construction, repair, or adaptation. In other words, it is not enough to have a "money person" on the committee. There must be someone who, in most cases, will at least know if a widget can be attached to a "whatsit" and if such action would make for a safer machine, piping system, electrical safeguard, or whatever. Also, include reps from all shifts.

Committees and Labor Unions

Everyone on the committee must be paid at their regular (or average) rate of compensation for their time spent in committee meetings or when doing authorized, related research, inspections, reports, or other work. If there is an NLRB-authorized union (organized labor), the labor-management contract must be scrutinized for references to committees. In fact, there are numerous aspects of the overall safety and health program that could relate to that contract. For instance, matters of discipline, personal protective equipment, and the involvement of safety personnel directly employed by the international union must be considered. Many plants have a safety steward. Larger plants have safety stewards for different areas and/or different shifts.

By way of collective bargaining, committees have often taken a role (ideally, earned a role) in finding solutions to safety and health disputes. The committee can take the point in examining the merits of all sides to the dispute in a fair and timely manner devoid of considerations of internal politics. In so doing, the committee can help cut through weeks, perhaps months, of inaction that creates a protection void. Such voids can result in injuries and illnesses, while more and more layers of arguments, memos, and hearings evolve.

Permit the committee to get to the heart of the matter and base its clear recommendations on pure safety and health factors. It is imperative that the committee, like the overall program, be concerned with the entire establishment—all of the buildings, grounds, equipment, etc. It is obvious that areas with greater hazard potential may be the topic of conversation and study more often, but no area or operation should be totally abandoned.

The Time and Place for Meetings

As for the actual meetings, there should be a regular schedule. Barring emergencies, it should not change. If meetings are scheduled for 9 A.M., they are to start at 9 A.M.. If you find that, for various reasons, members often do not drift in until 9:15 A.M., find out why. If there is no good reason, those members will just have to arrive on time, or they will no longer be members. There should not be a casual attitude about the starting time or the finishing time. Meetings must be controlled so that there is not a lot of wasted time or introduction of peripheral issues. Such issues may have value, but if they do not belong at a safety and health meeting, then refer the member who brought up the matter to the right party or department. It is possible, for instance, that the 9 A.M. meeting time was too early. If that turns out to be the case, it is far better to change the time to 9:30 than to keep scheduling a 9 o'clock start with the understanding that most members will start drifting in around 9:15 or thereabouts.

The meeting place must be comfortable and uncongested, with good lighting and adequate ventilation. Certainly, smoking should never be allowed. Interruptions should not be allowed either, whether in person or by telephone, unless the matter is critical (i.e., something akin to a personal or business emergency). I seriously recommend a small selection of beverages and snacks, the exact assortment being appropriate to the time of day. No, I am not advocating that all committee meetings be held at a resort! I am recommending that the situation be made pleasant. Such an environment can only be helpful, as long as the group concentrates on the purpose of the meeting. If, in fact, there is not a pleasant venue on site, then an off-site location may truly be a good idea, particularly when meetings are annual reviews or otherwise figure to consume a significant amount of time.

The meetings should be held at least quarterly. If the work establishment is quite large or if the potential hazards are many, it might be more appropriate to hold monthly meetings. In extreme situations, emergency meetings are not out of the question. However, this does not mean that, on a whim, a meeting should be held simply to tie up a loose end.

Topics to Consider

The meetings should be somewhat formal and have an agenda. It is preferable to publish the agenda several days before the meeting, although this may not be necessary if the agenda always consists of the same general categories. This does not mean that meetings are to be rigid and stuffy. Everyone should be at ease and not intimidated by confusing and overly strict rules of decorum. Yet, there must be a plan regarding the flow of the meeting. There should always be a part of the meeting that is used to address accidents that have occurred since the committee last met. There should be a time to discuss the progress of safety projects. Were they completed? If not, why not? Just what was done to effectuate the desired protections? New or proposed training programs should be discussed, as should any draft or pending plan related to safety or health. There may have been cases of progressive discipline related to safety and health. If so, they bear discussion. There should be a thorough, detailed review of inspections performed by OSHA, insurance carriers, consultants, the corporate office, or a committee team itself. It can be helpful to go over old business and new business, but there should not be a time that a speaker or subject is lost because someone did not speak at exactly the correct moment. How about lively, expert guest speakers occasionally?

When there is agreement that a safety project will be undertaken, accountability and responsibility must be assigned. This must be communicated with no possible ambiguity. Too often, at the next meeting, Kareem and Juanita understood that Frankie was going to take care of it, but Frankie thought that one of them had that job. When a project is set down to be done, the record must show who has overall and specific responsibility, when the project is to be completed, and basically how it is going to be done. Once such a project has been set in motion, it must be brought up at every single subsequent meeting until it has been successfully completed or until it has been dumped for an alternative, which must then be tracked until its completion. It should never be assumed that because something was planned, it was actually done. Follow it through. If it is not accomplished by the planned date, the reasons must be made clear. A new date must be set and scrutiny of the progress must be tightened. Tracking is all important. In a separate chapter, I address the issue of setting priorities.

Keeping Minutes

Concise but complete notes of each meeting must be printed as soon as reasonably possible, and copies of those notes should be disseminated to each member of the committee. I suggest that the notes be retained for at least 5 years. Highlights of meetings can be put into newsletters and posted on bulletin boards. There must be a particular person (e.g., a secretary) to compile, distribute, and maintain the notes. A dedicated file for all safety and health committee notes and related documents should be established. Besides the secretary, other titled positions are needed. A chairperson should be selected. There is some debate over whether that individual should be a member of management or not. I suggest that the chairperson be a member of nonmanagement.

Keep in mind that, unless a contract indicates otherwise, the employer does not necessarily have to be bound by the recommendations of the committee. The way it should play out in real life, however, is that the recommendations of the committee should be followed unless there was an absolutely incredible oversight in the decision-making process. Such faults should not be sought. Recommendations of the committee should not be analyzed in great depth unless the recommendations call for a major financial expenditure. There should not be an additional committee to vote on the recommendations of the labor-management safety and health committee. The whole point would be lost.

Legal Problems

Quite unfortunately, the existence of safety and health committees can cause special legal problems. This is most often a possibility where there is no authorized union. In certain situations, the National Labor Relations Board (NLRB) has determined that the committee fit the definition of a labor organization or union. The definition applies to entities that:

1. are based on employee participation;

2. exist at least in part for the purpose of "dealing with" employers;

3. address such topics as wages, hours, and conditions of employment when dealing with employers.

If the NLRB decides that the circumstances of the formation and administration of a committee meet this definition, and also finds that the committee is dominated by management, then a charge of committing an unfair labor practice can be leveled against the employer.

The NLRB finds little problem with committees whose brainstormed ideas are made available to the employer. In addition, the committee can collect or disseminate data or other information. According to the NLRB, the committee may plan safety and health educational programs. The interaction between management and labor is a key focal point in the interpretation of the legality of the committee. I have already stressed the importance of having more employees than management representatives on the committee. It appears that the higher percentage of employee members on the committee, the greater the chances that it will be allowed by the NLRB. You may choose to have no management personnel as actual members. Rather, they can serve as facilitators or merely as nonparticipating attendees. In all cases, the NLRB is less likely to find fault when the feelings and concepts of the committee are not offered as a proposal to the employer. The fine-line semantics involved can seem childish, but in the interpretation of the law, they can, for better or for worse, make a world of difference.

A central test of whether or not a safety and health committee is viewed by the NLRB as being a labor organization surrounds the question of employer domination. For instance, if the employer has delegated to the committee complete power to take measures to try to heal safety and health problems, there should be no finding of employer domination. In these circumstances, when the committee only contacts management to indicate its decisions as to resolving safety and health issues, the NLRB currently sees no remaining legal roadblock. There is no universal prohibition against management representatives cooperating with committees as a necessary part of the logistical process. The problem exists if the employer places pressures on a committee to act in a certain manner or to favor one aired option heavily over another. You need not stray from the principles of labor-management involvement, but do not form or permit the existence of a committee that is a puppet of the employer, regardless of the facade.

Committee Mission

Generally, safety and health committees are thought of as reacting to hazards or accidents. The committee can and should play a major role in helping to prevent the existence of hazards. Therefore, the committee should not just be an after-the-fact discussion group. Besides being involved in plans for incentive programs, training, and the like, the committee can deal with potential problems up front. Thus, the committee should be involved in discussing plans for plant expansion and for the introduction of new machinery, work processes, or chemicals.

A labor-management safety and health committee can be a significant part of any safety and health program, but only if its mission is clearly understood and if it remains active, focused, and committed. A well-oiled committee actually works as a team and plays a major role in establishing and perpetuating a safe and healthful work climate. It also makes the employer's job easier.

15

Accident Investigation

The main purpose of an accident investigation is to preclude or greatly reduce the chances of a recurrence. Thus, ironically, through comprehensive accident investigations as well as other elements of a safety and health program, accidents can be reasonably predicted and prevented in the great majority of potential cases. Unfortunately, although industry often refers to "accident" investigations, that usually plays out to mean "injury and illness" investigations. In reality, even if no one was hurt, but there was a near-miss incident, an investigation is called for. This also applies to fires of any size. Any noted unsafe behavior or condition should be investigated, but that is usually through less formal channels. There is still another situation that should initiate some sort of investigation. That is when damage to equipment or property has been discovered. Such damage may well have been just one step away from a situation resulting in an injury or illness.

Causes of Accidents

The initial step in conducting an accident investigation is to determine the causes of the accident. Note that I have used the plural—causes. In most cases, more than one cause can be identified, even if one or more causes played a greater role in the onset of the accident. Each cause must be analyzed, with the greatest emphasis being placed on the root cause(s). The investigation should not be considered complete just because some causes were determined. The investigation must keep digging until the root cause(s) are brought to the surface for thorough examination. Then a full accounting of the chain of events can be studied in a pragmatic manner.

In breaking down the basic sequence of events leading to the accident, the contributing causes, as well as the primary causes, must be sought out and evaluated. Contributing causes can involve such factors as the lack of adequate communication (crucial at shift change), inadequate skill, and/or poor attitude. Primary causes can involve such factors as unsafe conditions or procedures (one may lead to the other), unsafe acts, and natural phenomena. It is a cop-out to reflexively yell "unsafe act" whenever someone is hurt on the job. In almost every case, even if an unsafe act occurred, there was also an unsafe condition and/or an unsafe procedure (or lack of proper procedure) as well. On the other hand, even with a situation characterized by a clearly unsafe condition, behavior should be considered.

System Safety

All accidents should be viewed as representing defects in the overall system, whether or not specific deficiencies can be found in the behavior of the relevant individuals. An entire subfield exists within the accident investigation arena that includes several hazard identification and evaluation models. This subfield is generally called *system safety*. A detailed study of system safety can be quite lengthy, so I only briefly visit the subject here.

The models are often more effective in achieving their goals or the goals of the practitioners when used in combination. Some of the techniques target very particular branches of safety, while others have a broader application. Although they can be used for accident investigation purposes, the best principle remains to take a proactive, rather than reactive, stance. Aim first for prevention.

Thus, at least several of these systematic approaches should be considered by a company when trying to predict potential accidents and remove that potential. New techniques, including those that have arisen from existing techniques and those that are hybrids, continue to emerge. Some of the more widely acclaimed techniques involve critical path analysis, contingency analysis, critical incident technique, single point failure analysis, subsystem hazard analysis, flow analysis, failure modes and effects analysis, fault tree analysis, and job hazard analysis. A later chapter in this book addresses job hazard analysis in fair detail. Job hazard analysis is one of the easier to apply, simpler techniques. Others are often more sophisticated and can be used for complex situations, including catastrophic events. Nevertheless, job hazard analysis, with its elemental format, can produce excellent results.

Fault Tree Analysis

As an example of one of the other more popular, more intricate, and more sophisticated system safety techniques, there is fault tree analysis. I offer a glimpse at the fundamental tenets of the method. It is founded on the idea that every accident or potential accident results from interacting causes within a system and that every accident cause and interaction can be logically separated into basic or component failures. Further, the premise follows that solutions can be developed to control each of the failures and/or interactions. Even where there were numerous contributing events that can be listed as causes, the elimination of one of those causes might have been sufficient to break the chain of events that gave birth to the accident. Ideally, all of the causes should be identified and eliminated, but to reiterate, the breaking of even one critical causal link may be enough to preclude the undesired results.

Fault tree analysis seeks to identify and correct sources of failures and malfunctions within the system. The system generally consists of persons, machines/tools, and materials that operate within an environment to perform a specific task using prescribed methods. Systems are normally defined in terms of the task or function they perform (i.e., they are task oriented). The theory is that the components of the system (i.e., persons, machines/tools, materials, and methods) and its environment are interrelated, and a failure in any part can affect the other parts.

The fault tree is a logic diagram. The first step in constructing the fault tree is to select the undesired event (i.e., accident or potential accident). Each of the primary causes and how they interacted to produce the undesired event are identified. The causes are then broken down into the events that led to them. The logic process is continued until all potential causes have been identified.

Throughout the logic process, a tree diagram is used to record the events as they are identified and labeled. The undesired event is represented at the top of the tree. The primary causes are depicted immediately below the undesired event. The events that led to the primary causes are shown at the next lower level and so on. The tree branches are terminated when all of the events that could eventually have led to the undesired event are shown.

In constructing a fault tree, symbols are used to represent events and to describe the logical relationship between events. Different shapes are entered to represent different types of events. Symbols called *logic gates* show the manner in which events at one level of the tree combine to produce an event at the next higher level. So, you can see from this synopsis that following the process through every last branch and completing every portion of the diagram can be a rather arduous task. As an overview, fault tree analysis has five basic steps:

1. define the undesired event (i.e., accident or potential accident) to be studied;

2. acquire an understanding of the system;

3. construct the fault tree;

4. evaluate the fault tree;

5. eliminate or control the hazards identified.

Regardless of what system is used, very few companies have the good sense to include the victim of an accident on the team that conducts the investigation. The victim should definitely be involved with, but not lead, the investigation and be a member of that all-important team. It is advisable to, somehow, simulate a reenactment of the accident, *but only if that can be done with exceptional controls to guarantee that another injury or illness will not take place.* The victim should not be coerced to take part in that reenactment. However, if he or she is willing and psychologically positioned to go through with the valuable lesson, then so be it.

Natural Phenomena

When considering natural phenomena, if an employee was killed in a flood, by lightning, or was stung by a bee and then went into shock, it does not mean that the accident was unavoidable. Hardly any accidents are purely unavoidable. Some are more difficult to predict and some are more difficult to prevent than others, but it does not mean that they had to happen. Injury or illness resulting from natural phenomena are often predictable and avoidable through sensible precautions and contingency plans.

For example, if the work establishment is in a flood plain, it is incumbent upon the employer to exercise special diligence in monitoring the weather and in planning for evacuations. Further, the location of the structures should have been considered upon initial site selection. Employees should not be allowed to work outdoors when there is an electrical storm. If there are bees regularly visiting your establishment, inside or outside, steps should be taken to remove their presence. If there remains a fair possibility that occasional bees might still endanger employees, be sure to find out if any employees are subject to anaphylactic shock. At any rate, as dangerous as the bees can be, their sons are even more of a threat; you know how those "sons of bees" can be!

It is simply not an acceptable excuse to blame hazards such as these on the proverbial act of God. One more example that is worth mentioning concerns exit doors not fully opening due to accumulated snow or ice on the exterior of the building. Don't look for a divine excuse. Shovel the snow and chop the ice. Period!

Workplace Violence

There is a supplementary, nonconventional category of primary cause that has surfaced recently: workplace violence. It can be perpetrated by a spurned lover who enters the workplace to hunt down a former partner or by a recently fired employee who has returned to get even with his or her supervisor. In some establishments, although not usually of the factory type, it can be the work of a thief. An even scarier possibility with possibly catastrophic results is the work of a terrorist. These situations are a world away from the unguarded machine, the unavailability of eye protection for the grinder operator, and the adjacent storage of acid and cyanide. Unfortunately, workplace violence has reached a degree of prevalence whereby strategies must be developed to meet the problem head on, and I cover this topic in a separate section. For now, it is important to realize that if persons are hurt or killed as a result of violence, you should not feel that the situation is exempt from the need for a comprehensive investigation. Specific steps can be taken to avoid a recurrence. This issue is addressed in the Appendix Self-Inspection Checklist.

Categories of Accidents

If you wish to categorize accidents or potential accidents by types, the following list should cover almost every scenario:

- struck against
- struck by (falling, flying, shattering, sliding, moving)
- caught in, on, or between (in-running nip, pinching, shearing, rotating, reciprocating, punching, pulling and jerking, carrying actions)
- fall on same level
- fall on different level
- slip or overexertion (strain, hernia, etc.)
- gradual onset and ergonomic (cumulative trauma disorders, repetitive stress injuries)
- exposure to temperature extremes (burning, frostbite, heat exhaustion, sunstroke, hypothermia, scalding, freezing)
- inhalation, absorption, ingestion (asphyxiation, poisoning, drowning)
- noise
- radiation

- flash
- skin contact (acid burn, dermatitis)
- pressure (air, liquid)
- electric current
- transmission of bacteria, viruses, or parasites
- vegetation, animals, insects
- vehicles
- violent crime

Types of Unsafe Conditions and Procedures

If you wish to break down types of unsafe conditions and procedures, the following list should be helpful:

- no guard, barrier, or device; inadequate guard, barrier, or device; guard hazard
- defective equipment: rough, slippery, sharp; poor design or construction (consider ergonomics); low material strength, inferior composition, decayed, aged, worn, cracked, or frayed
- tool or equipment not suitable for the task or job
- hazardous arrangement or procedure: unsafe storing and stacking; congestion; unsecured loads; impeded access to aisles, egress, disconnects, extinguishers, hoses, and alarms;
- overloading; misalignment; unsafe layout (including for traffic); unsafe processes; incompatibles together; slipping and tripping hazards
- improper illumination: too much, not enough, glare, shadows, stroboscopic effect
- improper ventilation: insufficient air changes; unsuitable capacity, location, arrangement; impure air source; dangerous mix
- unsafe dress and apparel (none, defective, or unsuited): industrial safety glasses, side shields, goggles, face shields (for impact and/or chemicals); gloves (for chemicals, grip, sharp surfaces, slippery surfaces, vibration effect reduction, electrical insulation); aprons (for chemicals, knives); steel-toed shoes or equivalent, metal-bottomed shoes, metatarsals, boots (for chemicals, water, slippery surfaces); knee pads; spats; "cold gear;" "hot gear;" respirators; head protection (usually hard hat). Also consider the hazards of high heels, conductive clothing, loose hair, jewelry, and clothing (including gloves, in some applications)

- lack of fall protection: railings, toeboards, belts and harnesses, warning lines, monitors, nets
- lack of proper maintenance, inspection, and testing
- lack of proper training, communication, signage, policies, and rules (including lockout/tagout/ZES, permit-required confined spaces)
- lack of first-aid equipment and emergency procedures
- lack of sanitary conditions
- lack of vapor ignition controls
- lack of shoring, sheeting, or equivalent
- electrical hazards

Unsafe Acts

If you wish to break down unsafe acts, the following list should be helpful:

- causing or allowing unsafe conditions or procedures (as listed above)
- horseplay, distracting behavior, startling co-workers, throwing things
- walking or working on moving, dangerous equipment
- cleaning, lubricating, dressing, and adjusting moving equipment
- caulking or packing equipment under pressure
- taking unsafe positions and postures: under suspended loads, too close to hazardous openings or moving parts, improper lifting and carrying, exposure to falling or sliding objects, riding in unsafe position (including forks, crane hooks, backhoe shovels, tailgates), in vehicle right of way
- poor driving (including too slow, too fast, tailgating, not using directional signals, improper passing, and not obeying traffic signals, lights, and signs)
- dangerous mixing of substances (including acid and cyanide, pouring water into acid, pouring cold water into hot boiler or onto molten metal, oil and oxygen, flammable liquids and oxidizers)
- using tools and equipment incorrectly (including poor grip, unsafe pressure, volume, and speed)
- using hands instead of tool
- aborting safety features (including interlocks and breakers)
- running, rushing, and jamming parts into a machine or container

The Costs of Accidents

One more list can be of great value when attempting to convince upper management of the value of adept accident investigations. Properly conducted, these investigations should lead to a reduction in accidents and their related costs. Several of the items may intersect with each other. Not all costs can be immediately measured in direct fiscal output. The following list addresses several aspects of costs that stem from accidents:

- direct medical
- follow-up medical (including therapy)
- spoiled work
- loss of orders
- damaged equipment
- cleanup costs
- overtime
- accident investigation
- retraining
- new hire (training, personnel department, breaking in)
- downtime (including overhead, e.g., light and heat)
- loss of efficiency due to crew breakup
- co-worker time off (hospital, funeral, "shock")
- loss of efficiency of co-workers and returning victims (gun-shy)
- lawsuits
- legal fees
- economic effect on victim's family
- lost time of supervision
- spoilage (water, chemical, fire, explosion, spillage)
- workers' compensation
- other insurance
- time spent with enforcement agencies
- penalties from enforcement agencies
- poor public relations
- labor-management problems
- damaged morale

- general paperwork
- cost to society (including government support of family)

Conducting an Accident Investigation

There are many special tips to keep in mind when conducting a full-blown accident investigation. The following list should prove quite helpful:

- Be observant, curious, creative, and logical.
- Get it all, get it right, and make it clear.
- Do not insist on only one cause.
- Do not make assumptions or insist on preconceived theories.
- Do not delay the start of the investigation.
- Avoid moving or touching evidence until it is photographed, video-taped, sketched, diagramed, or similarly documented. *Note: There may be times, in extreme cases, that evidence must be moved at once to free a body or remove a source of ignition or fuel.*
- Consider the three Es: engineering, education, and enforcement.
- Establish who had responsibility and who had authority.
- Do not look for fault as a primary goal.
- Establish the existence and content of training, policy, rules, preventive maintenance, and job hazard analysis.
- Consider the possibility of inadequate or inaccurate technical data and/or operating instructions.
- Do not rush simply to apply the label *"unsafe act"*.
- Do not neglect weather, visibility, noise distractions, or terrain considerations.

Personal Handicaps

A significant and large array of personal handicaps may have played a role in the accident under investigation. These handicaps are not generally considered to be accident causes in themselves, but they may contribute to the creation of unsafe conditions and may relate to unsafe acts. The awareness of the following handicaps can reveal a need for corrective actions in the form of training, discipline, placement, and/or accommodation:

Lack of knowledge or skill

- unaware of safe practice
- lacks information
- does not realize danger
- lacks coordination
- lacks experience
- lacks general skill
- (language barrier or illiteracy could be a related problem)

Improper attitudes

- disregard of instruction
- disrespect of authority
- disregard of rules or customs
- showoff
- adventurous
- lazy
- disobedient
- defiant
- inattentive
- purposely injures self or others or causes damage (*very* rare)

More traditional handicaps.   In no way should the inclusion of this section be construed as an implication that persons affected with these handicaps should be avoided when hiring, promoting, or reassigning.

Physical

- blind or visually impaired, which can include lack of color perception or depth perception
- deaf or hearing impaired
- prosthesis
- deformity
- missing finger(s)
- weak heart
- generally weak

- fatigue
- extremely short or extremely tall
- extremely heavy or extremely thin
- epileptic
- trembling
- lack of coordination
- inability to speak
- stressed
- substance abuse (alcohol, drugs, inhalation)
- *properly* taken drugs with side effects
- other illnesses or conditions

Mental

- inability to understand
- retardation or slow thinker
- inability to concentrate
- dyslexia
- poor memory
- attention deficit disorder
- other learning disabilities

Emotional

- nervous
- excitable
- worried
- temperamental
- depressed
- hostile

Phobias (abnormal fear of...)

- heights
- enclosed places
- crowded conditions
- water

- animals
- the name Lenore
- germs

Interviews

Accident investigation interviews can be a very delicate task, particularly if an accident results in a serious injury, illness, or death. No one enjoys being interrogated. The interviewee may feel a threat to his or her personal or professional security if not approached in a tactful fashion. Fortunately, it is human nature to want to be helpful when accidents do occur. The victim(s) or witness(es) are usually your best sources. However, other persons who are not part of the accident scenario may also contribute valuable input. By using the proper interviewing techniques, your efforts may yield quicker and better results.

Here are several suggestions to help improve your interviewing savvy and remain in the driver's seat during the questioning process:

- The interview and the broader investigation should be conducted by a person (generally a line supervisor) who is close to the matter and clearly understands the pertinent job, task, process, and area. The person in charge of safety and health should coordinate oversight.

- Interview each person alone, unless an interpreter is needed.

- Make the person feel comfortable and at ease.

- Be courteous.

- Be in control.

- Assure that he or she understands that your objectives are to gain a better understanding of the accident, its causes, and ways to prevent a recurrence through the gathering of his or her input. Do not make your interviewees feel that you are "grilling" them.

- Speak at his or her level of comprehension.

- If the interviewee is not the victim, establish his or her relationship to the victim.

- Encourage the relating of facts and observations, even when the person hesitates and says that it may be insignificant. When you commit the statement to written form, you can limit the text to useful detail as long as nothing is unfairly taken out of context.

- Encourage an explanation of the clear sequence of events.

- Do not write immediately. First, try to listen awhile.
- Consider the five Ws: Who? What? Where? When? Why?... And How?
- Try to determine if normal procedures were followed.
- Try to determine if relevant persons were adequately trained.
- Try to determine the history of relevant employees (including related disciplinary action) and the equipment and process involved (including machinery malfunction).
- Ask if he or she can *safely* show you what happened (great care must be taken with any demonstrations).
- Do not lead.
- Write the person's statements in his or her own words. Include a clause similar to "to the best of my knowledge and belief." Read the statement back to verify accuracy and language. Have the person initial any changes, deletions, additions, or corrections. Take a signed statement, particularly from individuals who claim to know nothing.
- Ask the person to contact you if there is something to change, delete, add, or correct (perhaps his or her memory will improve).

Cop-out Excuses for Accidents

There are several poor clichés that are used as excuses or alleged reasons for accidents. They are, for the most part, really just cop-outs. As such, if these explanations are to be believed, then the inference can be drawn that no more than a cursory accident investigation should be conducted. That is the primary problem with offering these trite, ill-conceived rationalizations. The following sections offer a brief examination of those that are the most useless, offensive, and ignorant.

She's accident prone.

If a person has been hurt several times, then the underlying causes must be discerned. Has she been hurt several times because she lacks the necessary skill or knowledge? Is it because the machine, tool, or procedure is unsafe? Does she have a physical disability? Is she under the influence of narcotics? What exactly is going on? There certainly could be other reasons behind the frequent mishaps. "Accident prone" should never be the sole explanation for why an accident occurred. Get to the root of the problem if this worker has been injured on several other occasions.

He was careless.

How was he careless? Did he fail to replace a guard? Did he not wear the required personal protective equipment? Did he hold the tool in the wrong manner? Was he running through the aisles or down the steps? As with the "accident prone" cliché, "careless" should never be the only explanation written on the accident report. In some cases, an employee trips on a loose item on the floor and is injured. Someone in authority wants to invoke "carelessness" as the reason and let the matter rest at that. Others may wish to add, "He didn't look where he was going," an explanation that could have some merit. However, if an employee trips on a piece of pipe that was on the floor of a main aisle, other factors should be studied. For instance, how did that piece get on the floor? Was it dropped, and if so, did the person who dropped it realize it? Why was it not picked up by the person who dropped it? How many other persons walked by it or saw it and failed to pick it up? Did it fall from some structure above? Had it been carried or transported in an unsafe manner? Is a regular place provided to store such pieces?

Another case comes to mind. It is about an employee who sustained a serious eye injury when a bungie cord that he was handling snapped (it did not break). Most employers would simply suggest that the employee was careless and should be more careful, and let it go at that. There were at least three key, very specific aspects of the accident that called out for review. Why weren't safety glasses worn? Why wasn't the cord used and stretched in a position parallel to the victim's chest so that even if it snapped or recoiled it would project to the sides as opposed to toward the employee? Was there a different type of device that could have been used, the kind whereby the tension is relieved so that the uncontrolled springing action could not occur? To have simply labeled this injury and accident as being the result of carelessness would have shut the door on the step-by-step evaluation that resulted in the implementation of solid abatement methods.

That's the price of progress.

The apparent meaning of that gem reflects the misguided perception that, in order to progress in the industrial arena, there must be an understanding that there will be some losses (including human suffering through injuries or illnesses). I recall a rubber mill operator who was missing three fingers, each of which had been lost in the in-running rolls in separate incidents. He joked that losing three fingers in 30 years—one every 10 years—wasn't too bad! It was as if three amputations were a fair price to pay for continued employment at a decent wage. A factory job may present more inherent hazards than a desk job, but there absolutely should not be an automatic tolerance for pain and disability just because the person works with machinery.

It was just one of those things.

One of what things? Is this supposed to mean that something occurred that is unavoidable? That is the common inference. Apparently, the speaker wants to convey to the listener the ostensibly profound observation that such accidents are bound to happen. However, in the overwhelming majority of accident scenarios, a thorough investigation will bring to light two simple facts: that the accident was predictable and that more could have been done to preclude or significantly lessen the chances of the occurrence. This type of statement is sometimes used for clearly avoidable accidents, such as finger and hand injuries resulting from the hazards of moving machinery, electrical shock, burns from chemicals, and falls from heights. These are only a few examples. In other words, even where practical abatement methods should have been obvious, the assumptions seem to be that fate intervened and that fate cannot be controlled.

On the other hand, there are times when the statement is not totally absurd. For instance, an employee might turn or move a body part in just the *wrong* way. Nevertheless, what may be learned from the accident investigation still has value. This may simply be how to move and how not to move. If a person banged a hand against the frame of a machine, the quoted expression might be called into play. It is understood that it is quite difficult to assure that every single employee will move or steady his or her body appropriately at all times. Still, this is not the end of the issue. Is there a problem with the employee's perceptions of space and objects related to the job task? In a more focused vein, is the employee's work station large enough and properly configured? Can the employee be taught to utilize the space in a manner that decreases the chances of an injurious contact?

Accidents will happen.

The comments regarding "it was just one of those things" largely apply here. The simple counterpoint is, "Accidents will happen...if you let them."

It was a case of bad luck.

There is such a thing as luck. It is not a complete myth. Yet, it is paradoxical that humans often cause their own luck. The misfortune we leave behind is often a monument to our mistakes. It may have simply been a matter of unfortunate timing that employees were standing in a particular location when the load from the crane fell on them. It could be that, except for that one minute, there was no one under the load. Of course, no one should have been under the load for any length of time, and no load should have fallen. Period.

She was only doing it for 5 minutes.

This brings to mind the just described under-the-load situation. As another example, a contractor who did some work on my home greeted me one morning while he was wearing dark glasses. Knowing that I am a safety consultant, he was embarrassed. He said that he always wears his safety glasses when he uses the power saw. I awaited the *but,* and it arrived as scheduled. He said that he had worn the glasses while using the machine, "but" took them off to take a measurement, and did not don them again when he made the one additional cut. Maybe this is where luck does come in. He had to wear dark glasses for several days and went through a prescription of eye drops, but did not sustain any permanent eye damage.

There is a famous case of an employee who stood behind a person who was operating an abrasive wheel. The person behind caught a chip in the eye and blindness resulted. The victim was only in the dangerous position for a very short period. Examples are all too plentiful and relate to numerous types of hazards.

In many instances, employees have been severely injured when completing one last task just before clocking out or when they had only one specific, short, work task cycle to perform. It could be one more cut on a power shear (without checking the adjustment of the blade guard), one trip up a hastily set up A-frame ladder (without assuring that the spreader and supports were fully locked), one quick "ride" up on the bare forks of a forklift truck (without attaching the safety cage), one fast repair on a 220 VAC circuit (without locking out the power), or one transfer of a flammable liquid from a 55-gal drum to a smaller container (without installing grounding and bonding). As for impeding access to exits or extinguishers "for just 5 minutes", the phrase that says it all is, "Fire can't tell time." Good safety and health practice does not have a break time.

We've done it this way for 20 years without an accident.

The speaker probably meant that there had not been a serious injury or illness, but that there may have been accidents and possibly minor injuries or illnesses. I have conducted a large number of accident investigations where I was told that it was the first time that such an accident had happened on the machine or in the process. I have been told, in such cases, that it was the first time that the employee (sometimes more appropriately referred to as the deceased) had done such-and-such. Employees have worked on unsafe jobs and performed unsafe tasks for many, many years without an injury. Then, just a month or less before retirement, the employee sustains a major injury. Superstition aside, when hearing the referenced quote, I get nervous about tomorrow, or even today.

Some find comfort (really, false security) in the thought that the truth can be found in history. That sophisticated philosophical gem—meaning that if no one has been severely injured doing the particular task over the 20-year span of the operation, it must be safer than the worrywarts realize—is misguided at best and a dangerous gamble at worst. If there is an apparent OSHA violation and/or an evident risk involved, then no one should rely on history. In general, if several employees get hurt doing the same thing at different times, it is highly predictable that more will be injured unless abatement steps are taken.

On the other hand, a statistician might argue that the chances of an accident and resultant injury occurring *in a situation where there is a discernible hazard* actually increases as the noninjury duration time grows. Such reasoning may appear convoluted, but it should give pause to anyone believing that the longer an unsafe operation is run without injury, the more proof is sustainable that no one will ever get hurt on that operation. It makes more sense to apply the old cliché, "Play with fire long enough, and you'll get burned." In fact, employees can get too comfortable at a job that they have come to take for granted as being safe (despite the obvious safety concerns) simply because no one has been severely injured in all of those years. Complacency and being lulled into a state of misplaced trust can follow.

We can't be perfect.

This excuse is frequently used in a context to refer to a facility undergoing a comprehensive OSHA inspection and coming away with very few citations or violations. That is rare, but it is definitely possible. The excuse is also used when an employee is injured, particularly when there has been a low rate of injuries. In either case, you should at least strive to be perfect. It is inaccurate to state that a company can't be perfect. It can; it has that ability. The higher you reach, the higher you should attain, even if you don't attain perfection. Always shoot for a zero tolerance of injuries and illnesses and, better yet, a zero tolerance of accidents.

In the realistic industrial world, there are injuries and illnesses even in plants that have excellent safety programs and excellent follow-through. Hopefully, the frequency and severity are very low. Some specific types of businesses by standard industrial classification consistently reveal a higher frequency of injuries and illnesses than other types. A company may boast that its rate is much lower than the average in that particular standard industrial classification. That is fine, but the goal should be to have a facility that is injury free and illness free.

There is no totally acceptable rate of injury or illness. Even as the rate comes down and closely approaches zero, there remains work to be done. Just ask the one person who was hurt, particularly if the injury was severe, such as an arm amputation. The victim may appreciate the fact that no one else in the company was hurt over the past year. Why would he or she desire others to suffer? However, the victim cannot find much consolation in knowing that he or she is the only one missing that arm.

We lost only two people in 30 years.

This fits in with the acceptance of less than perfection. As I have framed the expression (and this type of statement has actually been made), we are looking at something even worse than that arm amputation. When management attempts to offer statistical data which include at least one serious injury or illness as an excuse in this context, the human factor is sorely lost. Even one person sustaining an arm amputation is a very serious matter, even if that was the only injury in several years. To reemphasize, if it was your arm or your child's arm, there is little solace in the fact that the company has a low rate of injuries.

The word "lost" is frequently used as a euphemism for "killed." It is an egregious breech of decency to even suggest that any life be undervalued. No death can be excused by the mere fact that the company operates in an industry that is traditionally and inherently dangerous. The word "only" has no place in such matters. A quarter of a century ago, the statistics on casualties in Vietnam were published and announced at least weekly. After months of hearing the numbers, many people in the country were either numbed or too ready to accept such awful indicators. Incredibly, receiving the heartbreaking input (i.e., the number of deaths) became too routine. In some weeks, the number of deaths went down from the previous week. That can only be viewed as preferable to the alternatives. The bottom line is that numbers do not have hearts. They are devoid of warmth and they are devoid of faces. The shock of Vietnam could best (or worst) be felt when the television screen showed the family of a victim (i.e., relatives of a human being from a specific town, with specific family and friends—a human being with an individual and unique identity).

Reviewing statistics, even when they reveal in some measurable way an improvement, is no match for the impact of a funeral.

You take a chance even when you cross the street.

It is amazing that this excuse is so frequently offered. Depending on the accident or potential accident in question, the implication is often that, if the employees exercise good judgment and informed caution, then no one should be injured. The translation may be, "There is no need for a guard on the machine, as long as the employees are careful." To take this poor logic further, why guard any machines, why install hand railings on staircases, and why wear impervious gloves when pouring acid? The supposed point of the excuse is that, if employees do everything exactly right, they will not get their hand in the danger area of the machine, will walk up or down the steps in a manner whereby they will not need to grab a railing, and will be so steady and accurate in pouring the acid that none will touch the hand.

Is it possible to be struck by a vehicle when crossing the street? Certainly. The parallels between the occupational setting (e.g., in a factory) and a person crossing a street are so weak that no lengthy direct discussion is required. One chief difference is that neither the pedestrian nor the pedestrian's family or employer has any authority or influence on the driver. In a plant, on the other hand, there is the opportunity to set and maintain controls through authority.

I offer brief comments not to really address the pedestrian-vehicle relationship but to make the argument that even in such situations where potential victims (or those who would want to protect them) are not the ones erecting traffic lights, setting speed limits, or controlling the speed or direction of the vehicle, there are still specific safety steps to be taken. It is fair to say that the person crossing should look both ways and gauge the speed of the traffic, the drivers' visibility, the weather, the condition of the road, and so forth. However, there is more. Where pertinent, only cross when the traffic light favors the pedestrian, and then only within the demarcated crosswalk. Still further, the controlling authorities (e.g., police, DOT, etc.) should have carefully set the appropriate speed limit and installed signs and/or traffic lights with due respect to the area.

I told him a thousand times.

All right. Perhaps you did tell him several times, but what specific, assertive, definitive steps did you take to modify or terminate the behavior? Telling an employee something is not, in itself, tantamount to causing that employee to act in a different way or to cease acting in the unsafe way. If a fatality or an extremely severe injury was sustained, you might lay in bed the following evening telling yourself, over and over, that you had instructed or warned the victim many times about the unsafe behavior. You might try to convince yourself that it just wasn't your fault, that you need not shoulder any blame, and that you could not have done more to see that the accident did not occur. You might try, until the wee hours, to absolve yourself of culpability. Yet, somewhere in the darkness or first light of morning, you will probably search deeper and more introspectively and confront the truth. Why did you not take positive actions to alter the dangerous behavior? You had the responsibility and the authority. You could have done something meaningful and decisive, but you failed to do so. I delve into this excuse later when I address the issues of willfulness, employee misconduct, and isolated incident.

It was a freak accident.

The media has embraced this all too frequent response to how it could have happened. It has become commonplace to hear on television and radio or read in newspapers and magazines about a "freak accident." In almost all cases, the expression is only a fast-track exit from reason. An accident may have occurred as a result of an unusual set of events. There may have been a tragically odd domino effect of things gone wrong. That is not the same as a "freak accident." When this excuse is to be believed, the idea is that the accident was not even remotely foreseeable. The next thought is that there is nothing that can be done to avoid a similar incident so there is no purpose in even conducting more than a cursory accident investigation. Therein lies the danger of the "freak accident" label. It is a way to increase the chances of another person suffering because no controls are brought into the picture.

One common dictionary definition refers to "freak" as "a sudden and apparently causeless change or turn of events." The important word there is "causeless." If the assumption of causeless is accepted when taking an incipient look at an accident, the issue is over and any chance of preventing another such incident has been lost. As with several of the excuses that I have reported, this one is very often spewed from the lips of a person who is ignorant, stupid (yes, there is a significant difference), basically incompetent, or lazy to a degree that can result in dire consequences.

Several years ago, a wire service news story reported that three persons drowned "in a freak accident" while watching a rock 'n' roll concert on a beach. The very next paragraph of the story specifically mentioned that the germane section of shore was known to have dangerous and unpredictable tidal and undercurrent effects. Then why would anyone arrange and promote a concert at such a precarious venue? It sounds like a case where the promoters and producers could be susceptible to legal action. What it does not sound like—and what it most assuredly is not—is a "freak accident."

Examples of similar cop-outs in occupational situations are common. They are so common, with this excuse being invoked for so many different types of conditions, that it is difficult to know where to start giving examples. I have heard this poor excuse even used in cases where the causes of an accident could be discovered quite easily and where a study of the sequence of events readily revealed that the discerning of those causes would not require a highly sophisticated and profoundly technical examination.

The excuse has been used time and time again for incidents involving very basic situations. This includes (but is in no way limited to) situations in which the following types of safeguards could have precluded the accident from happening: machine guarding, personal protective equipment, perimeter/fall protection, suitable and well-maintained electrical cords, grounding and bonding of flammable liquids during transfer, safe driving of and safe condition of a forklift truck, sufficient exhaust ventilation, limiting of weight to be lifted by one individual, the assurance of clear aisles, and lockout/tagout. The "freak accident" excuse is tossed around far too liberally.

Even if an employee was stung by a wasp and then banged into a machine, it might not be appropriate to rush to the judgment offered in this excuse. Quite possibly, it was known that wasps were in the area, but no means were provided or implemented to get rid of them. If so, the incident cannot be correctly classified as a "freak accident."

I did come across one case that comes closer to a category in which not much could reasonably have been done to avoid the accident. An employee was working in a large textile manufacturing room that had many looms. There was a large amount of loose cotton, but not to the degree of very poor housekeeping or of allowing respiratory hazards. She reported that a large ball of cotton came at her, she was startled, and then banged her arm against the frame of a machine when she jumped back in panic. She added that she later realized that the ball of cotton was actually a frog that had somehow become covered with the material. She literally meant a frog—a leaping, croaking, Kermit-type of creature.

It is one thing to deal with the harborage of flying insects or with rodent infestation, although it is not easy. This case was even more difficult. The plant was situated in a wooded area, some doors had been opened for ventilation, and no one supposedly had seen a frog near the building nor had reason to believe that one might pay a visit. Taking this to the nth degree, it is *possible* to lower the chances of a frog scare now that the potential for such an intrusion has been recognized. For one thing, the company could install sonic devices to convince the frogs to stay away. For another, ventilation could be obtained through screens or through openings higher than these frogs can jump.

This case is somewhat bizarre, and the injury sustained was not serious. It could have been serious, though. Despite the possible steps to avoid a recurrence, I can take a softer line and admit that this authentic case at least borders on being a "freak accident."

If two airplanes collide 20,000 ft above the ground, there are many possible causes. They can include poor maintenance, poor design, defective controls or gauges, ICBMs, pilot error, and so on. Thus, the referenced excuse has no place in the investigation relating to why the airplanes collided. However, if someone on the ground is struck by a falling airplane part, little can be offered for the victim as far as what he or she could have been doing differently. Of course, if a victim was standing too close to the area above which airplanes flew while watching them perform difficult and tight acrobatic maneuvers at an air show, that is another matter. In that situation, the persons who ran the show should have assured, by signs, barriers, and security/safety personnel, that unauthorized individuals could not enter the danger area. Although the victim (i.e., the spectator) was probably not engaged in employment at the time of the accident, the very point is that this serves to show how rare cases are where we could not have done much, if anything, to protect the victim (again, while still realizing that there were discernible causes for the collision itself and injury to the pilots).

I am not trying to suggest that the frog case and the airplane case are common occurrences. That would be absurd when you are concerned about OSHA-type hazards at the work facility. It is the *uncommon* aspects of these cases that serve my purpose by way of illustration.

The contractor left it that way.

I have been involved in OSHA cases in which the compliance safety and health officer observed an open, live electrical box. The employer blamed the electrical contractor who had left the box in that condition after repairing it. The company could have easily noticed that the box was open. An electrician was not needed to see this. The excuse does not hold any value. The longer a situation remains visible or otherwise perceivable, the more jeopardy the host company is in. This is not just in regard to OSHA; it goes to their very responsibility.

To start with, the company should select and hire a contractor that they have reason to believe is appropriate for the job. The contractor must have the proper experience, skill, and safety record. Licensing and insurance should also be considered. I have previously addressed the issue of a written program to cover any contractor entering your facility. That program is mainly for the protection of the employees of the contractor, but it could certainly include steps to assure that your employees are afforded protection as well. The scope and nature of the work must be clearly defined in the purchase order. I strongly recommend that the contractor be required to sign out with the safety director, plant engineer, or other person of relevant responsibility before departing the site.

However, it is quite another matter if a licensed electrician improperly wired the box, the company had no electricians of their own on staff, and the box was covered. In other words, there could be a hazard regarding a lack of grounding, but you might not be aware of it, even with reasonable diligence. Before the electrician leaves, you can ask to view a test with an instrument designed to show if the wiring is safe and proper. Even then, a mistake could have been made, which might show up later. So, even if minor shocks occur or sparks fly 3 days after the alleged repair, the contractor must be summoned.

There are times and resultant citation defenses relating to poor contractor work which may be difficult or impossible for you to detect. The more sophisticated and technical the work, and the more removed from a layperson's understanding and your line of expertise, the greater the chance of avoiding or working out of an OSHA citation. This is not just a matter of excuses to OSHA, though. This is about accident investigations. So one way or the other, if an accident occurs in any way relating to the work of a contractor, steps must be taken to avoid a recurrence, just as if your own personnel had been directly involved in the repair or maintenance. Get to the bottom of it and make it totally clear to your contractors. Be certain that they comprehend their role in an accident and that, if they are asked back, specific and sufficiently detailed precautions will be taken.

He must have been stupid to get his finger in there.

Not all employees are geniuses. In a different light, individuals may be quite intelligent but not be knowledgeable about the hazards that confront them. It is clearly management's responsibility to assure that all employees are adequately intelligent and adequately informed to perform their assigned duties safely. Management must make every attempt to foresee things that can go wrong with a machine or process. By so doing, they can instruct employees about what should be done when potentially dangerous situations arise. What are they to do if a part is jammed? What if the machine suddenly stops in midcycle? What if a power transmission chain requires lubrication and no one from the maintenance crew is available? The employee must also be made to understand the hazards and to recognize when new hazards are presented or old ones have become more of a threat. That is not automatically the case when an employee is simply told what to do and what not to do.

As previously addressed, it can make the critical difference to teach employees *why* certain sequences or precautions must be followed. Although the referenced excuse concerns a finger being injured in a machine, the notion that an employee was hurt because he or she was stupid can apply to other operations. What should an employee do if a certain chemical is spilled? Is it safe to walk in a particular area considering the floor load rating, powered industrial truck traffic, or movement of the crane load? How does one know if the electrical power on that box is energized, and can one make an assumption either way?

Employees have done things that they were told not to do. They have done things out of character with their regular performance. For the most part, if the right employees are hired, and they are trained and monitored correctly, the issue of stupidity should be irrelevant. Take the case of a young employee whose accident resulted in the amputation of most of three fingers. His fingers were caught in the power transmission chain of a commercial bakery machine. Did he have to be stupid to get caught in that spot? That would not be a just conclusion. So what could have transpired? Many negative factors could have come into play. The main hazard (and violation) was that the chain was unguarded. It was that simple. Still, there were those who contended that the victim had to be stupid to have his finger in the danger area when the machine was running. The situation is not so simple, and that conclusion is not fair.

Regardless of what other factors actually came into play, several cause-related issues remained and included questions such as the following: Was the training of the individual machine-specific, thorough, and clearly communicated? Were there controls or gauges in near proximity to the danger zone? Were any parts, chemicals, or supplies stored near that danger area? Was he required to work close to the danger area while operating another machine or monitoring this one? Was he attempting to release a jammed part, lubricate the chain, or knock or brush away bread dough? Did he have to reach over or near the danger area to access the product? Was the floor slippery or was there a tripping hazard (i.e., did he fall into or onto the chain)? Was the danger area properly illuminated and generally visible? Did he, by simple reflex, attempt to swat a fly away? Was the area congested so as to significantly impede his safe movement and not allow for a safe clearance distance from the chain whenever he could be in that area? Was he distracted by another employee harassing him or inadvertently startling him? Was he under the influence of alcohol or drugs (including legally prescribed drugs)? Was his attention deficient (and if so, why)? Could he have lost attention and been injured because his mind was on the money he lost on a ball game the night before, because of a disagreement he had with his girlfriend the day before, or because his mother is in the hospital?

Naturally, an employee should be fully focused on the job at hand. In the real world, that is easier said than done. In this case, the key is that there was no guard. It is not as if he removed the guard or climbed over a 7-ft-high guard. If the chain was fully guarded, then none of the other possible factors should have mattered. With the guard in place, even if he fell toward the danger zone, instinctively tried to remove dough with his hand, approached a button close to the chain, or did not fully concentrate on his work, he *would* not have sustained the injury because he *could* not have sustained the injury.

Sure they lose a finger once in a while, but they don't mind; they're craftsmen.

This is my all-time favorite. These exact words were said to me when I was employed by OSHA and had gone to a small wood cabinet manufacturing shop for the purpose of conducting an inspection. The shop was one large room with a small glass enclosure (a booth of sorts) that served as an office in the center. When I arrived, the owner was on the telephone in that office. He looked up at me in a quizzical way, so I quietly showed him my federal credentials and indicated that I did not want to interrupt him and that I would wait. I glanced around the shop without moving from the office. It was immediately apparent that there was a major dearth of required guarding. This included various types of saws and other typical woodworking machinery. I did not say one word and did not consciously display any particular emotion on my face or with body language.

When he got off of the telephone, I said, "Hi." Before I could continue my greeting, he uttered the incredible, unsolicited classic excuse without any introduction. He did not mention saws, what he thought I was looking at, point toward the employees, or even use the word "employees." He simply said the aforementioned words with no noticeable lack of sincerity. His view was that such permanent injuries were acceptable and were emblematic in the sense of some grotesque form of a badge of honor. Nothing more needs to be said about this one.

16

Injury and Illness Records and Reports

The last chapter addressed accident investigation in great detail. If any injury or occupational illness is sustained, no matter how minor, a record should be maintained. In many cases, there is not an OSHA requirement to do so. Nevertheless, I advise that there be a record, which may be as simple as a first-aid log. One reason for that is because nonrecordable first-aid cases may evolve into recordable cases. It is also good practice to record significant exposures (relative to permissible exposure limits) to toxic and hazardous substances, including when there is no apparent legal overexposure and when no symptoms have been reported. Those exposures may have been discerned through the analysis of sampling data. In so recording, medical surveillance can be brought into play, and a foundation has been laid for possible delayed or chronic adverse effects. Again, the case may evolve into a recordable one.

Spotting Trends

Case logs, whether required by OSHA or not, can serve as an excellent guide to hazard identification. Some trends can be spotted more easily than others. Look for them when there are multiple cases involving the same department, occupation, task, part of the body, diagnosis, or victim. I further suggest that your interest should be piqued when there is a preponderance of cases relating to the same supervisor, shift, or approximate time of day. An even more detailed study could show that employees are being injured early in the work week, late in the work week, or while working overtime. The occupational safety and health community has begun to look seriously at the theory of biorhythms and circadian cycles. It is apparent that these body-clock factors can be relevant components in the predictability and avoidability of accidents. After working very long hours, an employee's performance may well be significantly impaired. The diminished ability to function effectively can be earmarked by not only fatigue, but a lack of alertness, vigilance, and concentration as well. The employee may experience a reduced body temperature. For those working at hours of the day (times) to which they are not accustomed, the situation is often more dire. This may translate to "the middle of the night" for most individuals, as that is when they are in the lowest trough of the rhythm or cycle. However, some employees have adapted to extremely early morning hours and actually perform well. Others (giving credence to the term "night people") may thrive at those hours without the need for adaptation. The general issue is an important one. There should be no assumption that particular employees will easily adjust to long or unfamiliar hours simply because they are motivated to do so.

Some obvious trends can squarely highlight the need for process review, implementation of written procedures, personal protective equipment, guarding, ventilation, and so on. Some can boldly point to the need for employee retraining, sometimes for an individual and sometimes for all those who perform the same particular task. Are certain tasks regularly performed so as to allow for high risks? Keep in mind that this is not an implied indictment of the victims. By way of conditions or instructions, the results of their behavior may have been predestined for failure.

The discovery and evaluation of any or all of these types of trends can form the basis of a sophisticated risk location map. In turn, this can be an invaluable aid in the detection of various contributing causes that had been previously overlooked. Be inquisitive. Look deeply for trends whether the tie-in appears to be subtle or substantive. Information in the accident investigation chapter can be applied to help formulate the probing questions.

Some trends may call for a more profound analysis. Where there are several cases relating to sprains, strains, bumps, and bruises, is there a layout problem? Are more lefties being injured on a particular operation than are righties (or vice versa)? Is the enforcement of safe conditions and safe behavior particularly lax on a certain shift or when in the purview of a specific supervisor? Might fatigue be a significant factor? Do particular employees function at a safer level at certain times of the day or night?

The questions are only limited by your imagination. The hunt for answers is a challenge, but it is not a game. It is a very important mechanism for the prevention of accidents. Still, the study of injury and illness cases is only one weapon in the arsenal to prevent future accidents. In jobs, situations, and locations where no injuries or illnesses have occurred, the quest to uncover "accidents waiting to happen" must continue. If there is high absenteeism or turnover in a particular department or job, that may well serve as a tip-off. Perhaps, the very point is that the potential risks are what is keeping employees from regularly reporting to work.

OSHA Injury and Illness Reporting Requirements

OSHA has three major requirements for the reporting of injuries or illnesses. The first two are generally labeled *"regulatory violations"*. If there is a fatality that resulted from a work-related incident, it must be verbally reported to the relevant OSHA area office or by using the OSHA toll-free central telephone number within 8 hr following the employee's death. The second scenario is when there is the in-patient hospitalization of three or more employees as a result of a work-related incident. This is sometimes referred to by OSHA personnel as a *catastrophe*. The same reporting requirements are in effect. They apply when the fatality or hospitalizations occurred within 30 days of the incident.

If there is any doubt as to the work-related nature of the fatality, I strongly recommend that it be reported. In one case, an employee suffered a heart attack and fell off of a scaffold. The company did not report to OSHA, as they later claimed that the death was not work related. This may have been a close legal issue. OSHA was on solid ground in its desire to know about the incident. What doubt could there have been that it was not work related? For one thing, might the employee have been stricken after losing his balance? Was perimeter protection in place and was the victim wearing a personal fall protection device? Might he have died as a result of the fall, as opposed to the heart attack itself? Could he have been overcome by an airborne toxin or could such exposure have contributed to the heart attack? Might he have come in contact with a live electrical part that played a role in the onset of the heart attack? Might he have been working with a compressed air gun earlier when an air bubble was introduced into his bloodstream, resulting in an embolism?

There can be a case where an office worker just simply keels over and dies. He or she was not near any live electrical parts, liquid chemicals, or compressed air. What if vapors or fumes from a nearby ventilation duct were the culprits? What if the person fell earlier in the day, particularly if the head received a sharp jolt, and suffered an aneurysm? Both examples are not only presented as cautions regarding OSHA reporting requirements. They are also presented to illustrate the kinds of elements, hidden at first glance, that can be underlying causes.

The third OSHA injury (illnesses are not involved) reporting requirement is not regulatory. Thus, it is neglected more often than the other two. This requirement can be found in the mechanical power presses standards. It requires the reporting of all mechanical power press point of operation injuries that were sustained by operators or other employees. The employer must make the report within 30 days of the occurrence to either the director of the Office of Standards Development in the national office of OSHA or to the state agency administering the approved OSHA-equivalent plan. The standard lists numerous details that must be included in the report.

If you choose to convey the information to your OSHA area office, they may not require you to contact the standard-referenced office as well. I do not suggest that you take this route. If you do, there is a strong chance that an inspection will be initiated. If you report to the national office, there is less of a chance. The national office should send your report to the area office, but this will usually be by way of the regional office. So depending on the workload of the national office, it may be quite a while before the relevant enforcement office receives the information. When it is received, the person in authority may decide that too much time has passed to allow for a worthwhile inspection, particularly if your report (preferably written) spelled out how and when total abatement and compliance were achieved. He or she may not contact you at all or may call just to request a written or verbal update.

OSHA Injury and Illness Recordkeeping Requirements

OSHA enforces a set of requirements concerning occupational injury and illness recordkeeping. The employer must maintain an annual log of occupational injuries and illnesses and comply with record retention requirements. The employer is required to record information about every occupational fatality. With recordkeeping system revisions, all nonfatal occupational injuries and illnesses that involve any of the following must be recorded: loss of consciousness, day(s) of restricted work, job transfer, day(s) away from work, or medical treatment beyond first-aid. A copy of an annual summary must be posted in the place or places where notices to employees are customarily posted. This copy must be posted no later than February 1. The official responsible for the annual summary totals must certify that the totals are true and complete by signing the form.

In my chapter addressing specific programs required by OSHA, I did not detail each of the provisions of every standard. That is not the purpose of this book. Although I employ the same logic here, I do allude to common errors. When making entries regarding the description of the injury or illness, be specific and do not simply describe how the accident happened. There is no requirement to explain how the accident happened. Do not just enter "finger." Be as precise as you can be, although a diagnosis may be formalized or modified later, and the entry can be altered. Which hand was involved (right or left)? Which finger was involved? Can you address whether it was a particular knuckle, fingernail, or other specific joint or area? Was it fractured? Was there a contusion, laceration, or puncture? Was there an amputation? Was there a case of dermatitis or a thermal, chemical, or electrical burn (if so, to what degree)?

Do not forget that light duty is a form of restricted work. It must be recorded, even if no time at all was lost from work. On several occasions, OSHA has uncovered trends of unrecorded restricted work. In many cases, the restricted work followed day(s) away from work. Even if those day(s) away from work were recorded, the record must reflect the subsequent occurrence of restricted work. OSHA is very attuned to these situations. In many cases of noncompliance, often involving cumulative trauma disorders, OSHA has issued "willful" citations carrying extremely high proposed penalties. Too often, companies would post a sign (often in public view) proclaiming "one million man hours without a lost time accident" or something similar. Generally, those numbers did not account for light duty (restricted work) cases. There were sometimes several employees working in the plant who were barely in working condition. At times, employees were brought to work in a wheelchair, or skilled production workers were performing menial (if not meaningless) tasks.

As addressed in my chapter on medical case management and return to work, there is definitely a highly regarded place for benevolent, well-designed light duty programs. In all such cases, there must be assurance that the particular assignments are not started prematurely. However, the hiding of workplace injuries or illnesses that were sufficiently severe to limit an employee's ability and output is a very serious offense. This type of concealment can be immediately dangerous to the employees involved. It also removes the focus from accidents that should be held up to the light for thorough investigation. Further, it hampers OSHA's efforts to refine their targeting systems by analyzing accident frequencies and trends in various categories, including that of standard industrial classifications.

For every injury or illness required by OSHA to be entered in the log, additional information must be recorded on a supplementary record. That supplementary record describes how the injury or illness exposure occurred, lists the objects or substances involved, and indicates the nature of the injury or illness and the part(s) of the body affected. As addressed in my chapter on accident investigation, there is much additional, critical information that should be included on some form, even when not required by OSHA. Most companies first use the OSHA supplementary form, an equivalent workers' compensation or insurance form, or a slightly enhanced, customized form to gain compliance. Then, they use a detailed accident investigation form that goes much further. That form should seek to identify all of the primary and secondary causes that were a part of the chain of events leading to the accident. It should also include steps recommended for immediate action and steps recommended for long-term action to minimize the remaining risks.

OSHA, in consort with the Bureau of Labor Statistics, is continually researching ideas to simplify occupational injury and illness recordkeeping requirements, while making them more amenable to meaningful interpretation. The agencies publish explanations which are designed to help clarify the fine points of the standards. OSHA is particularly aware of the need to distinguish between cases that need not be recorded and those that must be recorded. Most of the gray areas have been removed, particularly in regard to what constitutes medical treatment beyond first-aid. Additionally, that distinction— between first-aid and medical treatment—is based solely on the specific treatment involved, not on the person (even if a health care professional) who provided it.

As with all OSHA technical standards, I strongly recommend that you keep apprised of modifications and updates in the recordkeeping standards. Perhaps your establishment is affected by changing exemptions regarding employers in certain SIC codes, as well as those with small numbers of employees. Be prepared to comply with the provisions regarding improved employee access to recordkeeping documents. Become closely acquainted with OSHA's special list of recordable, work-related health "conditions" that do not necessarily fit into the categories referenced earlier. The criteria for recordability, for some of these conditions, relate to the results of medical surveillance or testing.

Chapter

17

Setting Priorities

The Prioritizing Loop

There are very few companies that have the luxury of being able to start work on many safety projects immediately and simultaneously. Priorities must be set, whether they be set by a safety committee, by a safety director, or by another individual. Some safety projects might be handled at the actual "doing" stage by one department and other projects by a different department. Training could be the job of the safety department, the personnel department, or a department in larger companies dedicated to across-the-board training in itself. In very small companies, it is understood that such departments will not even exist. Certain safety projects must work through an engineering department or at least a person who serves as plant engineer. In other cases, it is the maintenance department, which may be only one individual in a very small shop, that must get the job done. It may be the purchasing department that has the responsibility for assuring that no chemicals are received at the company, unless accompanied by the proper hazard communication label and material safety data sheet. At times, an individual responsible for building and grounds will be the person to see a project through.

In situations where corrections are needed related to on-the-floor practices (e.g., the frequent blocking of aisles and exits and the inconsistent use of required personal protective equipment), it will probably be the job of line supervision to deal with the matter and to assure compliance. That is a nonconventional form of a project, but the concept is still applicable. If the projects that require attention will be addressed by numerous departments or individuals (i.e., one department or individual is independently working on one project, while another department or individual is independently working on another project), then there is not much concern as to which project receives the highest priority.

If the project is interactive and involves more than one department, coordination becomes a factor. This is the case because one of those departments may have to wait for the other one to be free from the high-priority project that it has been working on and is obligated to complete. Only an imminent danger or extremely urgent matter should interrupt an in-process project. When there are too many midstream projects, especially if work on them has been halted and resources have been redirected, coordination, paperwork, and the general values of continuity become severely complicated and compromised.

There will probably be times that outside contractor(s) will be called in to be involved with or to totally address a safety project. This is more common in smaller establishments or in establishments with fewer pertinent resources. This could relate to the construction and installation of perimeter protection, additional storage racks, or machine guarding. It can also be for the repair or preventive maintenance of cranes or highly technical equipment. Other possibilities include contracting out for spill control, neutralization, and reversal; however, this is generally out of the prioritizing loop in that it usually must be dealt with rapidly. In some cases, HVAC (heating, ventilation, and air conditioning) work will be done by a contractor. Again looking at very small companies, electrical and plumbing work could be performed by a contractor. Another category for contracting out involves private safety and health consultants, who perform such services as training, demonstrations, air sampling, simulated OSHA inspections, and similar activities.

Criteria for Priorities

The company should have a plan that makes clear what departments or individuals will generally be responsible for what types of safety projects. In some cases, this may call for the summoning of help from the corporate office. This type of plan should be in place prior to when there is a need to assign a priority rating to a project. If that is not done, the process can be slowed by that basic glitch. If it is done when the need to assign priority rating arises, the frustration of deciding with whom the project will kick off will be greatly alleviated. When setting priorities and concentrating primarily on when they must be tackled in a sequential manner (first one, then the other), what criteria should be established? The main criteria concern the following: If someone gets hurt or sick due to the unsafe or unhealthful situation, what is the reasonably predictable severity of injury or illness (always leaning toward the more severe)? What is the probability of a related accident resulting in injury or illness? How many persons are exposed to the hazards? To universally categorize these three criteria in order of importance is not only difficult, but it can lead to a flawed decision as well.

When OSHA considers whether an alleged violation is to be classified as "serious" or "other than serious," it is supposed to ignore the second two criteria that I have mentioned. In deciding whether or not to label the alleged violation as "serious," OSHA makes a determination as to whether there is a substantial probability that death or serious physical harm could result. (Note: The other part of the OSHA legal criteria for classifying a violation as "serious" is unrelated to the subject in the present context. It is whether or not the employer knew, or should have known, of the hazard. In our discussion of prioritizing, the supposition is that the employer has already recognized the hazard.)

The theory behind OSHA's criteria relating to severity has a sound base. However, it is not the end all. There are situations when an accident would most likely result in a very serious injury, but the chances of any injury occurring are very, very small. For example, there is a requirement stating in effect that mezzanine storage areas must be provided with a clear, visible indication of the floor load rating. This standard does not presuppose that the floor is overloaded. What accident can occur if there is not a sign posted? If the sign is not up, *and* the floor is overloaded (a separate standard and one that does not have be violated for the posting requirement to be cited), the concern is that there will be a floor collapse. Moving to the next step, if the floor collapses, there might not be anyone in a position to be injured. Yet, keeping in mind OSHA's criteria, if someone is injured (whether he or she was below the raised floor or came down with it), the most likely severity will be high—extremely high.

I suggest that, in pure consideration of priorities, the formulation and installation of the sign is not high on the list. If the floor is overloaded, that is certainly another matter. There are untold numbers and kinds of hazards that have a far greater probability of injury resulting. In many cases, they would carry a higher priority rating than would the lack of the load rating sign.

To set priorities is an intricate decision-making process. To consider it anything less would be a mistake. Nevertheless, the decision must be made promptly, and the plan for abatement must get on track. You do not need a committee to form another committee to call a meeting to assign a task force to establish a priority for a particular project! Take a logical approach, integrating the facts germane to the criteria which I have described, and go forward. Priorities can be labeled A, B, C, or with a similar system. Naturally, the As would be the highest priority, but considerations do not stop there. If you steadfastly refuse to take care of any Bs or Cs until all As are completed, you may never get to the Bs and Cs.

If the concern was worth listing at all, it must be addressed. The concept must hold: All deficiencies will be corrected. Some examples of when smaller and/or easier projects of lower priority can be addressed are the following: while awaiting the arrival of equipment or personnel for the As, when there is an open time slot of short duration where work on an A would not be practical, when several of those Bs or Cs could be grouped, and when a person who would not be involved in the resolution of an A could "knock off" a B or C in a satisfactory manner.

Even if your company would not normally use a contractor for a certain type of job, you might wish to reconsider when you are backed up with a significant accumulation of A priorities. Permit the contractor to help absorb and eliminate your backlog, as long as you keep close tabs on scheduling and progress. At times, you will be able to (and find it desirable to) include a time completion clause in a contract. This sort of clause could help motivate the contractor to complete the high-priority project in a timely fashion. That motivation comes in the form of potential back charges and fiscal rebates. I do not recommend an extra payment for completion ahead-of-schedule, unless you can be 100 percent certain that the integrity of the work has not suffered through haste and cutting corners. Not only will such "out-sourcing" efforts help you to achieve compliance with OSHA more quickly and afford your employees the protections to which they have a right, but as a result you will then be in a better position to hit those lower priority items, including those annoying ones that have cropped up inconsistently but frequently or those that have simply remained on your "to do" lists for a very long time.

When evaluating what steps must be taken to achieve compliance, do not neglect your role as a person of particular title or as a company in assuring the safety and health of employees. Yes, there is a difference. There are often conditions and practices that, while not in violation of the law, present hazardous situations. OSHA standards are minimal. It is a good idea to exceed them, not just for personal protective equipment (addressed elsewhere in more detail), but for such subjects as guarding of belts, gears, chains, couplings, fan blades, and similar moving parts. In those cases, OSHA requires guarding whenever the dangerous parts of the equipment are within 7 ft of the floor or working platform. I recommend the guarding at all heights. To begin with, the 7-ft rule was derived from consensus standards that were in place decades ago. The average height of the population in the United States has grown. Many persons are taller than 6 ft, and there are those who can easily reach to a height in excess of 7 ft. In another light, guarding is important because power transmission equipment and other dangers may be approached by ladder, scaffold, or vehicle for maintenance or repair or because they are near storage racks or the tops of tanks.

Similarly, OSHA requires perimeter protection when open-sided floors or platforms are 4 ft or more above the adjacent floor or ground level (height is irrelevant, by way of a distinct standard, when above dangerous equipment or tanks). I recommend that the need for perimeter protection be evaluated whenever there is the potential of a fall to another level.

It should be a given that certain states of continuing compliance are of a very high priority. As a kickoff point, this concept revisits the constant diligence required to assure unimpeded access to exits, extinguishers, and disconnects.

18

Training and Education

Several OSHA standards require that employees receive training. All employees should receive at least basic safety and health training whether or not obligated by those standards. That is the case even if they are not covered by the training requirements of a specific OSHA standard. (There are very few situations where employees are exempt from training under the hazard communication standard.) Most courses deal with function-specific training, whereas others serve as a base and deal with general awareness and familiarity. With all training concerns, from the orientation to the standard-specific and task-specific, a particular order of decision making should be considered. Training needs should be identified. This includes not only the contents, but what specific messages are to be transferred to the trainees and what exactly those trainees need to do with the information that they receive. This is often determined by OSHA. However, a true need can become known when employees themselves make it clear.

Employee Input

One idea rarely implemented, but one that can have great positive pay back, is to hand out short, simple survey forms periodically. These questionnaires are designed to gather input on what training the employees feel is necessary. The forms should not be signed. They are to be filled out on company time. Questions would address whether any aspects of their jobs frighten them, whether they are unsure about how to perform the tasks safely, and whether they feel there are unsafe conditions that cannot be brought to an acceptable level even when they perform their duties as directed. The questionnaire can also address whether employees have experienced any close calls or witnessed others having experienced close calls.

Training Needs

Training needs include not only the contents but what specific messages are to be transferred to the trainees and what exactly those trainees need to do with the information that they receive. The establishment of precise goals and objectives is essential. Keeping this in mind, the trainer and the enabling manager must determine how the trainees have to demonstrate or confirm that they have learned just what they were intended to learn. Learning does not just mean that they now have absorbed all of the facts that were taught. It means that they know how to benefit from the intake of those facts and that they know how to integrate the information, skills, and techniques that have been taught so that they can perform in a safe manner. Suitable learning activities must be built into the course. After the training is conducted, the course is evaluated for effectiveness. Finally, it is improved if analysis and feedback reveal a need.

The company's orientation program should include a well-structured session dealing with the fundamentals of safety and health on the job. From the first day of such training forward, employees should be convinced to have positive expectations about safety training. Thus, they will look forward to training sessions and feel that, as participants, they are an integral part and benefactors of an important process. Their interest and attention can be captured when they view training (as offered by/through their employer) as a fruitful and useful experience. It is a nice touch to have someone from upper management introduce the training to demonstrate "live and in person" his or her support. The session should then turn to the types of hazards particularly relevant to the work at the establishment. The trainees should have a secure feeling from the start that the training was designed with their well-being in mind. The training session is a serious matter and management should view it that way; it is not simply time set aside so that some government rule can be checked off. Then, employees should be trained in the hazards relating to their specific duties.

Do not lose sight of the basic fact that training, no matter how effectively done, cannot take the place of safe work conditions. For instance, training should not have to include a warning to work carefully around unguarded machinery or to walk carefully when descending stairs that lack railings. The machines should be guarded, and the stairs should have railings. If employees are to receive training related to a specific OSHA standard, that training must be tailored to the specific hazards that those employees may face. For instance, hazard communication training cannot simply consist of a general explanation of the dangers presented by chemicals. The training must describe the hazards, precautions, emergency responses, and so on directly tied into the specific chemicals with which those employees work or to which they might otherwise be exposed. Still further, the training must cover the particular uses and handling techniques of those chemicals. For control of hazardous energy, it is not sufficient to train employees (at least those considered to be "authorized" or "affected" under the standard) in the general hazards and methods of lockout/tagout. The training must address, in detail, exactly how and in what sequence to bring each particular piece of covered machinery into a condition of ZES (zero energy state).

For permit-required confined space training, the specific, actual steps to be taken for employee protection can differ for each pertinent confined space. Those steps must be taught, and it must be made clear that the steps taken for one confined space must not be assumed to be safe and adequate for another confined space. For one more example, if training is necessary under the standard for powered industrial trucks, it must include the safe driving methods and special hazards relating to specific vehicles. One type of forklift truck may operate differently, and hence require different special precautions, than another vehicle, even if the other one performs a similar function.

Your plans for training should be coordinated with the personnel department. It is advantageous to avoid staggered hiring, which results in two persons starting work on the 1st of the month, three on the 5th of that month, and then two more on the 8th. It is much easier and much more practical to have them all start at the same time so they can all be trained at once. Otherwise, there is a proven tendency to not train the persons hired on the 1st and 5th until the 8th, when the last of the group is brought on board. Then, they will all be trained together. Of course, that puts the new employees and possibly those that they might endanger in jeopardy until the 8th.

The forgotten side of the coin is that this often constitutes a violation of a particular OSHA standard. I worked with one company that received a visit from OSHA the very day that a particular employee began employment there. She worked in a machine shop, where she was likely to have her skin contact a coolant that was in a trough. She had not received hazard communication training regarding that chemical or any other to which she could be exposed. The company received a citation. As it turned out (and OSHA discerned this), the employee suffered from an existing skin condition.

Trainers and Communication

It is of paramount importance that the trainers do not fail to recognize language barriers. Then, they must be sure not to neglect or simply "write off" those barriers. In an earlier section, I listed various types of handicaps, many of which can interfere with proper training. This generally means that the trainee is handicapped in some way that will impede the success of conventional training. However, at times, the person assigned to conduct the training (whether it be in a formal or a nonformal setting) is not suitable. Sometimes this is because he or she does not speak clearly. It can also be that the trainer lacks patience, which will severely hamper the transfer of information.

As to language barriers, there is another way to view the problem. One might evaluate the trainer-trainee interface with an interesting twist: The trainer is handicapped in that he or she cannot speak the only language that the trainee comprehends. Either way, regardless of the subject or substance, if the communication is not totally understood, the training will fall short of the objectives and a hazard trap may lie in wait. It is bad enough if the training is not understood, and the trainer realizes that more must be done to complete the training before the trainee is released to start the job. It is far worse if the training is not understood, and the trainer is laboring under the false assumption that the trainee understands.

Good trainers are also good active listeners. They must make it clear that they are not talking "at" the trainees. While the trainers' expertise naturally places them in the lead role, they should also engage in conversations "with" their audience. Trainers should deliver information by making a connection with that audience. In so doing, everyone feels included, not simply as a body taking up space, but as a valuable individual. Each trainee should recognize his/her role as a part of a team. At the same time, individuals must be recognized by the trainer. For that reason the use of name tags should be considered and trainers should try as much as possible to address trainees by name.

Trainers must be absolutely certain that the trainees "get it." They must assure that the trainees understand the purpose and objective of the training and why and how it will be useful to them. When a particular OSHA standard is being discussed, the rationale for it must be explained. Trainees should not be left to think that the standard was simply dreamed up by someone in Washington. In fact, whenever the need for particular conditions or actions is discussed, the trainees must be told exactly why sound safety principles and practices (not just company or government officials) dictate that those conditions and actions are required.

For the most part, the order in which the material is presented should coincide with the order in which it will be used on the job. A major goal of the training is to leave employees with sufficient practical information so that they can practice and apply the newly acquired knowledge and skills when they return to or begin the related work tasks.

If there is a conventional language barrier, an interpreter is probably needed, but it must be a person who can properly translate the exact words or their equivalent. The interpreter must know how to deal with the vocabulary of your business. This might involve high-tech language, for instance. If interpreters are used (whether from within the plant or through a social services agency or similar source), training records (particularly those required by OSHA) should have an extra statement indicating that an interpreter was used. Both the trainee and the interpreter should sign off. There is a large, potential problem if there is not going to be someone else available in the plant to translate at all times that the trainee is working. Keep in mind that some English words and idioms have totally different meanings in other languages and cultures, even if the words per se are understood.

A different type of language barrier can actually be caused by the trainer. That is when the trainer uses language that is unnecessarily complex and beyond the educational level of the trainee. Say it as simply as possible. I am not suggesting a baby-talk tone. Yet, get the points across, no matter what it takes. You might be training an employee who is mentally retarded. That person may have the makings of a superior worker, but you may well have to adapt your vocabulary and method of training to suit the relationship between trainer and trainee.

Trainers must be more than technical whizzes, if that is necessary at all. Naturally, they must be intimately familiar with the material, but must also be somewhat personable. Trainers must be highly skilled communicators and motivators as well. Further, they must be good planners and organizers and must give the audience an accurate and sincere perception that they believe in the material and the point of the session. Trainers must enjoy their role and be comfortable with it, or the audience will be lost, to a large degree, from the onset. If the trainer is someone known to the employees, they had better have gained a respect for their trainer, particularly as related to the trainer's interest and example setting when it comes to safety. Again, if not, the battle is lost before it begins. Ideally, the trainer has some actual hands-on working experience regarding the subject, regardless of how much book knowledge he or she possesses. Trainees have added respect for a trainer "who has been there" and who can address far more than theory.

The trainer is an integral part of the overall training experience. Handouts and audiovisuals are not enough. One person may be fully competent to provide in-depth instruction regarding one technical subject but not be the correct choice to instruct on another subject. Therefore, in almost all cases, different trainers are needed for different courses.

Ideally, the trainees should be able to contact the trainer days or months after the course has ended. This can be more difficult, logistically, if an outside consultant or corporate person served as the trainer. One way or another, employees should feel that learning is not bounded by the time on the clock at the end of the day and that they can always revisit the material to gain clarification or expand the scope and depth of their working knowledge of the subject. Learning is a continuous process, in general. The company must not insist that all learning be limited to specific rooms at specific times. The very desire to learn should be encouraged. More specifically, the trainee may simply have lost the meaning of part of the imparted information or may have run across an unusual situation that is relevant to the course material but had not been addressed.

Reality-Based Training

The posing of questions and the soliciting of points of view and germane personal stories should not be overlooked. It is essential to hear from the floor the actual life stories that give true meaning to the material being communicated. Real-life horror stories do have a way of drawing attention. So do those stories that have happy endings in which tragedy was avoided due to the use of personal protective equipment or quick emergency response. Whether those stories (the closer to firsthand, the better) are related by the trainer or by trainees, they can illustrate a significant point very well. Another possibility is to bring in individuals who are or were in a job similar to that of the trainees and permit them to tell their stories.

In all of these cases, the trainees need to know that real accidents, real injuries, and real pain happen to real employees, including those to whom they can relate. Did anyone in the room have an accident, or know of someone who did, relating to the subject matter? The more that the trainees can be shown that the subject is applicable to their work, the better. They must learn the personal relevance of the material.

Employee Understanding and Listening Skills

Comprehension checks should be done regularly. Are the trainees learning? Is the trainer clear in his or her explanations? Is the pace and volume of the trainer's speech appropriate? These checks cannot, effectively, be in the narrow form of inquiries as to whether the information is understood or the trainer is going too fast. The questions must be designed to elicit specific responses that reflect an understanding of the material. Ask questions for which the correct answers are definite and have been aired during the session.

An excellent course, in itself, concerns how to develop and implement listening skills. Although it may not be practical in all companies, it would be beneficial to present such a course (even a short one) for all employees early in their employment life with the firm. The course would be designed not only to improve the absorption and understanding of material imparted at formal safety training sessions but also to improve the general listening skills of employees. To appreciate the need for such a course, it is necessary to understand that good listening skills are not automatic; they must be learned. Even would-be trainers can gain quite a lot from such a course, which would in turn benefit the trainees. A person can possess 10 professional degrees and be a technowizard, but still lack the talent to convey the information in understandable terms. This could relate to conversations on the factory floor.

The Time and Place for Meetings

As with safety committee meetings, all training meetings should have specific starting times, be in comfortable surroundings (although there should be times when the trainees are brought out to the floor to see the machines, processes, or other visible presence of the subject matter), be uninterrupted (except for real emergencies), and so on.

Always keep the number of trainees at one session to a manageable level. Be sure that they can see whatever audiovisuals or written material are to be shown and that they can hear everything that you have to say. As obvious as this seems, I have attended training sessions, as both trainer and trainee, where this was a very real problem. Sometimes it has been because the room was too big, sometimes because street noise was heard through an open window, and sometimes (in a slightly different vein) because the room was just too hot to allow for a concentration level adequate to gather input. You might find that certain days of the week and certain times of the day work better for training to be effective. There is also the question of whether the training should be at the beginning of the employee's shift or at the end. When at all feasible, it is apparent that it is more beneficial to hold sessions at the beginning of a shift.

An added consideration comes into play when a company operates on more than one shift. Coordination of training sessions can become a bit tricky in those cases. There is also the matter of whether training should be given just prior to vacation or the first thing upon return from vacation. There are many professional points of view on the matter of what days of the week and times of the day are best for training sessions. Most experienced trainers, who have the opportunity to assess what days are generally the most effective for training, prefer Mondays because the workweek slate is relatively clean.

Try to assure that no one is attempting to learn on an empty stomach. On the other hand, it is also well known that attempting to learn (even stay awake and focused) after a heavy meal is quite difficult. Many professionals feel that a good time to start a training session is in the relatively early morning, after a substantial breakfast has been enjoyed. Meal patterns are more difficult to predict regarding second and third shifts. Nevertheless, starting at the beginning of the shift appears to be preferable.

Setting the Proper Tone

The significance of the training as it relates to the safety and health of the employees and, to a lesser extent, as it relates to establishing compliance with OSHA standards must be established at the outset of each session. To communicate that importance, the trainer need not be dour and stiff. In fact, the trainer should put the trainees at ease. I continue to urge the relating of firsthand experiences and the depicting of what can go wrong and what the consequences can be. Some tactful coaxing might be needed, but the attention gained can be priceless. The trainees must understand that the training is not just offered as a cosmetic exercise to please some starched white-shirt bureaucrats in Washington, D.C. They must understand that employees get hurt, get sick, and sometimes die if they do not follow the instructions offered in the training sessions. You might also tell about high OSHA penalties and how that can wreak havoc in the company. Just be certain to explain that the safety and health of each employee is the number one purpose of the training. Compliance with OSHA is secondary, but mandatory.

Written Material

The benefit of training sessions can be enhanced if the trainees are given germane documents, including checklists, cautionary information, and standard operating procedures. Besides the copies that are given to individuals, master copies should always be available. Some trainers feel that it is distracting to distribute these documents at the beginning of a session. On the other hand, without the literature in front of them, the trainees will often, needlessly, take a lot of notes. Sometimes they cannot keep up with the information being imparted, and sometimes they simply do not write down the information as it was imparted. There are those that may put down the proper information, but cannot read their own writing later.

Looking at both views, I suggest that you explain at the beginning of the session that literature will be given out. As each topic or module is about to begin, disseminate copies of what you want the trainees to have in front of them during that portion of the session. Tell them that you will let them know when each segment of written material is being discussed. You do not want them to get ahead by glancing at pages that have not yet been addressed. When the discussion corresponds to a particular section of the written material, you will inform them. Explain that they need not take too many separate notes because the documents should pretty much cover it.

If they have a need to take notes, do not dissuade them. Their questions and note-taking should send the trainer a message. Do I need to build in more up-front information and/or add something to the written material the next time I give the course? In some cases, you might ask for the written material to be returned. Then, you will simply go to the word processor, make additions, deletions, and modifications (i.e., corrections or clarifications), and give the latest dated edition to the trainees. Be sure to get the old edition literally back in your hand before leaving the room and redoing the material.

When dealing with the printed word, whether it be on a chalkboard, a label, or the written material given out, you should be concerned about employees who are illiterate. Sadly, this is not as rare as management realizes or chooses to admit. If the trainee is illiterate, it is the company's responsibility to be sure that the information is understood. Frequently, management can be fooled by illiterate employees. They may get help from friends or just be extraordinarily good listeners with excellent memories. Certainly, there is an up side to those talents and abilities, but it is imperative that management learn if employees are illiterate. Then deal with it. More and more companies are sending employees to classes to learn how to read and write in English, whether that is their supposed native tongue or not.

Audiovisuals

Audiovisuals are an important tool in training sessions. Overheads (e.g., view graphs) and slides have great value. If you use videotapes, it is preferable to have machine features that will allow for clear freeze-framing and clear slow-speed advance.

There are numerous sources of training videotapes. Some are available in languages other than English. Keep in mind that videotapes and other audiovisuals are often available on free loan from your insurance carrier or OSHA. If the audiovisuals were totally produced by the United States Department of Labor, you have the right to reproduce them. The same is true for any literature or audiovisuals produced by any branch of the federal government. Sometimes, OSHA loans audiovisuals that were, at least in part, produced by another source, and they will generally make it clear that the audiovisual is protected by copyright and that you are not allowed to create copies.

It is a good idea to produce your own videotapes that show your own machines, processes, and so on. The cost can be very low. Interest is piqued when your own employees appear. The videotapes can show what is a safe operation. I do not recommend that you videotape or photograph any of your employees doing something unsafe. If you want to depict the incorrect, unsafe way to do something, it must be done in a totally controlled manner that does not actually expose anyone to hazards during the taping. On the videotape, it should be explained verbally that no employee was in danger when the unsafe situation was depicted. You can add a subtitle to assure that understanding. For these depictions, you might prefer to use management personnel.

Interactive training is a special method whereby trainees, often at a time most convenient for the individual trainee, will operate a videotape and/or computer (perhaps a CD-ROM "movie" with audio) and be given the opportunity to answer questions before moving on. This type of system more or less forces the participant to be involved. An entire section could be written on this emerging technology.

Yet, no matter how advanced and comprehensive the audiovisuals, their use should always be supplemented with the direct verbal interplay between experienced, highly knowledgeable trainers and the trainees. Thus, audiovisuals should never be the sole element or vehicle of the training. I choose to leave you with the brief reference and to suggest that such programs cannot take the place of a full-blown training session with a live trainer. The interactive programs can be used as refreshers or even precursors to the full course.

Compensation for Time

As with safety committee meetings, the participants should generally be compensated at their regular (or average) rate of pay for any time spent in training. This does not mean their base pay; it means that if their average includes a bonus, so be it. There are still companies that do not do this. They may be in jeopardy of violating various labor laws, even though OSHA does not generally spell out such a requirement. There may be times that training will be provided off-site by a private or governmental source. It may be in the form of a half-day or full-day seminar. It may be in the form of a formal college course, given in the evening over a period of several weeks. In that case, if the employee volunteered to participate outside of regular working hours, an exception to the compensation rule would be fair.

Training Experienced Employees

When training schedules are announced, some employees may claim that they do not need the training, due to their experience and, perhaps, education. They may explain that they have operated the equipment for several years in another plant, been classroom trained at another company or by a union, received a formal education in the process, or simply that they know all about the job and respectfully decline the training opportunity. The answer to this should never be to let that employee bypass the training. If such employees know what to do, then they will have an easier time at the sessions. They will attend the training so that you are certain that they fully understand exactly what the safety factors are, exactly how the work is done at your particular facility, and exactly what is expected of them. The process or equipment may be subtly different from how or what they worked with elsewhere, or even during an earlier stint at your plant. State-of-the-art advances may have caused the machine operation or process to have evolved so that earlier training and know-how are simply no longer sufficient. A machine may have the same name as the one used at another plant but be controlled in quite a different manner. Then there are those employees who are assigned to drive a forklift truck and say that they do not need training because they already know how to drive an automobile. No good. It is not the same. Everyone gets trained.

Buddies or mentors are recommended. When trainees demonstrate by specific behavior and observable actions that they can consistently fulfill their obligation to work safely without prodding, the buddy or mentor can notify the supervisor. Then, the trainee has graduated in a sense. For some, the training will be easier than it is for others. So be it. Do not make assumptions as to the level of knowledge or skill. Although training is usually thought of as being for new employees, do not neglect those being transferred or given new assignments.

Evaluating the Effectiveness of Training

A questionnaire should be given out at the close of training sessions. Again, it should not be signed. It is essentially an evaluation sheet. Were the setting and logistics conducive to a learning atmosphere? (You need not use such fancy words in the query.) Were the objectives met? Did the trainees have adequate time and encouragement to ask questions or make relevant contributions (this is essential)? Were all questions answered completely? Do the trainees now feel comfortable that they can apply the lessons to their jobs? What were the most valuable and least valuable aspects of the training?

At the conclusion of the training, there should be some reasonable degree of certainty that the individual trainees have reaped the desired benefits, not just in feeling a part of the safety culture, but in having learned the practical applications of what they have been taught. The effectiveness of the training must be evaluated. Feedback is key, and the more specific, the better. Quizzes can help evaluate the degree of success of the course. If a score (not to be compared with other trainees) is not adequate to reflect a suitable level of learning, then more work is needed with the individual. That work might directly follow the regular class, or it might come in a follow-up class or one-on-one session.

The bottom line is that a person who slept through the class cannot be considered to have been adequately trained. Trainees who have successfully completed the training sessions should be given an attractive certificate as proof. To further clarify, if the system works properly, this is not a certificate of attendance but rather a certificate of successful completion. This can represent the investment of the company's resources and be a source of pride for the recipient.

If trainees are asked if they understand what to do and what not to do, they may offer an honest "no." Once you have officially hired a person, you should make it very clear that you plan to train and put the person to work and that you understand that he or she may not fully understand the work assignments until after training. It is understood that, at times, companies are only seeking someone who is totally familiar with a job. That is a different story, but even if the new hire has apparently impeccable credentials and is to join your company in a highly technical and professional capacity, he or she still needs to go through an orientation.

Back to the earlier point about those who may not immediately and fully understand the work assignments. If the trainer is satisfied with the trainee simply claiming an understanding, that is a problem. Despite your assurance to the employee that questions are welcomed and the employee will not be fired just because he or she fails to fully grasp all relevant points immediately, human ego, fear of embarrassment, and fear that they will be let go still often remain. So, when you ask employees if they "get it," they may say "no" the first time and possibly the second. However, they might say "yes" after that, even if your points are still not fully grasped. That is why you must assure that the information that you have imparted has been completely received and absorbed.

Do not worry about seeming to be condescending. You would hate to look back, following a serious injury that resulted because of a lack of understanding, and realize that if you had just taken a little more time to guarantee the transfer of information, the injury would not have occurred.

You should be unambiguous in your urging that, if employees have questions, whether 1 hour after the training session or 3 months after the session, they should come forward and seek clarification. You might ask that the information be repeated by the trainee. You might ask that the trainee demonstrate (in a controlled situation, so as to avoid an injury) the correct way of carrying out the job assignment. The point is lost if the trainee properly demonstrates, simply to please or fool the trainer, and then performs unsafely when back on the job. So, feedback is required from the foreperson if a trainee who has completed the course fails to act as taught.

No amount of training in the classroom, even if there were hands-on segments, can substitute for training in the actual setting where work is to be performed. Employees must be shown exactly how they are to perform the job, subtask by subtask. There should never be an assumption that a step can be glossed over, even if it involves what you believe to be a task that an untrained layperson can safely perform. The employee must even be shown where to stand or sit, how to grip, and more.

Do not let an experienced employee—as an official buddy, mentor, or informal helper—be the one to teach and demonstrate unless you are 100 percent certain that the experienced employee is doing exactly what should be done. Take the time to show it all and, if necessary, more than once. Watch it performed and correct every specific deficiency in the employee's performance. Another examination of the effectiveness of the training (or subsequent resistance to it) can be undertaken by giving a quiz about 1 month after the training has been completed.

Still another form of feedback is evidenced in the increase or decrease of accidents and near misses relating to the training. If the number is increasing or is unsatisfactorily steady, why? Was there a gap in the contents or method of training? Was the information imparted not adequately concrete? Did an individual trainee simply forget or misunderstand? Course redesign or improvement may be necessary.

Role playing can have great value as a learning exercise. Employees get a good opportunity to serve as benevolent critics and may well see a bit of themselves in the errors of others. A few laughs may be heard, and that is fine. Training does not have to be dry, and it is not supposed to be a torturous exercise. Time spent in training should be viewed as a privilege and a job-satisfaction enhancer, not as a sentence. Being afforded the opportunity to receive training is another signal from management that the trainee has worth and is an asset to the company.

Ongoing Training

Training should never stop at the desk. There must be true on-the-job training, wherein employees are carefully monitored. If employees are observed engaging in unsafe behavior, they should be stopped as soon as practicable, but not by startling, and the correct behavior should be explained. All too often while in the course of an inspection, the safety director, foreperson, or other person witnesses unsafe behavior, but chooses to not bring it up until he or she writes a memo or until a meeting is held. It is essential to stop the unsafe act as soon as possible. If follow-up documentation is to be instituted, it can be dealt with later. In that case, and with any form of training, the trainer is burdened with the responsibility of convincing employees that complying with safety regulations is to their advantage. They must understand exactly why. The answer cannot lie in some general philosophical cloud. Employees must understand cause and effect.

The trainer's tone should be constructive, and he or she should be persuasive in a straightforward, but not authoritative, manner. Training should not be confrontational, even if it is actually retraining spawned by the observation of inappropriate (unsafe) behavior. It should be constructive. An effort should be made not to embarrass the employee who was spotted acting in an unsafe way. Taking the attitude of a coach, rather than a warden, is preferable. In most cases, that coaching is given by a person who has some basic comprehension of the values and workings of polite, aboveboard, tactful human relations and should be done one-on-one and privately with the employee.

Eventually, continued acts of resistance will lead to a harsher tone. This is surely the case if the progressive disciplinary process has begun. Even then, the concept of training should not be forgotten; it should not totally give way to the sanction process. Work with employees. Talk with them, but not at them. When they have listened and heeded your words, show your appreciation. Yes, safety and safe work habits are a condition of employment, but a few words of appreciation should not taint your position of authority. Tell the employee (still a trainee of sorts) that you are grateful for his or her attention, patience, effort, and modified behavior. After all, that modification will remove or reduce risk to that employee and possibly to fellow employees.

Employees may hasten to correct unsafe behavior when noticing management in the vicinity. If training is presented so that the benefits are made clear, then safe behavior should be a habit that is hard to break, a habit exercised independently of management's presence. If employees slip into unsafe behavior, they should catch themselves and correct the behavior by reflex.

Emergency Action

Looking to a particular type of training, if a situation calling for emergency action occurs, appropriate and immediate response by reflex is surely needed and, therefore, should be taught. If there is an imminent danger of catastrophe, the idea that the employee will have the time to consult a manual is ill-conceived at best. For emergency preparedness and response training, crisis planning is surely an indispensable ingredient. Training should also focus on how to stop an incident (possibly triggered by a malfunction or a procedural error) from ever becoming an emergency.

However, panic and confusion can cripple even the best of plans unless there are frequent hands-on practice exercises. Some can be carried out at special off-site training locations. As long as there is reliable oversight and monitoring, disaster and near disaster simulations can be safely executed on-site and should result in an acceptable level of crucial readiness. These simulations must be subjected to thorough, detailed evaluation. It is also well worth the research to study and analyze the pros and cons of response efforts utilized in actual emergencies, whether or not they involved your workplace. Plan, practice, evaluate, and refine. Then begin the cycle again.

Remember that fires are not the only emergency situations. Other potential disasters include the following, singularly or in combination: chemical releases (spills and vapors), collapse of building or large machinery, earthquakes, floods, tornadoes, bomb threats, workplace violence, nuclear radiation release, and explosions.

Positive Reinforcement

When walking through the plant, whether as part of an inspection (e.g., audit, survey) or specifically to gain feedback on the effectiveness of recent training, do not overlook the opportunity to offer praise. Tell employees how well they are performing. If the behavior is uncommonly diligent and especially safe, that is a time to go beyond the one-on-one approach. Laud the individual in front of others. Point to his or her actions as commendable and as setting a good example. Then, peer pressure should be positive. The lauded employee might be chided by peers in a good-natured way. That is a small price to pay (if considered any real price, at all) for the benefits to all.

Peer pressure is always a concern, even if it does not stem from how individuals react to training. There is plenty of room for an employee to be a team player and a safe employee at the same time. There should not be any perceived benefit from fooling management and, in turn, endangering oneself or others. Employees must know (not only believe) that bowing to negative peer pressure, which encourages unsafe behavior, is a supreme example of misplaced loyalty. It is not a matter of being loyal to management or to fellow workers in a mutually exclusive way. Where safety is the subject, there should not even be the foundation for an argument. Further, employees (again, in some respects always trainees, continuously reaffirming good habits and ideally even adding some) should be loyal to themselves.

In the proper safety climate, employees have been empowered to control their work habits, have been given the resources and thorough training to understand how to exercise that control, and have the capacity to do so. At times, unsafe actions will be noted. Yet, every effort should be made to run an achievement-oriented process. Look for examples where positive reinforcement is applicable. No classroom instruction can take the place of that real-life scenario.

Training Supervisors and Forepersons

Do not make the mistake of taking supervisors and forepersons out of the loop. It is too common for persons to have attained those positions almost totally due to their production skills. In some cases, leadership qualities and ability to motivate have been considered, as they certainly should be. Still, this does not mean that these midmanagement personnel have continued to receive training. In many companies, when safety training is planned, it is planned for line employees. The unfortunate thought is that the supervisors and forepersons do not really need the training (why not?) or that they are too busy to attend. This is a major roadblock to the success of the total safety program and its desired results. Do not assume that these employees, who some view as being close to employers, would not benefit from attendance at training sessions.

Teaching the Right Material

No matter who receives the training, it is not enough to maintain logs of training sessions, with attendees signing off, to prove attendance. The trainees must reap the benefits of the training to become competent in performing their job assignments. Consider, too, that safety societies and trade associations often conduct formal training sessions. I recommend that such courses be monitored or investigated to some degree so that you can feel comfortable with the content and structure of the program prior to sending employees. OSHA will not be legally satisfied merely to be shown a piece of paper attesting to the fact that employees were physically present at a training session. OSHA compliance officers are likely to ask some employees (and sometimes in a private conversation, at which management is not present) about the specific information imparted in training sessions and about their specific working knowledge of the subjects under discussion. These types of interviews often include questions regarding hazard communication and lockout/tagout. They can delve into several other training standards, including such subjects as permit-required confined spaces, electrical safety-related work practices, welding, respirator use, and the operation of powered industrial trucks.

This is not to say that the company is obligated by OSHA to assure that each employee is an absolute expert in, for instance, hazard communication. They will definitely need a firm knowledge of specific precautions; methods of handling; necessary personal protective equipment and how it is to be used; how chemicals can enter the body (i.e., routes of entry); what to do if there is a spill, a leak, or a splash on the skin or in the eyes; with which chemicals or articles of the subject chemical it is incompatible; previously existing medical conditions that can be aggravated; whether a substance is flammable; what hazards it presents to particular organs, tissues, and so on. (OSHA may be somewhat lenient on the acceptable depth or sophistication of response on this one); where the material safety data sheets are located; and similar factors.

Unless it is a very unusual case, employees do not have to know what vapor density is. The point is that there is no need to clutter their minds with unnecessary technical details that will hinder their retention of the truly important information. Line employees do not have to become chemists. (I am not a chemist, but my father said I was because I could turn money into manure!)

19

Progressive Discipline

Employees at all levels must be keenly aware that working safely is a condition of employment. They should be motivated because they have been made to understand that, if they do not work safely, injuries and illnesses can result. Naturally, the explanation should go into depth about how employees can sustain injuries and illnesses, what types (e.g., parts of the body, systems, severity), and how working in the prescribed manner can make a difference. Thus, they should want to work safely because they are protecting themselves and those around them. They should also understand the high cost of injuries and illnesses, as outlined earlier. This includes the fact that OSHA may pay a visit which can result in very high fiscal penalties. Then there is also the matter of simply following the company's rules.

Organized Labor

A documented, unambiguous program of progressive discipline should help convince those employees who still are not sufficiently motivated. You must always be sure to work within any union contract when considering such a program. No worthwhile union should totally reject the idea of a progressive discipline policy, particularly when it relates to safety and health. In fact, no union should be in a legal position to force a company to abandon plans to institute such a program or to shelve an existing program.

Work with the organized labor group, not against them. Although it may seem radical, the union officials should be brought into discussions concerning the details of the planned program. This is not to say that the union is granted a vote in the implementation of any or all of the elements of the program. Simply, I respectfully suggest that management be up front.

Some companies love to complain to OSHA that they will not or cannot enforce their own company rules because they are afraid of the union (i.e., they are afraid that they will lose their case when it is reviewed by an arbitrator). If the program is set forth so that the rules are clear and if they are enforced consistently and fairly, there should not be a problem. No, this is not naive. Companies that lose such cases have usually erred in the application of their own rules and/or have failed to provide adequate documentation.

Disciplinary Action

Employees must clearly understand what is expected of them by way of the written safety program. Then and only then can the system work. That system is not to be in place to punish, but to deter unsafe behavior, to encourage safe behavior, and to protect those employees who would be hurt by unsafe behavior. The system should, in the final analysis, serve as a constructive learning tool. If the contents of the rhetoric and paperwork that are parts of the disciplinary process cause employees to engage in self-examination and, with mature guidance, come to realize why they acted in an unsafe manner, then something worthwhile has occurred.

Management must be clear in showing concern for the well-being of the employee. The employee should be made to realize that the problem was not primarily a breach of regulations, but the surfacing of a behavior that allowed or added risk. The employee was capable of making the correct (safe) choice, but did not do so and endangered himself or herself and/or others.

If the disciplinary action is to have a lasting, tangible, positive effect, the employee must understand why he or she made the poor choice and what choice should have been made. Ideally, the disciplining manager will be certain to point out the specific safety concerns, while not putting the violator in a figurative corner. Show the violator that you care and do not take away his or her dignity. The employee has to be treated respectfully and at the same time must be reminded in indelible terms that safety is not negotiable. The employee must come away from the disciplinary action with an appreciation for the significance of the misdeeds, an understanding of the harm that resulted or could have resulted, a desire to take personal responsibility for the behavior, and a sincere commitment to be a key player in the solution (i.e., accident prevention). Employees must be encouraged to report unsafe conditions. They must know that they will not be disciplined, for instance, for reporting a stripped electrical cord or a broken machine guard, even if they themselves inadvertently caused the damage. On the other hand, I suggest that discipline be levied against an employee who knowingly uses such damaged equipment, regardless of who caused the damage.

Effective Program Characteristics

There are variations to different progressive discipline programs, but generally speaking, each program must set out a sequence of adverse actions to be applied when safety rules are broken. For example, in the first instance, the employee might receive a written warning. Some companies start with a verbal warning confirmed in writing. The second violation could result in 1 day off without pay. The third violation could result in 3 days off, and the next in separation from employment.

There should be a policy explanation that calls for the skipping of steps when the violation was egregious and when it could have easily (or did, in fact) lead to a very serious injury. It is a little difficult to cover all such potential situations when attempting to list them. The words must be selected carefully. There should be a clause indicating that not all of the particular actions or inactions that could lead to that step-skipping, which could even involve immediate firing, have been listed. In such important cases of major violations of safety rules, there may be a committee to examine the case.

There can be no favors for long-time employees or friends. No one should be exempt from abiding by the rules. Even members of management should be covered by the program. Of course, within the company rules, there may be certain employees who are authorized to undertake activities, while others are not. This is a part of job description and should be spelled out so that there is no misunderstanding. Too often, safety enforcement has been known to take a holiday on the third shift. That practice cannot be tolerated.

You might wish to implement a system by which violations are erased after a certain period. Then, the employee returns to a clean slate. As an alternative, you might consider that a clean slate can be reearned, but that there is a particular absolute maximum of violations.

Nevertheless, I do not recommend that your system be complex. It must be straightforward and easy to understand. Although I have addressed egregious violations, I do not recommend several levels of violation gravity. When employees return from time off as a result of a violation, they should be required to attend a short reentry meeting where they will be told what sanctions are next in the progression. That meeting can also serve as a brief training refresher course.

Keep accurate, complete records. A little extra time invested to assure the adequacy of those records will pay off handsomely if you are visited by OSHA. Those records could help you demonstrate to OSHA the depth and sincerity of your efforts to provide a safe and healthful workplace. Such records will help you to bolster a defense of "isolated incident" or "employee misconduct."

Handouts and Postings

Handouts

Even when not distributed at a training session, handouts can be very helpful. They can be given to employees at the end of the shift or used as envelope stuffers enclosed with the paycheck. In such cases, there is no assurance that they will be read. However, if they are kept concise, the chances of the employee reading them will be greatly increased. Employees should be told about these occasional handouts before they are disseminated. When a handout accompanies a check in an envelope (particularly if it is mailed), there is the possibility that the spouse may take a look at it and encourage the employee to read it. Just do not overwhelm the employee with handouts that are too long, too complex, or too frequently given out. As a backup, copies of handouts should be made available in a bin and posted on bulletin boards.

Postings

As for other postings in the plant, they can be in the form of special tips or be true posters, even of the cartoon or slogan variety. They should not be posted where they impede vision of exit, extinguisher, or other critical locations. They should be placed in areas, and at heights, where they are likely to be read. After a while, they become the equivalent of invisible if they remain in the same location. Therefore, their locations should be rotated for maximum effect.

A novel idea is to photograph plant employees engaged in safe activities and then make the photographs into large posters. This would usually be staged, but that is not a problem. Occasional graffiti is a minor concern, but a small price to pay for the benefits. Some possible scenes include the following: properly inserting ear plugs, wearing chemical goggles (and suitable gloves and apron) when working at a plating tank, manually lifting in the proper manner, holding a railing while descending a staircase, properly positioning an extension ladder, placing a cap on a compressed gas cylinder in storage, properly aiming an extinguisher at a small (staged and controlled) fire, properly driving a forklift truck, properly rigging a load with a sling, standing in the proper position while operating a radial saw, locking out and tagging out a machine, picking up a loose tripping hazard from the floor, and working at an abrasive wheel that has the work rests and the tongue guards properly adjusted. Posters could also show particular employees reciting in a quotation bubble some safety slogan or particular tip relating to an OSHA standard or company rule.

Placards and banners can be used to proclaim safety slogans. They can heighten general awareness, but after a while their impact can fade into the woodwork. Catchy or humorous phrases may prompt interest in the general safety program. However, this type of campaign symbol can never be a substitute for a sign that addresses specific task-related hazards and specific methods of eliminating or mitigating those hazards.

As for OSHA directly, do not neglect to post the OSHA Job Safety and Health Protection notice where all employees can readily see and read it. Depending on the size and configuration of your establishment, and whether or not all workers pass by a particular bulletin board, you may need to post more than one of the notices. Do not allow other notices to be placed over any portion of it.

One more thought. Where can you pretty much be assured a poster will be at least looked at and probably read? Right, the bathroom stall.

21

Signs, Labels, and Tags

Exits and Fire Extinguishers

Exit signs are always the first concern. All exits should be clearly marked. In technical terms, the one door out of a small office could be called an exit, but OSHA is not likely to cite for that situation. So paradoxically, I do not have in mind such layouts when I refer to "all."

Do not make the assumption that all employees know where the exits are located. Beware of the same trap when considering whether signs are needed for fire extinguishers. There are times that extinguisher signs are not really needed. If there is a red extinguisher that is clearly visible against a white wall from all reasonable approaches and from a long distance, then the sign should not be needed. To place a sign 1 ft above such an extinguisher does not generally add to its conspicuousness. Nevertheless, it may take more time deciding where extinguisher locations are not needed than it would simply to figure on at least one sign per extinguisher! I have seen some extinguishers that were quite obvious from 25 yd away, but were not evident when the employee was in close proximity to the extinguisher. This was due to visual obstructions presented by stacks of material, racks, columns, and sometimes even the fact that the sign was so high that its benefit was lost when the person was too close. In such cases, an employee could be within 5 ft of an extinguisher, fail to see it, and waste a lot of critical time looking for an extinguisher in another location. Some companies place red bands high up on columns to indicate that an extinguisher is below. Of course, this can also be done with a worded sign.

There may be a need for arrows pointing toward the extinguisher location. Arrows are most often needed when the path to an exit is through a mazelike layout, when there are high storage racks, or if the establishment is very large. Triangular signs (with two sides showing) help draw attention to exits and extinguishers from various angles. If a door, archway, or similar opening could reasonably be mistaken for an exit or a way to an exit, even though it is not, then there should be a sign indicating *Not an Exit* or indicating the actual character of the opening (*Boiler Room, Supply Closet,* etc.).

It is recommended (and sometimes required) that exit signs be internally illuminated. A useful supplementary system is to install additional signs or arrows (preferably internally illuminated) near the floor, beneath the potential blanket of smoke. Another novel tactic is to provide glow-in-the-dark lines and/or arrows (above walking surfaces) leading toward exits.

Many other standards and factors relate to exits, extinguishers, and their locations, but they are beyond the scope of this section.

Characteristics of Signs, Labels, and Tags

In the case of any sign, be sure that the background contrasts sufficiently with the words so that the sign can be easily read. Also be sure that the sign itself contrasts with the wall on which it is posted.

In an earlier chapter, I addressed the labeling of machines, rooms, electrical disconnects, ladders, and so on. Some labels, particularly those used to meet compliance with the hazard communication standards, may be easily destroyed or obliterated if special precautions are not taken. When signs or labels are susceptible to chemical splash, excessive heat or cold, or the adverse effects of various elements of weather, you may need to use special materials, laminate the sign, or otherwise assure that the message will be readable and that the sign will remain in good condition.

Signs are often used to remind employees of the requirement to wear personal protective equipment in a certain area. Failure to post such signs is seldom cited, but I certainly recommend posting them. The problem comes when plants are overloaded with signs and their messages are not enforced. If you really do not require everyone in the department to wear hearing protection—and if it is not required by OSHA—then take down the sign. Otherwise, a "boy who cried wolf" syndrome can develop.

When do employees take the signs and labels seriously? If management did not really mean what the sign's message indicated regarding hearing protection, then how about the other signs? Similarly, do not place a *"flammable"* label on a liquid that is combustible but by definition not flammable. Also, do not allow a label to indicate *"inflammable"*; it means "flammable," but many persons think it means "nonflammable." Further, the "boy who cried wolf" syndrome can result when "DANGER" signs, effectively overstating hazards, are encountered at every turn. The signal word "Danger" implies a higher degree of concern than "Warning" or "Caution."

OSHA does have a loose standard by which the agency can cite for the lack of danger signs or caution signs. Do not make signs verbose. If the message can be conveyed in a few, well-selected words, then that is advisable. Too many signs can be annoying to employees and cause them to sour on the whole concept of reading or even heeding the wording. Place the signs logistically, at proper heights and angles, assuring that the wording "cuts to the chase." As with posters and handouts, there may be a great benefit to printing some signs in languages other than English. I am not referring to exit or extinguisher signs, but to warning and instruction signs and labels. In many cases, symbols can go a long way to convey the message. Be sure to move or replace signs and labels when they are no longer relevant or accurate. For instance, it often occurs that fire extinguishers have been relocated, but signs remain in the old locations.

It is understood that, unless a company has earned a variance from OSHA and local authorities have concurred, exit signs are to be red. Red is also used for danger (including flammable liquids), emergency stops, stoplights, and miscellaneous fire equipment and apparatus. Start buttons are green, as are first-aid kits and other medical items and locations, personal protective equipment storage areas, signs for eye fountains and deluge showers, and safety instruction signs. Yellow is the basic color for caution signs and for calling attention to physical hazards such as striking against, falling, tripping, and so on. Some companies prefer to highlight such areas with orange. While OSHA is not likely to cite in that case, the standards do refer to yellow. Yellow is also the color of choice for aisle markings, but OSHA is generally satisfied with any clear contrast. Orange is preferred for drawing attention to dangerous parts of machines, such as sections of moving parts that are under guards (for when the guards are removed). Purple is used for signs or markings related to radioactive hazards. Fluorescent orange, or orange-red, should be used for biohazards (including bloodborne-related). When I have referred to sign colors, this means that the referenced color is to be the predominant one. There are other types of hazards for which certain colors, symbols, etc. are either required or recommended.

OSHA and ANSI standards should be checked in regard to the more specific use of color, size, and shape of signs, labels, and tags. The purpose here is to draw your attention to the importance of these indicators as tools of a sort. ANSI specifically assigns color schemes to the labeling of pipes. A yellow background with black characters is used for hazardous materials. This includes material that may be toxic, chemically reactive, explosive, flammable, and so on. Green background with white characters is used for liquid hazards that are a low level of concern. Blue background with white characters applies to low-level hazards that are in a gaseous state. Red background with white characters should be used for fire extinguishing/quenching substances, including water, foam, CO_2, and halon (although halon is being phased out).

Further, near the label wording, arrows should point to indicate the direction of flow through the pipes. It makes sense to include labels and arrows at all critical sections, such as at valves, where pipes turn, and at the entrance and discharge from each and every wall, ceiling, and partition. It is also wise to assure that the words and arrows can be easily read by employees from all pertinent angles and distances. This includes from directly below and from catwalks and similar platforms from which work on the pipes might be performed. I have referred to different types of sign media, but for pipes with troublesome surfaces that make sign adhesion or anchorage difficult to maintain (possibly due to humidity or similar factors), total wraparound labels are quite valuable. These can be affixed or removed with snap devices. For less problematic surfaces and elements, pressure-sensitive (i.e., self-adhesive) labels are commonly used.

It is essential that any tampering (including removal or modification of the words) with signs, labels, or tags be dealt with in the harshest of terms. Changing one word can totally alter the meaning of a safety instruction sign and result in a very dangerous chemical reaction, for instance. The removal of a lockout-related tag can obviously lead to a tragedy. Under the lockout/tagout standards, there are several points made about such tags. Certainly, lockout-related tags and other out-of-service tags must be durable and must not be easily removable. Other types of signs are not of the flat, traditional kind. They may be in the form of brightly colored cones or warning tape and flags used to warn of a temporary danger involving construction work in a plant. There are hinged A-frame markers commonly used to warn against the hazards of wet or slippery floors. The cones, tape and flags, and A-frames should be affixed with adequate precautionary wording. Do not simply rely on the assumption that their physical presence sends the intended message. Fluorescent tape or paint on the floor and near light switches or exits serves as another category of signs.

One special twist on signs relates to those used to indicate that chocks must be used on truck wheels at loading docks. Besides the obvious type of sign, I recommend that signs worded backward for viewing in a rear-view mirror be installed. This is a stock sign that can easily be read by the driver who is backing the vehicle to the dock. In fact, stock signs are readily available to address virtually all safety communication needs. Nevertheless, there may be special cases, particularly regarding safety inspection signs where customized wording and/or configuration is required.

I suggest that you label each machine guard that is frequently removed during servicing or similar activities. An indelible marker could do the trick, but a neater, more formal designation is a better idea. Too often, time is wasted searching through a bank of removed guards. Further, sometimes a guard is placed on a noncorresponding machine or on the wrong section of the correct machine. In such cases, it is common that there will be a poor fit. This not only causes the guard to be poorly mounted and easily displaced, but also leaves a physical gap, resulting in exposure to danger area(s).

With all uses of signs, labels, and tags, keep focused on the ultimate goal of conveying very specific information in unambiguous terms. Be 100 percent sure that the words, pictures, diagrams, schematics, symbols, or codes are fully comprehensible to everyone who has a need to gain the knowledge intended to be imparted. The caution, instructions, or other category of message should be clear without the need for consulting another person.

However, all employees must know that they can (and are absolutely expected to) contact the proper personnel if they do not totally understand the meaning of a sign, label, or tag. Where warnings are involved, the type and degree of hazards should come across. There are many cases where there is a need to include the hazards associated with misuse and improper handling. There must be no doubt left as to what the hazard is and why it is a hazard. For example, a simple written communication (even when accurate) indicating that there should be no smoking or open flames near an area or product is generally insufficient. The message should convey what the adverse result can be. I do not suggest that the addition of the words "*vapor ignition*" is, in itself, satisfactory. The word "*explosion*" ought to be a part of the warning. In a related vein, I recall a fatality that occurred in a permit-required confined space. A sign near the entrance to a large pipe indicated words strikingly similar to "WARNING: ARGON PURGE". Those words do not suitably command the attention of the reader. The warning should have clearly conveyed the message that there was a grave danger of oxygen deficiency.

When several bits of information are needed on a sign, there is a logical, preferred order of message components. They are generally best conveyed when laid out from top to bottom; occasionally, a left-to-right orientation is proper. The signal word (such as "Danger") should be first. Then, wording should describe what actions to take or to avoid. Next is the hazard description, followed by the description of adverse health and safety consequences (injuries or illnesses) to be encountered if warnings and/or instructions are not heeded. In some cases, emergency contact information, including a telephone number, is appropriate.

22

Inspections

Inspections constitute the heart of any occupational safety and health program. In this chapter, I am generally addressing "in-house" inspections—those physical surveys performed by company personnel. Such site examinations conducted by professional safety consultants are also recommended. Keep in mind that "in-house" inspectors may well have job titles and responsibilities for areas other than safety. However, full-time professional safety consultants are dedicated to the recognition and abatement of hazards. As with the overall program, the inspections are specifically and directly conducted to recognize hazards, to analyze those hazards, and to result in actions that will remove those hazards or at least reduce the potential for injury and illness as much as feasible.

One gauge of the adequacy of the actions is to attain compliance with OSHA standards, with the agency deeming that you have achieved abatement. Periodic inspections are needed even if, at some particular time, it appears that the facility is in compliance. Subsequent inspections should, on the one hand, focus on the efforts that were made to remedy the hazards detected during earlier inspection(s). Thus, there is no assumption that, just because a bad situation was recognized and a plan of abatement was set forth, the matter has been adequately addressed. In that light it is imperative that maintenance and other personnel assigned to accomplish abatement fully understand the objective of the measures to be taken. Too often, through poor communication, hard work only results in a well-made guard that does not cover the danger areas. The welders, carpenters, and so on, must know exactly what end is to be achieved. On the other hand, I stress that successive inspections should not just focus on previously discerned hazards. There may have been hazards inadvertently overlooked during earlier inspections. There may have been operations that were down so that hazards were not noticeable. There may simply be new or altered processes or machinery that have not been thoroughly monitored while in service.

Inspecting New Operations

Any new operations should certainly be scrutinized during initial start-up and through several repetitions or cycles. Further, they should have been field tested in simulated true-service modes and scenarios before they were actually placed into use. Inspections of recent abatement methods may point to the insufficiency of those efforts and be cause for reevaluation.

It is extremely wise to seek the input of those employees who regularly work with or at operations where abatement methods are still being evaluated. They may know that under particular or unusual circumstances the abatement methods have gaps (not necessarily physical) that allow for dangers to surface. In other words, whoever inspects the operations may not have witnessed a certain part of a cycle, a repair, or some similar aspect of the process or operation wherein the abatement strategy simply failed to take the resultant hazards into account. It might even be something that is done on the machine only once per week. It could relate only to when a certain part is worked or only to when a particular succession of events occurs, for instance. Certain operations may only take place on the third shift, and the relevant inspection personnel will have to be there at that time.

Hazards and Violations

When conducting inspections, as I have referenced in an earlier chapter of this book, it is absolutely essential that everywhere and everything be inspected. If a safety director did not know about a certain machine or had never been in a particular room, a red flag should go up immediately. Inspections should concentrate on hazards and violations. What is the difference? Some hazards may not relate to OSHA standards. If you do not possess a substantial working knowledge of the OSHA standards, you can easily miss violations, particularly those that are subtle or just not readily evident to the eye. You have to know what to look for. However, if you seek and find hazards, you will uncover the vast majority of OSHA violations. The argument can be made that you really have to know the specifics of OSHA's paperwork and training standards. If you do not, you might be able to run a safe shop, but not be in full technical compliance. Therefore, when conducting a comprehensive inspection, you should not just try to find violations of OSHA standards. In broader terms, you should be concerned about where, when, and how employees could become injured or could become ill. In the appendix of this book, there is a lengthy checklist to guide you in determining what to look for. It is a most valuable tool, but must be used with caution. Do not become focused only on checklist items and OSHA standards.

To conduct a worthwhile inspection, you should always have in mind "what would happen if?" You should want to speculate. What if his finger gets too close to that point? What if she walks over there? What if she drops that container of liquid? What if he missteps while descending those stairs? What if there is a fire and they have a need to evacuate the area? A more subtle potential hazard could be considered when asking: What if that window is opened? (can change ventilation patterns). The questions go on and on. Another category of questions concerns tags, signs, and labels. Is there conformance with the instructions and warnings? For instance, are guards in place; is personal protective equipment being worn; is there assurance of specific pressure, volume, rpms, and pHs; and is placement and use of ladders in compliance with specifications and capacities? Be curious. Do not assume that all employees will function exactly as they were trained. Do understand that employees are not robots. Even if they are not supposed to be near that unguarded power transmission chain, what if an employee trips and falls toward it? What if, out of mere curiosity, somebody places a finger on or near the bare copper wire (energized electrical parts)?

Inspecting Machinery

Conduct a physical inspection. Certainly, you do not want to place yourself in a position where you can be easily hurt. You do not want to place yourself in a position where you would be creating a violative condition. However, you do want to look up, down, back, and under. To conduct a reliable inspection, be systematic in your plan of attack. When you are about to leave a room and you look back, you may see hazards that were overlooked on your initial pass simply because your visual perspective is significantly different. You are catching things from another angle.

You will probably have to get dirty. If you are conducting a detailed, full-blown factory inspection, do not wear a jacket, tie, and white shirt. That mode of dress might be all right for counting sprinkler heads, but it is not appropriate when a complete inspection involves dark areas, dirty areas, machines with moving parts close to the floor, ladders, old basements, flammable liquids storage rooms, and so on.

Wear a jump suit or at least old clothes. OSHA compliance officers generally get "down and dirty." Be sure to wear all of the required personal protective equipment. If an area is truly too dangerous to enter or if an operation is out of service, note it so that you can return to it later. Do not permit such situations to slip through the cracks.

You may have to watch certain operations for quite a while. When you first come upon a machine, it may be performing a job wherein not all of the sections are operating. You will want to see the machine when everything is moving. Through employee interviews and even simple conversations with forepersons, you may learn that the other cycle and/or operation may not be run for hours. If so, make every effort to return. If there is any aspect of the operation that you do not see in use, make a note. You will need to see it at some time. When viewing a nonworking machine, do not assume that you fully understand how it operates, including what parts move and in what direction. See it operate. There is no substitute.

Inspecting Task Performance

When observing the performance of tasks for which there is a fair-to-strong potential for ergonomic risk, you may have to watch different employees working at the same job. They may differ slightly in their postures, hand and arm manipulations, and similar movements. In fact, even with non-ergonomic concerns, there is a value in observing different employees performing the same job. They may (for instance) operate the machinery differently, in one case not following the SOP (standard operating procedure) or JHA (Job Hazard Analysis). You might also find that one employee goes through a different sequence of steps to achieve the same result. In so doing, he or she may be at greater risk. Although two employees performing a task together is frequently safer than the work being done by an individual, that is not always the case. For example, if two employees are working on one machine simultaneously, there may be an increased probability of an injury. This is basically related to the fact that one employee may be capable of activating moving parts of the machine, while the other is in a dangerous position.

Inspecting and observing for ergonomic hazards can be difficult in that you do not want to ask an employee continually to repeat a movement. Your very asking may significantly affect the manner in which the self-conscious employee does his or her job.

It is often difficult to analyze and find the words to describe the movements. In turn, it is difficult to describe in notes just what you are watching. It helps to know words (and their working definitions) such as ulnar deviation, radial deviation, flexion, extension, and others often used to describe movements.

For all of these reasons, the evaluation of ergonomic considerations, particularly those that relate to cumulative trauma disorders or repeated stress injuries, is well suited to videotaping. Later, the ergonomic task performances can be reviewed over and over. They can be watched in slow motion, and you can easily fast forward to certain critical portions. On a related note, if you take still photographs with flash, be sure that you do not startle employees. With such equipment, as well as certain testing apparatus including that which is designed to check for electrical hazards, be extremely cautious to avoid any chance of vapor ignition.

Although there may be times when inspections are announced, there is no question that many inspections must be conducted as surprises. In virtually all cases, OSHA inspections are unannounced and it is best to simulate OSHA inspections for the most part. Do not give employees the opportunity to neaten up for an inspection, to replace guards, to scurry to don personal protective equipment, or to put out cigarettes. See the shop the way it normally runs. Period. Vary the route and order of the tour. This will help remove the predictability of when a certain area will be entered for inspection. Do give the employees ample opportunity to ask questions and to offer comments. Put the employees at ease in these conversations. Let them know that you are there to learn. The stages of the inspection that bring employee-inspector interface into the fold present a marvelous opportunity to enter the world of reality. Forget how things are purported to be. Although you definitely want to see jobs performed as they normally are, I do not suggest that your inspection should be clandestine. Rather, when employees see that they are being watched, tell them what is occurring and what the objectives are. Explain that you are not conducting a time-rate study and that you do not want them to speed up. Tied into this consideration, piece work is generally considered to be a negative in the safety business. Playing "beat the clock" has inherent dangers. That is an understatement.

At times, safety committee members, or other well-trained employees should be asked to perform inspections of the peer review type. This is not to be an "authority trip," but rather a constructive tool used to identify unsafe procedures (unsafe conditions may be noted as well). The goal is definitely *not* to seek disciplinary action. It is to find out if inappropriate behavior is occurring, if so why, and to correct that behavior promptly and permanently. Right or wrong, employees are customarily more receptive to criticism from peers than from managers or supervisors. This perceived level of comfort and security can be used to everyone's advantage. As long as the peer inspection findings are communicated in an empathetic, supportive manner, the results will be rewarding for all.

Other Inspection Considerations

You may come upon circumstances where facilities, processes, materials, and equipment are in the course of being added or renovated. In some instances, they may simply be in the planning stage. These situations should not be ignored and bypassed merely because there is no exposure yet. Once equipment is put into place—possibly with major construction involved—it may be quite difficult to move it or add certain types of ventilation, physical safeguards, and so on. Analysis should be conducted while such matters are in the planning stage. In fact, with the help of the engineering and purchasing staff (other departments, e.g., maintenance, may well be pertinent as well), no equipment should be allowed into the plant until there is an assurance that it is essentially safe for the particular intended use and location, or that a firm plan is laid out to make it safe before being put into service.

As a component of inspections, industrial hygiene or similar sampling and testing might be performed. Earlier, I made references to the specific types of hazards for which such testing and sampling are most pertinent. As a crucial reminder, the use of testing and sampling equipment need not—and in many cases should not—wait for the formal walkaround inspection process. For various hazards, the equipment should be used much more frequently for such purposes as monitoring to determine if oxygen deficiency or IDLH (immediately dangerous to life and health) levels of toxic vapors or fumes are present. There are also operations such as those involving permit-required confined spaces where testing is mandatory whenever there is potential exposure. Always be sure that sampling and testing equipment are calibrated as needed. Understand the limitations of the specific pieces of equipment and the attachments. If cassettes or similar collection devices are left unattended (which is not preferable) or are clipped to employees' clothing, be sure that all employees in the area understand the purpose of the devices and how to treat or not treat them. Clearly explain the negatives involved in tampering with these items.

When considering (even if peripherally as compared to the main purpose of the inspection, which is to recognize hazards) possible avenues to abatement, the best approach is to engineer out overexposures. For instance, as referenced earlier from a slightly different perspective, try to improve exhaust ventilation and noise reduction at the source so that personal protective equipment is not needed. Frequently, the personal protective equipment cannot be avoided. Another option is to implement administrative controls so that exposure time is reduced. With most chemical hazards and violations, overexposures are not based on pure numbers; they are based on concentrations (usually parts per million or milligrams per cubic meter) over certain durations. Therefore, a higher level of concentration is permitted for a shorter period. Similarly, for noise standards, more decibels are permitted for a shorter period. The concept is to shorten the amount of time that the employees are exposed. This can be accomplished by way of more breaks, by rotating jobs, and by a combination of the two strategies.

Written Report

Depending on the depth and scope of the inspection, a written report may be generated. It might take several days to produce such a report. In the great majority of situations, it would be a waste of time to reference OSHA standards in the report. One section could highlight trends. The majority of the report should be formulated in a bulleted format. If there are particularly important nuances to the hazard and possible violations, they can be briefly summarized.

The person(s) who conducted the inspection will not necessarily be the same ones who will devise all of the abatement plans. If there is any doubt as to whether a certain approach toward abatement will leave or place the company in a violative condition, then further research into OSHA standards is necessary. The maintenance staff, for instance, must understand what they are trying to accomplish in their work to address the uncovered hazard. Keep in mind that with reports (whether as the result of an in-house inspection or an inspection by a consultant, insurance representative, or other noncompany individual or team), there is added OSHA enforcement jeopardy if that report is seen by OSHA and identified concerns have not been fully satisfied.

Before a report is written, then, there must be a commitment that it will be taken seriously and that action will commence to correct the noted discrepancies in an expeditious manner. As with safety committee notes and similar documents, do not offer them to OSHA if you will be handing them a record that will serve to point out your apparent indifference to observed and documented hazards and violations. It is understood that, if the inspection took place just days before OSHA's visit, then not all of the items on a long list could have been abated. Still, use caution to avoid the insertion of large feet in large mouths. In my section addressing "willful" OSHA violations, I delve further into these concerns.

I am not an attorney, but you might consider consulting with suitable professional counsel regarding the matter of disclosure of such reports. If an attorney issues a written request for an inspection to be conducted and for the total results to be given directly to him or her, you may have a form of insulation. The attorney should indicate in the request that its purpose is to help in providing legal advice. Then, the attorney might be in the position to assert the existence of a client-attorney privilege regarding the results of the inspection (i.e., the written record). I again caution you to seek appropriate legal advice on this specific point.

Another type of written report that is much shorter and does not necessarily address all deficiencies can be produced on the spot. Such a report would be neatly written on carbonized forms. Before leaving the department or area, one copy is separated and left with the foreperson or supervisor. One copy is retained by the inspecting individual or team. Depending on the gravity and complexity of the noted hazards, a copy might be instantly given to the maintenance chief or plant engineer. I do not recommend too many more copies. The copies must all be coordinated, and in the end, there must be accounting for each copy.

The Authority of Inspection Personnel

Authorized inspection personnel should carry *Danger: Do Not Operate* or similar tags, as addressed in a previous chapter. Unless there is a very real question of whether or not the shutting of an operation can cause a major hazard, then the sound judgment of the inspector should override any other arguments against the halting of the operation. There are some cases where the shut-down of an operation can add or increase risk. Some examples involve ventilation, electrical power (affecting the power supply in other operations), chemical reactions, and machines that when stopped in midcycle cause material to fly or dangerously accumulate and overflow. Barring such scenarios, the tag goes on and the proper persons are summoned. The tag is only removed if the relevant persons are convinced by a chemist, plant engineer, or other person with sufficient knowledge that the operation is safe to resume. Naturally, some situations dictate lockout, disassembly, or equally effective methods to assure that the machine or process cannot resume.

Where imminent dangers or extremely hazardous situations are detected, abatement or job cessation must be effectuated at once. In such cases, at least one inspection person must remain near the hazard as long as there is no current exposure. Even with the "imminent danger" term, this usually, but not always, means that such danger exists when an operation is running. Another employee would immediately seek out person(s) with the expertise to confirm the initial determination. In the case of a supposedly essential operation or the type of imminent danger independent of whether or not employee(s) are actually working on a task, that person could explain what emergency procedures should be undertaken to end the crisis.

Creative Approaches to Abatement

In an earlier chapter, I discussed the importance of brainstorming and looking beyond traditional problem-solving concepts. These approaches should be embraced when seeking abatement remedies. When machine hazards noted during inspections appear to be very difficult to abate, it may be feasible to simply eliminate the machine. Can the work be done on another kind of in-house machine? (I am *not* referring to newer or better guarded models of the same machine.) What if several operations (different work on the same parts, or the running of substantially different parts) are performed on each of a series of machines, and it is a significant undertaking to provide abatement for *all* operations by *one* method? Perhaps certain machines could be dedicated to certain operations, allowing for easier abatement on a machine-by-machine basis. In another scenario, it might be a complex job to guard against the hazards presented by particular rotating protrusions. Maybe those protrusions can be removed, as they no longer perform a function. If the placement, movement, or other factors regarding a treadle, used to put machine parts into motion, prove extremely troublesome, the use of hand controls should be considered. Thus, the need for the treadle could be totally negated.

The idea of substituting, in place of taking highly burdensome abatement measures, does not just apply to machines. Substituting a flammable chemical with a water-based, nonflammable one can remove or substantially mitigate the hazards associated with vapor ignition. Similarly, it may be possible to replace a chemical that is highly corrosive with one that is both far less injurious to human skin, and nevertheless possesses properties that allow it to satisfactorily produce the desired product effect.

In exceptional situations, the best tactic might be to eliminate an operation from the facility and to contract out the work. Such operations might include chemical dipping processes and spray finishing work. Further, there can be specialized machine operations, for which the cost and exacting nature of abatement are impractical. In fact, the machine may be seldom used, arduous to maintain, and large enough to occupy a good deal of precious plant square footage. If so, there are additional incentives to go the contract route.

Keep in mind, too, that by contracting out, employers can essentially "max out" of entire OSHA standards that require training and written programs. This can move the company out of the legal scope of the standard. Some of the most relevant examples are permit-required confined spaces, hazardous waste operations and emergency response, bloodborne pathogens, and electrical safety-related work practices. Many employers are, in effect, not covered by one or more of those standards because their employees lack exposure to the related hazards. Even if a company has permit-required confined spaces, it might elect to have all work performed in those areas done by outside contractors. If a company is initially subject to the stringent requirements for hazardous waste operations and emergency response, because of the severity and size of potential chemical spill or release (*not* specifically defined in the standard), it can choose a course by which no employees will be involved in said operations. In case of emergency, the documented plan would dictate evacuation by all employees and bringing in a specialist firm to deal with the spill or release. For bloodborne pathogens, a company can decide to have no employees who could be reasonably anticipated, *as the result of performing their job duties,* to face contact with blood or other potentially infectious materials. In a factory-type setting, the company would have to take into account not only the potential hazards involved with assisting injured or sick victims but also the hazards presented by the clean-up of facilities, equipment, and other surfaces, following (for instance) an injury that resulted in a severe loss of blood. The electrical safety-related work practices standard addresses the requirements for work performed on or near exposed energized and de-energized parts of electric equipment; the use of electrical protective equipment; and the use of electric equipment. The standard is intended to protect employees from the electrical hazards that they may be exposed to, even though the equipment itself may be in compliance with OSHA's electrical *installation* requirements. Thus, in many facilities, no employees have exposure, as defined by this standard. In those facilities, any work that would fall within the scope of the standard, is awarded to an electrical contractor.

There is a distinct twist to the OSHA standard addressing process safety management of highly hazardous chemicals. The standard applies to a process which contains a threshold quantity or greater amount of toxic or reactive highly hazardous chemicals that are listed in the standard's appendix. A specific threshold quantity is assigned to each of the chemicals. The standard also applies to 10,000 pounds or greater amounts of flammable liquids and gases and to the process activity of manufacturing explosives and pyrotechnics. As a form of abatement nullifying the applicability of the standard, an employer whose process had fallen within the scope of the standard, may decrease the amount of chemical(s) in the process to less than the designated quantity.

The bottom line is that it is good business practice to avoid being locked into the confines of conventional approaches to risk reduction. Innovation and alternative thinking fall right into this philosophy. This is not to say that the most effective and efficient means of abatement must be something never before attempted. Clearly, customary, field-tested solutions may provide the answers. However, it is paramount that all aired ideas be studied. When faced with problematic abatement circumstances, do not fail to seek help from the employees who have hands-on knowledge of the machine, process, or other germane facility features. Generally, this translates to a hands-on "line employee," such as a machine operator or assembly worker. Individuals may even know little about other facets of the operation, but be in the best position (even better than the plant manager) to comment on hazards and safety considerations concerning the job tasks that they perform.

An employee may propose a unique suggestion that may not be pragmatic exactly as explained, but perhaps peers or others in the organization can build upon it. Do not be too quick to dismiss what might first appear to be unworkable or just plainly absurd. Do not scoff at any proposal offered as a serious response to the question at hand. Accept all input. Occasionally, even a suggestion offered in jest may have merit. Keep an open mind. Excellent remedies may grow from an off-hand remark or from a recommendation that is seemingly unsophisticated or nontechnical in nature. Sometimes, finding an abatement method that protects employees to a legal sufficiency, while not significantly hampering production output, can be a challenging ordeal. Imagination and creativity know no boundaries. Although problem solving is not a mere parlor game, it can be a stimulating exercise that will teach those involved much about the critical decision-making process. Considering the possible legal consequences of noncompliance, along with the risks of injury or illness that remain when solutions are not identified, it is a pursuit at which failure is not a reasonable option.

23

Incentives

Systems that reward prosafety activities or indicators are used in many facilities. These incentive and premium programs should be designed to enhance existing overall safety programs. Such programs can then help propel the overall program further in the right direction. They should only be put into place after a reasonably effective safety culture, with positive results, has been established and nourished. If the overall program is not yielding clear and definite signs of success, then the incentive idea must wait.

First get everyone on track and assure that they know how to work safely. No incentive program can serve as the major part of an overall safety program. However, if formulated and implemented within the proper framework, it can shore up gaps that allowed risks to remain or to surface. It can aid in the philosophy of employee self-regulation and show a most worthwhile return on investment. The employees have an added personal stake in working safely. They must understand that they are in the strongest positions to exert control over the program. It is their conduct that defines the success or failure of the program.

Planning the Program

In any type of incentive program, it is beneficial to solicit employee participation from the incipient stages of planning. Permit and encourage employee input into the design of the program elements, the rules, and the rewards. First line supervisors and forepersons should also be involved from inception. They, along with line employees, may even be able to help fine tune the drafts of the proposed programs. The rules should be clearly stated, and the goals or landmarks that are to be attained for the earning of rewards and recognition must be reasonably attainable. The system must not be so confusing that employees lack an understanding of the basic purpose and objectives. When programs involve evaluations of success over time periods, the periods should not be so short that it becomes difficult to administer the program. On the other hand, particular campaigns or cycles should have scheduled ending dates that are not so extended that employees are hindered from seeing the light at the end of the tunnel.

The Purpose of Incentives

Never lose sight of the all important fact that the purpose of incentives is to realize improvement in actual observable and measurable safe work habits; it is not simply to be able to point to improved statistics. There is a marked difference. Strive for a system that rewards safer conditions and safer work practices as opposed to a reduction in a certain classification of injuries and illnesses. The former will eventually take care of the latter.

Incentives for Accident Avoidance

The most common form of incentive program reaps rewards for certain time periods during which there was no lost time injury, or recordable injury or illness, or similar gauged data. The first concern should always be the question of the stifling of incident reporting. This is more of a concern when a whole department or large group would lose out on a tangible gift because one employee was hurt. Negative peer pressure to not report can have a devastating effect. The employee who was hurt or who became ill may place too much emphasis on "not wanting to be the bad guy." He or she may not report an incident because that would mean that no one in the department would receive their turkey or gift certificate. So, any program of this kind must strictly require reporting of all injuries and illnesses. When cases are not reported, an injury can become infected, chronic health problems can beset the victim, and the hazard can remain unabated and possibly become more of a risk endangering additional employees. As a part of the overall safety and health program, the reporting of injuries and illnesses is not optional.

No employee, whether involved in an incentive program or not, should ever be disciplined because he or she sustained an injury or illness. If that injury or illness was the result of unsafe behavior, that is another matter entirely. Employees who hide injuries or illnesses, or influence others to do so, should receive firm sanctions. If unreported

injuries or illnesses are somehow uncovered, the cases should be publicized. I do not recommend using the name of the victim, although
that name may become known. The object is not to humiliate. The
object is to point out the very hazards of not reporting and how that
can adversely affect the victim and other employees who are potential
victims of the same or similar hazardous circumstances.

Incentives for Safe Work Performance

More meaningful incentive programs center on good work performance. In such cases, the danger of not reporting is irrelevant.
Employees are rewarded for positive actions as opposed to the lack of
negative results. Criteria for milestones include how many days have
passed since an employee was noted violating a safety rule or simply
operating in an unsafe fashion. In those systems, peer pressure can
have an enormously positive impact. Co-workers can play a large role
in motivating each other to work safely. If the program's purpose was
stated clearly and the program is driven to generate interest in avoiding risks (as opposed to just earning gifts), then success should be
forthcoming. Employees can show support for each others' diligence in
trying to be safety conscious not just in response to the program, but
because they understand the full range of benefits to be enjoyed by
safe work practice in a safe environment. The extra incentive is there,
but safe practice becomes ingrained and habitual. After a while, to
perform in any other manner seems awkward, and to witness others
performing in any unsafe manner becomes an annoyance. Something
appears out of sync, and positive peer pressure takes over.

Do not forget about over-the-road drivers. Although clear wisdom
dictates that accidents, even with no resultant injury or illness, are
always a concern, this subject finds added relevance for the drivers.
So, build in incentives for safe drivers. On one hand, you can issue
rewards (or opportunities for such, where those included for special
prize drawings are from the "safe population") for when a driver did
not experience any motor vehicle accidents for a certain time period.
More in line with my theme, though, the incentives can be awarded
when, along with a lack of accidents, the driver has been praised by
the public with calls to a telephone number appearing on the vehicle's
bumper sticker.

Size of Groups

If the incentive program is set up so that the whole plant is considered to be the population of the incentive pool, there may be a blueprint for failure. With such large numbers of employees and potential

actions being considered, it is too easy for one act of one employee to, in effect, penalize many, many others. When the program addresses individuals, those who deserve the rewards will receive them, and they will not lose out due to the poor performance of one individual in another department.

There is a middle ground that has practical value. Let the criteria relate to work teams, cell groups, or small departments or subdepartments. Now the program becomes fair and manageable, while still encouraging positive peer pressure and camaraderie. Certain jobs and tasks do carry more or less potential for risk than do others. If, for instance, the office staff is included in the program, their chances for success are fairly high. Perhaps they can be involved in a different type of program, or they can be required to pass more risk-free time prior to being eligible for a reward.

Employee Suggestions

Be sure to remind employees—by bulletin board postings, newsletters, and so on—of how the rewards can be earned in this program as well as others. Keep the enthusiasm high, while always assuring that the specific ways to work safely are addressed and reinforced. Enthusiasm can easily be awakened by another type of incentive program: one that focuses on suggestions. Usually, simple suggestion forms are made available, and there is a specific box or place to put them after they have been filled out. The forms should be carbonized so that employees making suggestions always have a record of what they submitted. There should not be any general requirement that the forms be signed. However, they obviously must be signed if the submitting party is to be eligible for a reward. Every suggestion that is even half-way serious (i.e., just about anything short of ideas unarguably presented solely as expressions of sarcasm) should trigger a written response. Do not forget that some pretty far-out ideas may become viable with a little rework.

Rewards would be given when and if suggestions are translated into actions that will truly yield a benefit to the system and therefore to hazard reduction. Another reward category can be for those who have submitted at least a certain number of serious suggestions, even if none resulted in action. You want to continually stimulate and inspire employees to come up with ideas. Even if several ideas did not bear fruit, the next one might produce an orchard.

If an employee detects a significant safety hazard, he or she must not be required to take the time to commit the findings to writing just to be considered for a reward. With that in mind, it would be a shame, and possibly lead to a tragedy, if an employee waited until break time, wrote up a suggestion on a particular form, turned it in, and the form was not reviewed until the following day. Permit the system to allow for immediate notification of the foreperson. Then, the matter could later be documented. This exception, again, is for critical matters. Otherwise, the forepersons could later be listening to ideas and transcribing them all day.

Comparing OSHA Inspection Results

A distinct type of program can have criteria that are easy to quantify. It involves actual OSHA inspections. Of course, in almost all cases, it is illegal for even OSHA itself to give advanced notice of an inspection. Then there is the problem of trying to compare apples to apples. In other words, compare the results of one comprehensive, wall-to-wall OSHA inspection with the results of the last OSHA inspection of that scope. Many OSHA inspections involve only limited portions of the facility. When the results of an OSHA inspection are finalized, record the total proposed penalty, the number of standards allegedly violated, and the number of instances of alleged violation. Choose which category or categories are to be graphed and compare the numbers to the last inspection. If this does not work, due to an apples to oranges situation, find a way to grade the inspection results versus an absolute or versus reasonable figures. Do this in a way that does not condone the citing of any number of items, no matter how low. Do not forget that your goal in that field is no citation. However, be fair and leave room for looking to the positive side of the inspection.

Contests

Still another type of incentive program is more like a contest, although rewards are involved. Distribute photographs of actual hazardous situations or cartoon-type mock-ups of hazardous situations. The contest involves as many specific hazards as possible. Employees should be allowed to work on the problem with their families. Families can also be brought into the fold with picture-drawing or poster contests. Use specific criteria for the size of the submittal, and insist that safety messages be specific. Do not accept those posters, for instance, that show persons smiling, along with a caption to *Think Safety*. There could be different age groups in the contest if children do the work.

Rewards

For all of the types of incentive programs, the rewards can be akin to production bonuses. As for the choice of rewards given, one slant is to give out merchandise that is directly related to safety. Some examples are the following: smoke alarms, carbon monoxide alarms, first-aid kits, automobile emergency kits, fire extinguishers (always with instructions and precautions), personal flotation devices, personal protective equipment for bicycle riders and in-line skaters, and flashlights. Popular items include clothing, such as wind breakers and sweat shirts. With those items, as well as coolers, thermoses, paperweights, mugs, and even watches, a safety message can be imprinted. A special touch is for the message to begin with a "Thanks." In that way the item serves as a constant, visible reminder of why it was earned. Depending on the exact message and where the item would normally be used, the reminder can also influence employees and their family to be conscious of off-the-job safety. If you include the company name or logo, there may be an additional benefit.

With items that are used in recreational settings, there is a nice tie-in drawn between the concept of working safely and the opportunity to enjoy good health when pursuing activities with family and friends. With gifts that can be enjoyed by the family, the spouse or other members become interested in the incentive program and, in turn, gain heightened interest in the employee working safely. They want more bonds and more merchandise. How about gift certificates to restaurants, sporting events, amusement parks, and movie theaters? There is no company message imprinted, but this type of gift is usually very desirable and something to strive for.

Straight cash is a dubious choice. It will probably be welcomed and appreciated. Nevertheless, it can be spent on anything (even a six-pack on the way home from work). Then it is gone. There is no solid reminder of what was earned or why it was earned. Savings bonds have a similar value, and are not likely to be spent too quickly. If a few of these are earned, they can form the basis for a particular, planned savings account.

With the bonds and other items, a letter of appreciation to the employee is recommended. It should include the name of the individual employee who has earned it, and not be to "Dear Employee." If accompanying a bond, the letter could respectfully suggest that it be put toward a child's education. This will not always apply to the specific recipient and his or her family, so the wording might have to be adjusted.

I have listed several good choices for rewards, but the affected employees should be provided with an opportunity to offer their own recommendations. What do they feel is appropriate? What would they like to earn? It is most suitable to provide this opportunity when the program is still in the planning stages. The choices of gifts should be broad so that employees remain interested. Although savings bonds and gift certificates are a different matter, how many of the same item can an employee use? An excellent approach is to provide catalogs in the workplace or mail them to the employees' homes. An item can be selected, ordered, and received at the employee's home. Once again, the family feels a special part of the program, even if the item is somewhat personal. The whole family had the opportunity to be involved. Paradoxically, although the item was not handed to the employee, the place to which it was delivered heightens the personal nature of the recognition.

Whenever rewards are garnered, whether by groups or individuals, the results should be publicized in the company newsletter and on the bulletin board. The results can be highlighted at an annual company picnic or similar function that is often attended by the families of the employees. Make a big deal out of it. It is a cause for celebration.

An excellent form of individual recognition is the plaque. As with letters of appreciation, plaques must be personalized. These could be for special "beyond the call of duty" safety acts or for significantly sustained periods of good safety performance. One caution here. Do not get caught in the trap where employees are risking their lives unnecessarily to earn a plaque. As an example, an employee seriously threatens his or her own life attempting to perform an unsafe rescue in a permit-required confined space—a rescue attempt that was performed contrary to company regulations. Of course, in the final analysis, this is an unsafe act in itself as just stated.

An additional type of reward, best suited for groups, departments, or in cases of exceptional achievement an entire workforce, is to have special catered lunches. In a different light, some companies use bingo-type systems and drawings. Cards and numbers are assigned to those who have earned them by attaining milestones. This generally does apply to everyone in a large work area. It is a fun activity and gets everyone talking, but the individuals who win may not be the most deserving. In addition, the rules governing the games sometimes get a bit confusing. They must be kept simple.

Whatever the type of incentive and premium program, it is paramount that the employees are empowered to affect their own fate. Provide them with safe working conditions, a healthful environment, and excellent training. Then, let them show what they can do. Continually welcome feedback from employees as to whether or not they like the programs and what might be done to improve them. Periodically, disseminate survey forms seeking that information.

If the overall safety and health program is sending the correct message, employees will find an internal motivation to work safely. The absence of injuries and illnesses and the accompanying feeling of pride and accomplishment will serve as substantial rewards in themselves. To maintain one's good health also rewards family and friends (and ideally, fellow-workers) who care about the well being of the individual. Incentives have a place in the fostering of safe work habits, but a good attitude cannot be bought. If your incentive program is properly integrated, employees will continue to work safely, even in times when tangible gifts are not being offered as inducements.

24

Employee Assistance Programs and Wellness Programs

Participation in Employee Assistance Programs

Employee assistance programs are usually known as EAPs. An EAP is a program designed to help employees cope with personal problems and pressures that adversely impact on job performance and general well-being. That adverse impact can most certainly be manifest in preoccupations, attitudes, and behaviors that undercut safe performance. A successful EAP intervention can result in the employee being less apt to be the victim of a workplace accident. This can occur through heightened concentration and attention, increased self-esteem, a decrease in distractions, the removal of the judgment and coordination inhibitors previously caused by substance abuse, an improved attitude that symbolizes a recognition of the value of good health, and a reinvigorated interest in one's work tasks, work surroundings, and co-workers. The employee is also less prone to absenteeism and generally becomes more productive. Surely, catching a problem early on can successfully put a halt to the employee's emotional, mental, and possibly physical deterioration.

The next step is to reverse the effect of that deterioration and build up the inner strength, confidence, sense of worth, and wholeness of the individual. In so doing, individuals can overcome their frustrations or at least hold them in a proper balance. In very real terms, this can preserve a most valuable asset of the company: a skilled employee. In more humane terms, it can replace calamity and pain with reason and joy. Occasionally (but increasingly), the early intervention can literally save a life—either the life of the employee or the life of someone else that the employee had considered taking.

EAPs have the responsibility to identify and assess the problems that are causing interference with employees' ability to do their jobs correctly and safely. They should be formulated and run as cost-effective, humanitarian, job-based strategies that assist in conserving human resources, while operating within a nonpunitive framework. They provide counseling to help resolve conflicts, freeing employees to deal more effectively with their surroundings and daily frustrations. The goal is to cause an improvement in employees' ability to be successful at work and to lead generally happier lives.

EAP Referrals

In many cases, the EAP staff finds it fitting to refer the employee to a social service agency, substance abuse counselor or program, medical doctor, psychologist, marriage counselor, financial counselor, attorney, or support group. EAPs often serve to intervene promptly in a time of crisis or when the onset of a crisis stage is at hand. Early intervention is of the utmost importance. The system must not be allowed to bog down in miscommunication, paperwork, or backlog. It is essential that the right resources be identified for referral so that the appropriate treatment, care, or other service can begin as soon as possible. For this reason, the EAP staff must be highly trained to recognize which specialized resources fit the particular person and case. The employees should not have to be filtered through levels of generalists down to specialists and then subspecialists. The referral should be as accurate as possible and take place as soon as possible. All employees seeking EAP services must be confident in the knowledge that their interaction with the program will be held in confidence.

Once the employees have the information that details the referral system and the intake process, they can opt to make a start with a self-referral or a referral through a supervisor. With most EAP-employer relationships, the supervisor of the employee is to refrain from interrogating the employee about the specific perceived need for contact with the EAP. Supervisors are not to probe, but are to be supportive and sympathetic. They are to follow up with the particular system in a prudent, tactful, and efficient manner. All requests for entry into the program should be viewed as serious. You may find it difficult to believe that an employee who you thought of as being the last person to have an EAP-type problem is serious about needing and seeking help. Do not make matters worse by trying to convince such individuals that they are not legitimate candidates and that bona fide EAP candidates are not as "well-adjusted" as they are. You may, in effect, be embarrassing individuals looking for help and pushing them away from professional assistance that is sorely needed. I urge you not to place employees in the awkward position of having to emphasize just how serious their problem (or perceived problem) is.

Employees that make known their desire to enter an EAP may experience some degree of relief as soon as they know that there is a company-sponsored avenue toward professional, competent, and ideally, subsidized assistance. (Sponsorship can also be a joint venture, with the union and/or insurance carrier bearing some of the cost and responsibility.) That, in itself, may open the door for a repair and healing process to begin, but it does not necessarily simplify matters. Some participation in an EAP may span several months before substantial progress can be legitimately claimed. Yet, entry into the program symbolizes that employees recognize a need for intervention and are ready to help take part in their own recovery or in the resolution of problems that weigh heavily on their ability to function daily in a safe and healthful manner.

The Cost of EAPs

EAPs are designed to take the employee as far as is practicable at no cost and then seek reasonable costs for further participation. The EAP must keep close tabs on the case and assure that proper treatment is occurring as needed. All too often, participants find themselves in a financial position that seems too dire to allow payment for the full course of services. In such cases, the EAP should make every effort to find a method mutually satisfactory to the employee and the resource for resolving this added pressure. In other cases, the participant simply becomes impatient and withdraws from the program prematurely. The program must be mindful of the necessity to follow through on every case, with an eye to the continued care and complete recovery of employees who have given their trust to the EAP.

Particularly as related to the cost of services, consider that some larger unions have become involved in EAPs, in some cases overseeing Membership Assistance Programs or the equivalent. How about the employer attempting to contractually secure not just the cooperation of the union with regard to the EAP but its assistance in the development, implementation, maintenance, oversight, and financial subsidizing of the program?

Publicizing EAPs

EAPs should be in a position to address (at least by way of referral) concerns including the following: drug abuse, alcohol abuse, tobacco addiction, depression, hostility, gambling addiction, eating disorders, paranoia, nervousness, grief, depression, apathy, chronic fatigue, chronic tardiness, financial problems, legal problems, family conflicts, phobias, tension (whether externally applied or self-induced), and other significant barriers to the maintenance of a reasonably stress-free and happy life. Employees who work second or third shifts are generally more susceptible to problems relating to poor diet and lack of restful sleep. Individual biorhythms may play a part, but it is often more difficult to arrange and assure healthful sleep, relaxation, and nutrition when work schedules are not in sync with much of the rest of the culture. This is particularly the case when the worker changes (rotates) shifts, basically confusing sleep patterns and meal types and times. This type of schedule can also play havoc with one's family and social relationships. The EAP may be called in for assistance, and the wellness program should also take into account these unique concerns. EAPs must be well publicized within a company. Preferably, a copy of the literature describing and explaining the program will be given and/or mailed to each individual. If copies are simply made available, employees might feel too embarrassed to be seen taking them. The contact channels must be inconspicuous.

Participation in Wellness Programs

The emergence of comprehensive wellness programs, run by numerous employers, has proven to be a great source of pride for all involved. For our purposes and interests, the important fact is that employees who are robust, or who at least enjoy a strong total health profile, are generally less apt to experience injury and illness on the job. Yes, there are exceptions, but the relationship is powerful. An effective and efficient wellness program should result in the enhancement of both physical and mental health, which is surely a bonus for all employees. It can also contribute to improvements in morale, human dynamics, alertness, and productivity while at work. In no way does this suggest that employees with disabilities are less capable of performing their assigned tasks. In fact, those with disabilities should certainly be given the same opportunities as all employees to participate in wellness programs. Although a particular company's program may have certain segments of the workforce in mind—as defined by departments, those who are perceived to be in positions of greater stress, those who are obese, and so on—all employees should be viewed as worthy participants.

The physical fitness segment of the program should include the following: medical clearance, informed consent, physical fitness assessment, nutritional analysis, exercise prescription, health and fitness education, and follow-up assessment and monitoring. The medical clearance stage must be anchored with a broad-based screening process. The screening should include, at a minimum, vision, hearing, cardiorespiratory, neuromuscular, and musculoskeletal history and function.

The health and fitness education aspects should be ongoing. The material should be aimed toward the development of a keen awareness of the benefits of leading a healthy lifestyle. Subjects should include nutrition, hypertension and cholesterol control, relaxation, weight control, smoking cessation, substance abuse, and injury/illness prevention. Employees should be provided with the means to evaluate their lifestyle and related health status. They should be guided toward activities that will, in fact, measurably improve their health, including through lifestyle modification.

When they are at work performing their job tasks, employees must feel that there is no hypocrisy. Their workplace must be a total environment that fosters safety and health values consistent with those learned in the wellness program. Those values must be unambiguously supported by all levels of management.

Good health is not merely the absence of disease, ailment, injury, or illness. From a holistic perspective, it is also a state of mind and body that yields a feeling of well-being. From this broad perspective, a truly healthy person consistently approaches life with a spirit and vigor that cause self-contentment to thrive. When this self-contentment is coupled with a comprehensive knowledge of specific risk avoidance measures, it is even more crucial to safe behavior than is the complete freedom from physical discomfort. The key is to maintain a balance so that needs, desires, and satisfactions, including those in social and work life, are in reasonable harmony. Another important balance to promote is that, although job, family, civic responsibility, and general interpersonal relationships should be taken seriously, there is a strong need for smiling, laughing, and simply unwinding.

Wellness programs should not be as concerned with quick turn-around regarding health status as they should be with a long-range approach to good health. Rather than simply "fixing" employees, the objective is to help the employees sincerely desire to fix themselves, to provide them with the information, guidance, and facilities to allow for that change, and to instill in them the confidence and desire to take charge of their own lives. Thus, the employee will engage in a personal preventive maintenance regimen of sorts that will be self-regenerating. The "machine" will automatically service itself (i.e., the employee), and have the knowledge to make an informed decision as to when an "outside contractor" (i.e., physician, therapist, etc.) should be consulted. Although ailments can even strike those who take good care of themselves, veteran active wellness program participants will, as a group, require fewer outside contractors.

Putting aside all of the sophisticated medical terms, a well-rounded wellness program can lead to improvement in the participant's endurance, stamina, strength, flexibility, and even attractiveness (as relative as that is). That just covers obvious physical attributes. The psychological advantages are plentiful. That is not just a matter of "look good, feel good." Vigorous, controlled physical exercise releases natural endorphins, which cause a positive "high." My explanation is purposely offered in simplified terms, and the point is that the relationship between the exercise and the feeling of well-being is not casual. It is physiologically fueled. To simplify further, within suitable, case-by-case guidelines, active persons are happier, healthier, and less prone to be victims of accidents over which they had any reasonable degree of control than are those persons with sedentary lifestyles. Might there be a person who is quite pleased simply to sit and read all day? Sure, but the axiom remains valid and statistically significant.

Dedicated participation in any wellness program of substantial worth should result in an increased resistance to life stressors. The handling of external stress, as well as self-induced stress, will be improved. However, wellness programs should also include seminars, workshops, and similar presentations specifically designed to address stress management. EAPs offer similar presentations, but they are usually structured in a more formalized mode, including those with support groups. Besides the physical and psychological advantages, there should be a clear improvement in mental acuity and sustained, focused concentration. As for further comments on the psychological factors bearing on recovery, I address those when discussing medical management.

Private Health Clubs

More and more forward-thinking companies subsidize employee memberships in health clinics, fitness centers, and gyms. Although there is less opportunity for monitoring the level of meaningful participation than there is with company-sponsored or company-managed programs, the benefits can still be notable. However, many gyms are limited in their objectives or offerings. They seldom provide the wide range of services associated with multifaceted wellness programs.

On the other hand, employees may prefer such establishments just because they are not directly related to work. Additionally, such health clubs often have excellent, state-of-the-art facilities that would be a financial burden on any relatively small wellness program. There are few exercises that receive as much praise as swimming. Many clubs have conventional swimming pools and some even have against-the-current lap pools. Whirlpool spas are common and exceedingly popular. They are not exercise facilities per se, but can certainly serve as stress reducers.

Whether it be at a private facility of this sort or in a small room set aside at the company, various types of exercise equipment, including steppers, riders, rowers, bikes, striders, ski machines, treadmills, and others can serve as part of an overall plan to increase good health. The use of a heavy bag (with suitable gloves) can be a great stress-releaser, an excellent way to work off job and private frustrations. Proper instruction is mandatory, and there are always liability concerns, particularly if free weights are used. It is preferable to choose resistance equipment. Supervised aerobics have gained popularity and can be a wonderful, equipmentless avenue to cardiopulmonary function improvement.

Incentives

An interesting and novel incentive and premium type of program, far different than those discussed in an earlier section, could involve measurable health improvement. For example, baseline readings could be established for individuals, noting their blood pressure, cholesterol level (and HDL/LDL ratio), triglycerides, body fat percentage, pulse rates, and weight. Then, after a given time period (at least 3 months), readings would be taken again. Those who show the most improvement by numbers or percentage would be rewarded with gifts. If such gifts are to be given out liberally, then gym bags or warm-up suits are logical choices. For major campaigns and major accomplishments, an interesting concept would be to waive an insurance deductible or further subsidize an insurance policy or membership to the gym or its equivalent.

This is a doable type of contest or promotion, but it absolutely requires special handling. Many employees may not want their numbers to be known by others. That is fine, and they need not participate in the friendly competition. The wellness program is voluntary but strongly encouraged. The incentive program is voluntary but participation should not be pushed too fervently. Then again, it would be possible to have an administrator keep their actual numbers confidential and compute the number or percentage differential that reflects the level of improvement.

An additional category for the incentives could be actual exercise units or duration. How many situps or pullups can a person perform?

How many miles can he or she row, pedal, or ski? In the exercise category, beware of participants who might purposely underachieve for the baseline readings so that their later performance will seem miraculous. I think that it would be less appropriate to include pure weight lifting as a category.

Rewards should never be given for the best raw numbers. They should, again, be given for personal improvement. It should be obvious, but it's worth emphasizing, that medical monitoring must be performed. Employees must be adequately educated so that no one endangers their health by working too hard or too long. Some program participants might have numbers that are so good that there is little room for improvement. They may be the type of individuals who see those numbers as rewards in themselves. Eventually, everyone should enjoy that rich perspective. It is necessary to stress (if you pardon my choosing that term) again the need for medical monitoring. Professional fitness trainers can also be instrumental in keeping an eye on contest participants to assure that there is no ironic impairment to health while in the pursuit of improved health. This can be a critical concern with weight loss, for instance. If the shedding is too rapid or comes at the expense of good nutrition, the results can be catastrophic.

Literature and Videotapes

Wellness programs should regularly make available brochures, pamphlets, and various other forms of educational literature. A nice idea is to have a videotape machine available and a large selection of health-related videos. Literature and videotapes may be available for free or for very reasonable costs through your insurance carrier, the federal or state government, or various health-related charitable and educational organizations. The educational portion of the program should not neglect hazards that are generally associated with home or recreational activities. Whether it be by way of literature, videotapes, audiotapes (employees can even listen in the car), or speakers, valuable and practical safety tips should be communicated. Subjects could include topics such as the following: boating, swimming, defensive driving and the use of seat belts, shoveling snow, power lines, contracting sexually transmitted diseases, avoiding carjackings (and personal safety in general), lawn mowers and other outdoor home maintenance equipment, heat stress, and hypothermia. Various topics involve concerns inside the home, such as frayed electrical cords, items left on staircases, bathtubs lacking nonslip footing surfaces, kitchen fires (and evacuation plans in general), carbon monoxide dangers from improperly used or maintained fireplaces or from furnaces or heaters (or from vehicles left warming up in closed garages), household chemicals (accessible to children and general hazards of intermixing), legal drugs accessible to children, and firearms storage and safety.

For medical response, it is preferable to not rely on literature or audiovisuals alone; live instructors should be utilized for courses in basic first-aid, CPR, or the Heimlich maneuver (or similar system). Just remember that improper or incomplete instruction could result in a tragedy and/or liability.

Other Wellness Practices

Even if there is not a highly structured, integrated wellness program, most of the referenced elements could be introduced into a plant in some fashion. More companies are setting aside several minutes (at the start of shift and, possibly, one other time) for planned, coordinated light calisthenics and stretching exercises (i.e., limbering up). To be effective, these must be led by an assigned individual.

How about serving nutritious meals in the cafeteria? At least offer them as an alternative and provide calorie, cholesterol, and fat contents information. Occasional, preannounced lunch time speakers can be brought in. Heed the cliché about the cost of prevention versus the cost of cure. With even this type of limited program, medical personnel should be consulted, and the human resources staff and safety director should be involved.

Whether as a part of a wellness program or not, preemployment physicals have become more sophisticated and in-depth. I have already alluded to most of the categories that should serve as targets for medical evaluation. To be more specific, the back is an area that deserves very serious attention (including as related to obesity). Vision and hearing should certainly be checked. For the eyes, the concerns go beyond general close and far vision. Depth perception should be checked. Testing to determine the presence of eye disease is very important. Complete health histories should be obtained. Numerous baselines should be determined. When relating to OSHA standards and periodic medical surveillance (whereby comparisons can be drawn), hearing, lung capacity (and pulmonary function), and heavy metal content are among the most critical.

Substance Abuse

Drug (substance) screening has gained great favor in recent years. It is a sad fact that such testing has become so important. The subject is arguable, and I will not pass unequivocal judgment on the general concept. Permit me to simply submit that my longstanding (even if

well meant, but misguided) liberal aversion to such testing has started to bend toward grim reality. In most cases, drug testing is legal, if
performed fairly, within accepted professional and ethical practices,
and without favor. Do not embark on a testing program unless you
are on firm legal ground to do so. If there is a union, it should be
brought on board without a doubt.

A separate matter involves the implementation of a workplace substance abuse program. The program should include five basic steps: a
written substance abuse policy, a supervisory training program, an
employee education and awareness program, access to an EAP, and a
drug testing program where appropriate. Honing in on the supervisory training, the relevant personnel are to be responsible for observing
and documenting unsatisfactory work performance or behavior and
talking to employees about work problems and what needs to be done
about them. Supervisors must understand the policy, be able to
explain it to employees, know how to look for signs of substance
abuse, and know what to do when they find them. Their education
should include information on specific drugs, updated as necessary,
and methods of detecting probable drug and alcohol use. Supervisors
are not responsible for diagnosing substance abuse problems or for
treating substance abuse problems.

With substance abuse, most persons immediately think of illegal
drugs. Substance abuse can involve alcohol or legal drugs used in
excess. It can also involve the combination of alcohol and legal drugs.
As for alcohol, the obvious concern relates to drinking on the job.
Studies strongly suggest that alcohol abuse, even when not occurring
while at work, can increase the risk of occupational illness from
chemicals at the workplace. I caution against premature, public accu-
sation, but maintain the need for drug abuse awareness.

So much has been written about the adverse effects of drug abuse
on and in the workplace that no extended discussion is needed here.
Specific drugs produce different symptoms. As a quick reminder, drug
abuse in general can cause a large variety of specific deleterious
actions and inactions. There can be clouded judgment resulting in the
lack of ability to predict or even consider the likely harmful outcome.
There can be a marked slowing of reflexes. There can be a loss of concentration at a critical moment. Drug abuse can result in a deterioration of coordination, leading to dangerously impaired balance and
motor skills. It can also result in numerous manifestations of diminished eyesight. Other adverse reactions of particular concern at the
workplace include the following: drowsiness, confusion, anxiety, dizziness, distortion of space or time, panic, confusion, and tremors.

As with other unsafe behavior, the concern is not only for the subject employee, but also for others threatened by his or her behavior.

Suffice to say that the adverse effect on the maintenance of occupational safety and health can be devastating. It is far more than a legal or moral issue. Substance abuse is frequently a causal link to serious workplace injuries and illnesses.

Do not forget the possibility of EAP referrals when dealing with alcoholism. Misuse of legally obtained, legal drugs (even if by design) may still fit into the EAP concept. Chronic conventional drug abuse may be another matter concerning which an attorney should be consulted. The ADA may give a level of protection to alcoholics, viewing their addiction as a disease. There have already been claims, without a clear, binding, consistent case law foundation, that drug addicts should be afforded the same protections under the law. Once more, stay current and consult with an attorney who has specific expertise in these areas.

Tobacco Smoking

A rash of local and state laws has limited the places where tobacco smoking is permitted. The myriad, dire consequences of tobacco smoke have finally gained wide recognition. A detailed recounting of the salient statistics is not needed here. Yet, tobacco smoking (usually cigarettes) is still ignored or condoned (by implication) in workplaces all over the country. For that reason, it is important to note comments concerning smoking in the workplace in particular with just a few reminders of the more general concerns and dangers.

In 1994, OSHA published its proposed rule to regulate environmental tobacco smoke (ETS, which is essentially second-hand smoke) in the workplace. If the standards become law (and the odds strongly favor the passing of at least similar standards), they will apply to all of the environments—industrial and nonindustrial—under OSHA's jurisdiction. Employers would be required to prohibit smoking or to designate nonworking smoking areas that are separate, enclosed, and exhausted directly to the outside. Cleaning and maintenance work in designated smoking areas would be conducted only when no smoking is taking place. Employers would be prohibited from requiring employees to enter those areas in the performance of normal work activities.

At the core of the proposed rule are the numerous serious health hazards associated with tobacco smoke (including ETS). They include heart disease, stroke, cancer of the lungs and pancreas, decreases in pulmonary function, a large number of birth defects, and other illnesses and diseases. According to the EPA, cigarette smoke contains many carcinogens, with the most dangerous hazards being found in the fumes rising from the tips of lighted cigarettes as opposed to the smoke that users exhale. In any case, irritation to the eyes, nose, and throat is a problem. There are also numerous adverse synergistic health effects when associated with other occupational respiratory hazards.

Another major hazard of smoking involves vapor ignition, usually relating to solvents and similar products. In those cases, the event may begin with an explosion. Of course, any type of fire can result in an explosion when it reaches flammable vapors or liquids. Consider more conventional types of fires, such as those starting with paper, corrugated material, rags, or similar items. Further, there is the hazard of oil (i.e., oil-spread or oil-fed) fires. Although oil does not ignite easily, it can be extremely difficult to extinguish once it is ablaze.

There is the problem of lost work task attention due to an employee dropping a cigarette or concentrating on the use of the weed. In these situations, various types of severe injuries besides those directly related to fire and smoke can occur. I have seen employees smoking while operating crane pendant controls, forklift trucks, portable tools, and various machines.

I urge that no one (including contractors and other visitors) be allowed to smoke in any area of your facility. It is not merely a matter of safety, health, productivity, economics, or ethics. Within the near future, federal and/or state laws will almost definitely require you to prohibit or substantially limit smoking virtually all of the time and in virtually all of your facility. It is time to start to wean employees from smoking tobacco. I recommend that you incorporate smoking cessation into your wellness program and your EAP. This could ease enforcement and compliance. It could also put you in a better position to retain the services of your long-term, skilled employees who currently smoke.

I am not a medical doctor. In no way should my comments on the suggested contents of preemployment physicals, medical screening, or surveys be considered complete. My goal was merely to emphasize some of the more important considerations for work-related medical examinations. Surely, most of the contents of workplace physicals are standard and obvious, and I have not attempted to list them all. However, no one should assume that all such examinations are equal or sufficient. I strongly recommend an in-depth knowledge and understanding of the Americans with Disabilities Act so that no laws are broken relating to the exams or the posing of questions. You will find that the timing of the exam and of the questioning does matter as to when certain information can legally be requested or sought.

25

Medical Case Management and Return to Work, Including Ergonomics

Medical case management is not an event or a series of events. It is a smooth-running, well-integrated process. By way of this process, individualized patient care is planned and accomplished in a careful, systematic manner that is consistent with high quality medical practice, while at the same time being cost effective. The theories of managed care are brought to bear to help assure the smooth transition of the victim from one stage of progress to the next logical stage of progress until he or she has attained maximum possible recovery.

A vast array of health professionals are considered for consultation or direct involvement in each case. Depending on the exact circumstances, as well as the wide-angle view of the specific case and the specific patient, individuals are selected to play a role in the recovery. Some of those considered include the following: physicians, physician assistants, chiropractors, physical therapists, occupational therapists, nurses (including nurse practitioners), rehabilitation specialists, psychologists, and social workers. Other potential providers include managed care programs, insurance companies, wellness programs (including from a reactive perspective), and clinics.

All cases of injury and illness must be tracked from detection, through the return to work, and ideally to the return to original health status. However, the suggested introduction of in-depth medical management of the case can be triggered by certain flags, including the following: an obvious need for long-term treatment, wide options for treatment, significantly differing professional opinions as to the most appropriate types of treatment, questionable diagnosis or prognosis, highly unusual cases, and the professional recommendation of a course of treatment, which can include surgical procedures, that would be extremely expensive.

At first blush, the mentioning of cost in any context related to injury (even response to injury) seems incongruous. I simply reference cost to help illuminate the real-world use of medical case management. In no way does this imply that valid medical or collateral treatment should be withheld due to price of service. The company does have a vested interest in maintaining control of the case. One way to do this is to control the gate to the community of service providers. Can proper action be taken in house or through those whose delivery of services are in the company's normal sphere? There is no suggestion that treatment should be prohibited or delayed. It would be helpful and more manageable if the company has relevant personnel on staff and/or has a regular list of sources that fully comprehend the company's business.

Occupational Health Problems

It is troubling that not all medical doctors and those in closely related fields are trained and experienced in occupational health problems, including those related to cumulative trauma disorders. Some employers feel that all physicians should be able to handle workplace medical concerns and conditions without special training. Those without special training will at least need input from the employer if they are to have the necessary facts that will allow them to evaluate any return to work considerations.

However, there is a legitimate specialty in occupational medicine. Although it generally includes the study of such things as acute trauma associated with machine-related injuries, some main concentrations of the specialty are occupational disease and cumulative trauma disorders.

Many physicians lack a two-way relationship with the employer. The physician often renders a diagnosis and recommends treatment which is generally agreed to and then provided. End of story. This limited approach is often the fault of the employer. With situations unique to a work environment and/or concerning return to work considerations, the employer and physician must have two-way communications.

Return to Work and Light Duty

This communications gap is particularly problematic with ergonomic concerns, including range of motion and lifting capacity, but is frequently pertinent at other times. Anticipated postures should be noted. The providers (including the physician, occupational health nurse, and physical therapist) should be supplied with sufficient information to make reasonable assessments regarding the employee's return to work. This is especially important when light or restricted duty or vocational rehabilitation programs are realistic options, and this is often the case. Provide the physician (and others, as needed) with the employee's job description. If it is not appropriately specific to allow for meaningful interpretation and recommendations by the medical professional, then supplement it with the required detail. The response should not be that he or she "is a factory worker." Be function specific. What tasks are undertaken? What movements are necessary? What physical limitations are germane? (do not lose sight of ADA here)

The physician should be told that the company desires to return the employee to work in whatever limited fashion is medically suitable as soon as possible. You will want to receive periodic progress reports in simple English from the physician so that a close eye will remain on the case and, more important, on the victim. Again, keeping in mind what specific functions the employee will be allowed to perform, when a return to work date becomes somewhat predictable, the employer must know promptly. Then light duty or restricted duty jobs can be considered. It is preferable for forepersons or supervisors to be the point people in the design of such jobs. I suggest that you embrace the "medically suitable" part of the first sentence of this paragraph. I expound on the values of a quick return to work later in this section.

It would certainly be counterproductive, for numerous reasons, to return an employee to perform unsuitable tasks that will cause the medical restrictions to be exceeded. Premature return, whether to full original duty or to modified assignments, can defeat the whole purpose. It could even result in the citing of an OSHA "willful" violation. Whether the situation is related to an ergonomics and medical management standard or to the general duty clause of the Occupational Safety and Health Act, the company may be in great jeopardy of a huge proposed penalty.

What does the employee actually do at work? Some companies perform functional capacity assessments. The chief goal is to figure out the normal, task-specific, required ranges of motion (i.e., movement). Diagrams with pertinent degrees, angles, and linear measurements are completed and often computerized. Weights of items to be manipulated can be incorporated into the formulas. One instrument used for this purpose is a goniometer, which measures angles of solid bodies. Dynamometers are used to calculate forces exerted on areas, including the wrist. There are instruments to quantify the forces of vibration on the hand and arm. The field is blossoming. It is technically quite sophisticated. The idea is to conduct testing and then log and graph the results. There will then be a record of what forces, ranges, and so on are required for particular tasks.

If someone has suffered an ergonomic-type impairment and return to work becomes viable, these records are scrutinized. Testing on the particular employee will be conducted. The physical limitations will be compared to the limitations that were determined earlier for the tasks. Therefore, a suitable return-to-work job (considering all of the tasks involved in that job) can be selected for the individual.

Too many companies use the all or nothing principle and only return employees to work if they can work a full shift, at full capacity, on the job that they performed prior to the injury or illness. This is a mistake. Can work hardening take place in the company, instead of the enlistment of (or shortening the use of) a rehabilitation center? Work hardening involves getting employees back into mental, emotional, and physical condition so that eventually they can fully resume their original jobs, if possible. This is done in steps. The gradual and safe buildup of musculature is paramount. There is also the matter of the reconstitution of the employee's confidence in his or her ability to perform the job in a safe and productive way. Work hardening includes healing and reconditioning. How and when this is done depends on a wealth of factors.

A major priority is to take all reasonable steps, as specifically outlined by the appropriate medical personnel who have stayed in close professional contact with the employee, to assure that the medical impairments are not aggravated. It is one thing to return to work from a period most notably marked by rest; it's another thing to return after an extended round of therapy sessions. Other medical concerns become involved if surgery was involved. Premature or overly aggressive attempts to reverse impairment of tendons, ligaments, or nerves can cause inflammation and have dire consequences. The healing and reconditioning must be part of a logical, step-by-step progression. This is often accomplished by the modified or light duty slowly working up to the original schedule, if prudent. The job could even involve the same work and the same hours, but with more break times.

Some workers may require permanent reassignment. It is also understood that some workers will not return to employment status. Nevertheless, the return to full work status is a worthy goal for all parties. Other means to ease return to work include flex time and job sharing.

Other Company Concerns for the Victim

In a snapshot look at medical case management, the whole topic may appear to be an unworthy tie-in to an occupational safety and health program. However, the relationship is a natural one once the need for medical case management has, unfortunately, been made necessary. The main objectives of a safety program have been examined in detail. If, through whatever shortcomings or failures, a gap existed and an injury or illness resulted, then an efficient, effective, compassionate, and totally responsive effort must be made to make the victim whole. Besides the ethical foundations for this effort, there is the concern that a returning employee who is not fully adjusted to the resumption of job duties is more likely to be involved in another accident. He or she must be totally prepared. There must be no added distractions.

Where in-depth medical case management is to be utilized, there should be early intervention into the case. Management's involvement must be visible from the start. That involvement begins with a demonstrated concern for the well-being of the victim. Management should stay in touch with the victim and the victim's family. Unless unmistakably pushed away, contacts should be maintained. The message of caring can come through loud and clear, in any case, without being a nuisance. It is a simple human truth that, for the most part, interest shown in the recovery of an employee can in itself aid in the recovery process. The employee has apparently been disabled. The obvious manifestations of the disability are in physical limitations and pain. The disability extends to the psyche as well.

Telephone calls and cards are helpful, but visits (even of rather short duration) are very well received. Just do not convey a message that the desire for total case cost reduction lurks behind that basket of fruit that you brought to the employee's home. Psychological and social factors certainly play a role in the degree to which the employee feels motivated to want to get better and return to work. Additionally, the good feelings engendered by the positive psychological and social interactions often stimulate physiological recovery. Happier, more vibrant, and exuberant employees have fewer roadblocks to recovery. If they feel the gentle and sincere good will of the company and not a harsh angry hand trying to force them back to work, then they are much more apt to distance themselves from their infirmities more quickly and more completely.

This is not to place too much pressure on a person who has sustained an injury or illness at work. It is not to say that every person who does not get well quickly or, tragically, who does not get well at all bears a major responsibility. They have not failed. Was the injury or illness simply too severe? Had medical science not yet evolved to a point where effective treatment existed? Did the medical personnel, even with the proper tools available, fail to perform with the required excellence? It can be down right cruel to blame an employee because he or she is not recovering at a pace that meets the expectations of the employer. Nevertheless, strong will and desire do have a place in the recovery process.

Make the employee feel valuable. Remove as much stress as possible from the entire situation. Instill the employee with optimism, if such optimism is even in a small way grounded in the facts of the condition and the outlook. The employee must want to heal. He or she must feel that there is a someplace to go from here. One place to go is back to work. Employees must have a sense of self-worth and of control over their lives and futures. In a benevolent way, do not allow them to feel too dependent. The longer employees are out of work, the more chance for the accumulation of deep-seated feelings of depression and helplessness. Show them that there is a light at the end of the tunnel.

It's a fact that events such as injuries and illnesses can cause stress. More important, however, is how victims perceive the event. The level of stress experienced by employees is largely determined by their perception of the event. Do not lie to your employees; rather, guide them in the direction of hope.

It will help if the employees feel that they can and will return to work. Better yet is if employees look forward to returning to work because they will be welcomed back into a hospitable workplace that offers job challenge and low stress at the same time. Despite conventional wisdom, those concepts are not mutually exclusive. Try to mollify employees' fears concerning their return to work. They may be afraid that they will be hurt again or will aggravate the existing medical problem. What has been done to preclude or significantly reduce the chances of additional accidents? Factors surrounding ergonomic-type cases are varied and complex. Yet, ergonomic stressors can be identified and mitigated, if not removed.

Employees might also fear returning to work as a subordinate to their regular supervisor. That fear must be driven away. Injuries and cumulative trauma disorders, particularly those in which there is obvious subjectivity involved with the level of pain and difficulty of movement, are less apt to be reported by employees who are satisfied with their jobs. This is notable with cases involving the back. In turn, job satisfaction is largely tied to satisfaction with the immediate supervisor. So, it behooves the supervisor to keep that in mind. This is not to say that any employee should be allowed to be unproductive or that sensitive employees should be permitted to accomplish less and exert less effort. It does mean that there is a human relations aspect to high-quality supervision. Employees are assets, but they are not androids.

Looking at the other side of the employee-supervisor relationship, supervisors should not influence employees with whom they have a pleasant relationship to hold back on reporting injuries or illnesses. If they feel that the employee is holding back, even as a misguided favor to management, that attitude and behavior must be corrected. The employer should be an asset to recovery efforts and not a barrier. If there are serious concerns about fraudulent claims or exaggerated effects of the impairment, those situations must be investigated. Often, the insurance carrier has personnel available for that purpose. There are criteria that can set off doubts, but they are too tangential to the direct subject at hand. The number of actual fakers is quite small. Again, where there is job satisfaction, that number is further diminished. Do not assume the worst. Be supportive instead of suspicious. Do not confuse the emotional complications associated with injuries and illnesses with malingering. Management plays a key part in easing the employee's transition back into the workplace, and then toward full-time, full-load assignments. However, management is not to take the place of medical personnel.

The medical case management program must be administered by a physician or occupational health nurse. There is certainly a place for second opinions of medical personnel, but those second opinions are also to be rendered by medical personnel. Their knowledge of medical case management and of advances concerning treatment of cumulative trauma disorders must not be allowed to stagnate. They must keep abreast of new developments and cutting-edge theories and measures. In so doing, the delivery system will be not be bogged down in outdated concepts. In addition, the service providers will stay current with the best ideas and methods for assuring effective and efficient recovery.

Cumulative Trauma Disorder

I have frequently introduced ergonomic concerns into this discussion. Although medical case management has other applications, it is well-suited to being an important component of a comprehensive ergonomic program. You can find a reference to several ergonomic-related actions that should be avoided in this book's lengthy hazard checklist appendix. At the risk of tackling a major subject in a limited forum, I offer just a few additional comments about ergonomics and cumulative trauma disorders. An entire thesis could be written about this broad and complicated subject. Ergonomics involves the study of biotechnology. It refers to the interface between the workplace environment and the worker. It is more generally thought of as relating to movements of the employees, although there are many other elements of ergonomics. Depending on one's definition of the scope of conditions encompassed by ergonomics, these can include light, ambient temperature and humidity, eye strain (including from video display terminals), noise, and many other factors.

For ergonomic concerns, break down each job into sequential tasks. Then analyze each finite element of each task. The following thoughts are most suitable to the conventional applications of workplace ergonomics. Teach exercises (preceded by stretching) and efficient, low-stress element-specific movements. However, do not rely solely on calisthenics and the instruction and demonstration of proper reaching, lifting, and carrying techniques. Consider the use of hoists, conveyors, carts, lift tables, pneumatic lifters, fatigue mats (also consider proper footwear and medically prescribed shoe inserts), ergonomically designed tools, multiadjustable chairs, sit-stand structures, smaller, lighter unit loads, and new or redesigned, repositioned grips for items to be moved. Do not overlook the value of a more logical placement of the work, parts, gauges, and/or controls. From a broader perspective, it may be necessary to reevaluate the work station design.

As for ergonomically designed tools, manufacturers are constantly introducing more vibration attenuating tools (also consider antivibratory gloves), optimal hand diameters, torque control features, improved balance, more suitable angles and lengths, and adjustable products (e.g., chairs and other furniture). Tools are available that require much less pressure to operate, particularly when the palm, index finger, and thumb work as a group or individually. Ergonomically designed keyboards are widely available. Some consist of opposing half boards designed for striking in the hand-wrist neutral position.

For all ergonomic concerns, I urge that various suppliers be contacted to determine the availability of state-of-the-art products. Tell them exactly what your problems are or seem to be. The field is growing rapidly. If you make your needs known, it can result in the design, construction, and sales of products that will aid in compliance with OSHA and reduce ergonomic risk factors. Abatement might include more teamwork, job rotation, or slowing down the pace of a task or the speed of a conveyor belt (over which employees should have some control).

Lifting

The National Institute for Occupational Safety and Health (NIOSH) makes available a lifting guide. In fact, it has been revised to take into account additional factors in the equations that relate to injury risks, most notably of the back. The formula addresses such confusing terms as *"asymmetry factor," "coupling factor,"* and *"load constant."* The bottom line is that lifting concerns are not to be judged by the weight of the lifted object alone. In fact, the word *"lifting"* is a problem in itself. Items might be moved or manipulated in a way, or from position to position, that does not exactly fit the definition of lifting. Following the integrated NIOSH formula can certainly be helpful in the avoidance of injuries and even in the avoidance of an OSHA citation. I recommend that you obtain the literature and gauge/slide rule or equivalent.

Taking into consideration the NIOSH formula and adding several other factors that are extremely difficult to quantify, major considerations for predicting lifting hazards and related object movement hazards include the following: weight of object, repetitions per time period (i.e., frequency), horizontal distance of object from the body, heights from which the object will be retrieved and then abandoned (this includes how much stooping and extending upward is required), number of persons who will conduct the lift (hint of abatement implied), stability of the object (added concern if more than one object is lifted at once), center of gravity of the object, wet or slippery object surfaces, sharp corners or edges of the object, and ease of material portion of the object shifting (e.g., powder in a bag or liquid in a partially filled large container).

Additional factors include the following: distance to be carried, impediments along the route (e.g., doors, tripping hazards, slippery or wet floors, thin aisles, etc.), time required to sustain the lift while placement is considered or landing area is prepared (should be predetermined and set), elevations to be negotiated while holding the object, necessity of turning the body while feet are stationary (a poor practice), adequacy of gripping surfaces (including placement of handles or areas likely to be grabbed), and even obstruction of vision while carrying the object.

Never forget the specific person who is to be considered for performing the lift. That is a critical and constant concern. With an encore ADA warning, what is the particular employee's capacity to perform the lift safely? Consider flexibility (being limber), not just strength. Does the employee fully understand the proper lifting techniques?

Other Ergonomic Stressors

I remind you to become familiar with the ergonomic section of the checklist. It will help with carpal tunnel syndrome concerns. Further, keep in mind that other parts of the body, such as the neck and legs, can suffer the results of ergonomic stressors. I have stated that employees are not robots. True, but employees' bodies do function as a sort of working tool. It is necessary to know how to fix them when they are broken, but the goal must be to care for them and avoid a need for repair. These tools should receive preventive maintenance, and they should be used properly and with adequate periods of rest. Remember that they are equipped with hearts. Even if the bodies are in excellent working order, employees may become dysfunctional if they lack the knowledge, skill, and proper attitude to perform in a safe and productive fashion.

I do not suggest that all employees should be treated with kid gloves or that every workplace should be a country club. Just remember that a major part of an employee's waking hours are spent at work. It is not enough for supervision to consciously avoid adding unnecessary stress. That can be viewed as a neutral gear. Watch the dynamics. Are employees frequently arguing with each other? Is it apparent that grudges are harbored? Are certain employees the targets of harassment (whether sexual, racial, or other)? These are real concerns.

I have already discussed the significance of certain criteria for job satisfaction. Another slant is that if employees are unhappy, particularly to the point of distraction, disillusionment, or fear, they are more likely to become the victims of accidents or cause others to be victimized, even if inadvertently. If they do become disabled and cannot work, for authentic or ostensible reasons, they are more likely to want to postpone their return to work.

All employees, including new ones, on jobs that have likely ergonomic stressors must be carefully monitored during the break-in period. This period can run from several days to several weeks depending on the individual and the particular job. Medical concerns should be triggered by the surfacing of symptoms including swelling, tingling, numbness, weakness, stiffness, and soreness. The degree of resilience of muscle-tendon groups serves as an early indicator of whether the employee will be successful at performing the work satisfactorily. Regarding all of these symptoms, it should not be inferred that any less than perfect score indicates unsuitability for the job. Employees may experience some adverse symptomology during regular adaptation periods. This can be akin to acclimatization. Are there early warning signs of the onset of cumulative trauma disorders? Properly trained medical personnel (i.e., not just any physician) should be in a position to answer this question as long as they understand the finite tasks performed. Professional ergonomists with formal training and practical experience can be well versed in responding as well. They cannot substitute for physicians, but having given that disclaimer, I can add that excellent ergonomists should be able to respond more reliably than most physicians.

Ergonomic Design

Professional ergonomists should be involved in the design of jobs, tools, work stations, and workplaces. They should monitor performance as well. It is not sufficient to view a work station and be told what goes on. They have a need to observe and assess actual human interface with all elements of the specific tasks. Computer simulations can help, but even they are not good enough. At first observation, several employees may appear to be performing the tasks in the same manner. A closer look, which may require watching, videotaping, and analyzing dozens of performance cycles, may reveal something different. Even fine differences in the angle of a shoulder, the rotation of the hand, or the extension (supported or unsupported) of the leg may hold the answer to why a particular employee is experiencing difficulties, which are heading him or her toward a cumulative trauma disorder. Then, a little education and hands-on instruction can go a very long way. Break the habit, explain the significance, and correct the behavior. Contract the services of a professional to perform this type of assessment.

Discerning other variables may aid in the prediction of cumulative trauma disorders. What hobbies does the employee have? In what recreational and athletic activities does he or she regularly engage? I do not recommend a rash judgment with a hasty finger pointing to the outside activity as the cause. There can be a relationship, though. If the hobby did not cause the cumulative trauma disorder, did it aggravate it? With employees returning to work, the same series of considerations addressed earlier still apply. You may now discover how and why the cumulative trauma disorder had developed.

A professional ergonomist may desire to perform the actual job tasks. This can only be done under stringent supervision. If the ergonomist does not actually work the parts, for instance, then he or she can at least shadow the employee's movements. The closer the simulation is to reality, the more valuable the assessment. It is difficult to judge by doing or simulating the tasks just a few times. The number of hours, cycles, and/or repetitions must be considered as well as the number and duration of work breaks.

The ergonomist can use quantitative methods and generally reports instrument-collected data, but it is a different matter when creativity is needed for problem solving. Then qualitative methods are required. This calls for a strong knowledge in physiology, medical anatomy, and the type of engineering principles applicable to the design and redesign of work stations, processes, positions, postures, methods, tasks, tools, and equipment. For example, quantitative methods can result in the recording of ranges of motion, task frequency, and force magnitude. Then the ergonomist can help select state-of-the-art furniture for the keyboard operators. This can include chairs, articulating keyboards, armrests, pointing devices (e.g., the mouse), footrests, and monitor (i.e., screen) elevators. The ergonomist can provide training in the proper use and adjustment of this furniture and equipment.

What if operator(s) still experience pain in specific, describable areas of the body? Note that this is not always the hands, fingers, or wrists. It may be the back, shoulders, elbows, or thighs. In response, ergonomists operating in a qualitative mode will attempt to identify exactly how and why the discomfort is occurring. They will want to observe the normal work habits of each affected individual closely. Then a determination can be made as to exactly what each individual is doing or not doing that is contributing to the pain. Practices may differ among even those operators seemingly afflicted with the same ailments. This can relate to working positions and/or equipment adjustment.

Testing Physical Ability

An ergonomist or other professional may test physical ability when there are real doubts as to whether all individuals can perform a job task in a reasonably safe manner. The testing must employ an objective, discernibly reliable method that is not just a general strength evaluation, but one that can address specific movements necessary for job-task performance. One way to do this is to have the employees lift a stable weight through the actual range of motions required by the particular tasks. For instance, there is certainly a difference in whether a weight is to be lifted straight up and down from point A to point B versus attaining the heights while pivoting. Thus, this is not a situation whereby an employer is screening out employees due to a lack of general strength, even though that gauge is irrelevant to a particular job. There is also the possibility that an individual can handle the test, but would not fare well over a normal shift. Endurance and deterioration of ability can be telling factors.

ADA is a major consideration, and accommodations may be feasible. Anyone who takes the test, but does not meet the minimum requirements (interpreted as a floor for safe performance), should be afforded the identical opportunity to take the test again. The idea is to match employees fairly with jobs that they can physically perform safely. I strongly urge that you familiarize yourself with ADA regulations as well as court decisions, which will help form the basis for consistent legal interpretation of the regulations, where they were written in somewhat vague language. I caution you not to assume that an individual who appears to have or actually has a limited physical capacity in one or more ways should be automatically judged unsuitable for a particular job. The results of a fairly administered and 100 percent relevant test may surprise you. Further, the legal implications of your hasty decision to exclude an individual must be considered. Yet, especially when there are strong ergonomic considerations, there is good reason to seek a reasonable assurance that the job and the employee constitute a suitable fit.

Other Ergonomic Considerations

OSHA vigorously pursues ergonomic hazard recognition. They do not prescribe exact, limited abatement methods. They perceive corrective measures as performance oriented. OSHA leaves options and wants to see positive results. The Agency places a large burden on the employer to conduct dialogues with employees and disseminate information to them. Regular reviews of all potential indicators of musculoskeletal disorders must be conducted. OSHA injury and illness logs are one source. Employers must evaluate jobs that meet specific descriptions relating to frequency and duration of different specific types of movements. Such movements include (but are not limited to) the following: lifting, use of vibrating or impact tools, particularly troubling postures, fast paces out of the exposed employee's control, and highly repetitive movements with possible emphasis on fine motion.

Risk factor checklists, including the type of "red flag" situations addressed in my overall hazard checklist, should be used to aid in the identification of the very specific gross and fine movements that give rise to the ergonomic-related ailments. The job hazard analysis system can also be valuable in providing a path to ergonomic risk reduction.

Always keep in sharp focus the fact that ergonomics problems should be viewed from a human engineering perspective. To some, human engineering even serves as the very definition of the word. The training regarding lifting, reaching, and other movements is necessary, but do not fall into the too often welcomed trap of blaming ergonomic-related ailments on unsafe acts. The correct and logical approach to lowering risk factors is predicated on the concept of designing and constructing the equipment and workplace around the needs of the employee. Start with safe conditions. This is always the obligatory foundation into which safe employees can be incorporated. This is far different than attempting to fit employees into inferior layouts or requiring them to undertake a series of tasks for which there are inherent dangers. The inherent dangers are there because the process was developed and implemented without ergonomics in mind and without the human operator in mind. Now it is too late to correct it. Perhaps the hazards can be minimized within the range of those inherent dangers. However, if the employees were the last variable thrown into the equation, they can only be forced to make the formula work. The human-machine relationship should have been considered at the inception of design. That was the time to generate sensible, pragmatic, brainstormed ideas, fully inoculated with sound safety principles.

Minimizing Human Error

In addition to the obvious goals of avoiding conventional ergonomic risks (i.e., stressors), be progressive and seek to minimize opportunities for employees to become fatigued, confused, frustrated, uncomfortable, and so on.

Realize the importance of good illumination, not just to see the work, but for alertness and mood elevation (in a healthful way). Bright lighting should be consistent and uniform, while not nerve-rackingly intense, or stroboscopic (like flickering fluorescents). Vividly colored walls or accents can add a positive element as long as the hues and patterns are not dizzying or hostility-provoking. It is preferable to avoid incessant or otherwise annoying equipment noises, such as low frequency hums. Perhaps, music (at a controlled volume) can be used to lift employees' spirits and to camouflage the monotonous tones. It can be rather hard to more-or-less satisfy everyone's taste in music. However, it is worth a try even if style compromises or alternative types of music on alternating days must be worked out. Certainly, loud noise in general presents an assortment of hazards, but that is an obvious issue addressed in the OSHA standards. Additionally, room temperature and humidity are ergonomic elements to which attention should be paid. Workplaces need not be palaces, yet they should be designed and maintained so that the ambient environment is hospitable, while not anesthetizing.

Strive for maximum depletion of windows (opportunities) for human error. Assure that controls, displays, and gauges are located for optimum efficiency. Some of the concerns fit more properly into a discussion of machine safeguarding. This includes any discussion of interlocks, constant pressure controls, antirepeat, fail safe, required prior action of one control to allow for the operation of another, redundancy, and the like.

For our discussion at hand, assure that all emergency buttons, switches, pedals, cables, bars, and similar devices are easy to reach and operate. As mentioned in an earlier chapter, red is the anticipated color for "stop" and green is the anticipated color for "start" in American culture. These colors may not be universally recognized as such. I have been in plants where stop buttons (including those used for emergencies) were green, and start buttons were red. These situations presented tragedies waiting to happen. If an employee has been caught in a machine and is being drawn in further, an employee striking the incorrect button (e.g., the red start button) would contribute to an increase in the severity of the injury. Even if that employee was an excellent worker who always operated the proper controls during normal circumstances, he or she could easily err in a panic because primal instinct dictates going for the red. In a lockout/tagout situation, the lock and tag could be applied to a control in the "on" position if the coloring was the opposite of what it should be.

Assure that all employees, particularly those from other countries, understand the general expectations that are shared in the American workplace. At the same time, assure that no assumptions are made. Stop buttons should be of the mushroom type so they can be actuated easily. Start buttons should be located and arranged to avoid accidental activation. Recessing and shielding are but two common methods. Similarly, start pedals (i.e., treadles) should be guarded to avoid accidental activation, mostly by the foot but also by dropped objects. In addition, levers can be set in detents or offsets.

Proper operating sequences should be easy to understand so that mistakes can be avoided. A catastrophe could occur, for instance, if the wrong order of valves and switches were operated, allowing a dangerous release of chemicals. Some other of the myriad cultural expectations that should be adhered to, wherever possible, include the following:

hot water faucet on the left and cold water faucet on the right (or hot water drawn by counterclockwise turn and cold water drawn by clockwise turn or lever shifted toward left for hot water and right for cold water);

knobs (particularly for electrical equipment) turned clockwise to start and counterclockwise to shut;

electrical disconnects pulled down to shut and pushed up to start (preferable for regular light-type switches, but not the case with three-way switches);

steering wheels turned clockwise to turn vehicle to the right and steering wheels turned counterclockwise to turn vehicle to the left;

numbers arranged in a round (like a clock face) should increase in a clockwise direction;

numbers arranged in a vertical line should increase as they go up the line.

Besides color, other means to differentiate the purpose of controls and to avoid the wrong one being operated include shape, size, feel (tactile), and location relative to controlled operation, where feasible.

26

Job Hazard Analysis

By breaking down jobs into basic tasks, a plan for risk reduction can be created. The end product of such a plan is a job hazard analysis (often called a job safety analysis). Each step of the job is studied very carefully, and this is recorded in simple language. The word *job* in this context refers to a particular procedure (perhaps the use of one specific machine to manufacture specific parts) rather than the general, overall tasks that an employee performs in line with his or her occupation. The record is usually kept in a multicolumn format. In some companies, there are loose-leaf books of laminated or similarly protected job hazard analyses for virtually every job in the plant.

When setting priorities for which jobs should be selected first for the analyses, consider where the frequency and severity of injuries and illnesses are greatest. In addition, consider a job as a high-priority candidate if there have been many near misses. Revisiting comments offered relating to ergonomics, it is best to perform job hazard analyses when the jobs are in the design and planning stages. Once the job is put into a real factory setting and there is less control over the actions of the employees the analyses may have to be modified. The original simulated conditions and trials may not match the conditions in which the job will be performed by line employees day in and day out.

Employee Participation

Employees should, by all means, be brought into the analysis process. It must be made clear that the analysis involves studying the job; this is not a time-work study or a performance evaluation. As a byproduct of employee participation, they should automatically increase their level of safety awareness. Employees should become greater advocates of safety as soon as they understand that their involvement translates into the opportunity to help alter their own destinies. Once again, there is the value of employee empowerment and employee control. If the job has been conducted for quite some time, which is usually the case when the analysis is proposed, who can better explain and demonstrate the way it is performed? It is also a good idea to solicit contributions from employees who have run the job, even though they no longer do so. You may learn from employees or from injury and illness logs about more illusive hazards associated with the operation. I am referring to the more subtle types of scenarios that can result in injuries or illnesses. By definition, these are more difficult to predict. The discovery of these hazards is another benefit of an increased level of open communications.

Where actions are needed, as indicated by the tallying of the comprehensive analysis facts, they may result in the improvement of the physical conditions and environment affecting safe performance. Whether or not these modifications are necessary, step-by-step safe procedures should be established. Those procedures can form the heart of the relevant training program. Prior to conducting the analysis, all generally considered physical and environmental factors should be considered and addressed. As with ergonomics, it may help to videotape the procedures and the area. Do not look only at the actual machine involved. Look, for example, at the floor and surrounding area. Consider noise, light, ventilation, congestion, and so on.

The Job Hazard Analysis Form

The job hazard analysis form should list, near the top, the job title and job location. The job location might have to indicate more than one place. Also at the top, list the title and/or occupation of those employees who perform the job. The date of the analysis must be given on the form. The significance is that, if modifications to the machine, process, or operation have been made, one should always be able to compare when the most recent analysis was completed. The analysis should be compared against the job whenever modifications are done. Even if there ends up being no need to change the analysis, it should be given the new date. That way, there is no question of whether or not a review took place.

Reviews are also necessary after accidents and near misses. In all of these cases, employees affected in any way, as well as those injured or involved in the near miss, should be consulted. They may provide the most meaningful input. Further, this always enriches the partnership alliance between management and line employees. This is another pragmatic example of a company fulfilling its goal to be coactive as well as proactive.

The first column of the analysis should address the basic steps (i.e., tasks) within the job as they are performed in normal sequence. List all discernible steps, even if some of them are omitted for certain products run on the machine. Perhaps a particular groove or finish is not needed in some cases, therefore enabling the employee to skip work purposefully on a secondary station. Some keys for deciding where and when a step ends and another begins include changes in activity (including switching from one hand tool to another), changes in type of employee motion, and changes in direction of the employee motion.

The second column to the right should address potential hazards. The entries should be horizontally aligned with the steps in the first column. Do not only think about hazards that can obviously surface at each step when the job is performed perfectly (paradox noted). Think creatively. What hazards could arise out of a slight miscalculation, reflex action, or unusual series of what would otherwise be individual glitches? Do not just concentrate on probable hazards. What hazards are possible? When identifying the potential hazards, always consider the need for personal protective equipment as well as the need to avoid items such as long hair, unless it is restrained up and close to the head, and items of clothing or jewelry that can be pulled into moving parts of equipment. Other factors to consider include (but are not limited to) the following: fixed objects such as sharp edges, unguarded (or inadequately guarded) moving parts, varied ergonomic concerns (see the preceding chapter), falls on the same level or to another level, electrical shock, and toxic substances (by way of ingestion, absorption, or inhalation), noise, welding rays, flying parts (including sparks), radiation, heat, and suspended loads. There is a lengthy list in the accident investigation chapter that references just about every conceivable type of accident. Therefore, that list should be examined.

My short list in the preceding paragraph should not be interpreted to imply that careful work behavior can take the place of guards, open pits, poor layout, and so on. The point is to remove or greatly mitigate any hazards identified, certainly to at least the point of compliance with OSHA standards. Subsequently, devise or revise the job steps to preclude or further minimize any remaining risks. However, do not be too hasty in giving up on the idea of removing the hazards. First try—and then try again and again—to remove the source(s) and/or cause(s) of the potential hazard(s).

If this cannot be done while keeping the same equipment, for instance, it is time to think about other ways to do the job. Will a simple equipment redesign or retrofit be helpful? If a particular piece of machinery cannot be guarded, and still allow for adequate production, is there another piece of machinery that will? If you cannot come up with a safe way of working with a particular tool, is there another tool or another method that can be safely substituted to complete that portion of the operation? If it is extremely difficult to avoid the use of a respirator, even if the employee works very carefully, why not add suitable exhaust ventilation? If there is a major concern about ignition of the vapors of a particular chemical and if the removal or isolation of all ignition sources will be extremely difficult, then is there a nonflammable chemical that can get the job done? Examples abound.

Other considerations include altering the sequence of steps and/or combining steps. If employees are still exposed to hazards of any significant degree, it is time to decide if the job should be performed at all. If it is really necessary, then reduce the frequency as much as possible. The ideal bottom-line mind-set should be that there is no acceptable excuse for allowing the continuation or commencement of a job that is known to expose employees to significant risks.

To help evaluate each of the potential hazards listed in the second column, the third column should list cause(s). This virtually forces the evaluator to invest the required time to look deeply into the potential hazard. Without this column, the evaluator might have a tendency to generalize the recommended action or procedure (i.e., preventive measure) that is the subject of the fourth and final column. In the fourth column, the preventive measures must be specific. General warnings that caution must be exercised are not sufficient. Start each concise sentence or sentence fragment with a verb.

When the documented analysis has been completed, it will consist of four distinct columns with a clear layout to facilitate horizontal tracking from step-task, to potential hazard, to cause(s), to preventive measures. Then the preventive measures must be acted on with an eye toward engineering out potential hazards. Once the engineering is completed, determine if there is still a need for the wearing of personal protective equipment. Then, redo the analysis, as fewer potential hazards should remain. For each of those hazards (and recalling the earlier comments regarding substitution of equipment and similar approaches), be explicit in describing the precise preventive measures still required for the job under consideration.

27

OSHA Visit Action Plan

Most companies lack a contingency plan for dealing with the arrival of OSHA. Accordingly, matters get off on the wrong foot and trepidation sets in. It is of paramount importance that top management develops a simple plan of action and assures that all relevant personnel are well-versed in its elements. Otherwise, even if the initial chaos gives way to some atmosphere of order, the ensuing experience can be disquieting.

As a natural preface to your OSHA visit action plan, or as an integral part of it, you will have taken all of the measures that have been outlined in the previous chapters of this book. Every assurance will have been made that the facility is in compliance with OSHA standards and with accepted occupational safety and health practices. Special note will have been taken of any and all OSHA citations on file to be extraordinarily careful to avoid "willful," "repeated," or "failure-to-correct" citations. All of the paperwork, particularly regarding required programs, training, inspections, certifications, and injury-illness records, will be readily accessible and available to show to the compliance officer.

Filming the Visit

As a part of the plan, I advise that you have a still camera and a camcorder on hand. Be sure to have a plentiful supply of videotape and a minimum of two fully charged batteries for the camcorder. Have the equivalents for the still camera. If you have someone on staff who is trained and trusted to conduct industrial hygiene sampling and testing, be sure that the equipment is properly calibrated and available for use. As an alternative, you may wish to call a consultant with whom you already have a working relationship to be available to conduct that sampling and testing. Whether it be for audiovisuals, industrial hygiene sampling and testing, or general accompaniment on the inspection, you might well have a predetermined individual from the corporate office on your list for immediate notification.

You will be in a position of readiness to take audiovisuals unobtrusively of whatever the compliance officer notes as an apparent or potential violation. Most likely, the compliance officer will visually record every item for which photographs or videotape can have any practical value. Be prepared to take your shots from the same angles and distances used by the compliance officer. As a word of caution, remember that the compliance officer's camcorder has ears as well as eyes. If the compliance officer performs sampling and/or testing, you or the person that you have designated can perform side-by-side sampling and/or testing.

Greeting the Compliance Officer

All receptionists should be aware of exactly what steps are to be taken when an OSHA compliance safety and health officer enters the premises. There is no reason to suspect that Mr. or Ms. OSHA would attempt to enter through an unintended door unless your signs or layout are poor (I might also question your security plan). Yet, if your particular situation gives you reason to believe that this might happen, then the management representatives in that area should also be aware of the initial steps. In fact, you might have a general rule that anyone entering your establishment must pass through the main lobby and sign in. Then, no one is to go forward to wander through the establishment without an escort. As an aside, you might make special concessions and arrangements for contractors who are regularly in the plant.

The receptionist must know exactly where the compliance officer should wait, which should be very close by. There is nothing to be gained by inviting the compliance officer to roam freely. Then, the proper management official must be contacted at once. It must be understood that the official may be interrupted for the purpose of informing him or her of OSHA's presence. That official or representative must be intimately acquainted with the basic OSHA inspection process and with the company's overall safety program. It is also desirable that this individual has a solid grasp of the OSHA standards. However, if the official's knowledge of the standards is more casual, then the plan would dictate that he or she immediately contact the predetermined individual who is most knowledgeable. If there is a safety director, he or she should certainly be involved from start to finish. The hierarchal order for greeting the compliance officer should be listed on paper.

The management person (preferably two) should be polite, while requesting to see the credentials before departing from the lobby. After the identity and position of the compliance officer(s) have been verified, the group should all adjourn to a comfortable, quiet office. The walk to that office should not cause the group to go through, or even near, the manufacturing areas. Additional management staff may be notified, as long as that is done in a composed demeanor and there is no implication that you have sounded the alarm to scramble the troops.

Note my earlier comments about consultants and corporate individuals. Do not get carried away. Do not employ obvious stalling tactics. Be courteous, but be prepared to get down to business. The compliance officer will want to explain the purpose of the visit promptly. Once this is done, there is nothing wrong with offering a cup of coffee, tea, juice, or a similar beverage.

I return to this topic when detailing the OSHA inspection process in Part II.

Responding to OSHA

28

The OSHA Visit: Arrival and Entry

Do not panic. It is time to implement your OSHA visit action plan. That plan was discussed in an earlier chapter and should be reviewed. In this chapter, I briefly touch on the main points and offer some additional comments.

The receptionist will be cordial, yet businesslike. He or she will have the compliance officer wait in a comfortable, public area, while the predetermined employer representatives are summoned. The compliance officer will not sign, now or at any time, any form or release or agree to any waiver. This includes any employer forms concerned with trade secret information. Do not be alarmed. Your right to halt an inspection at any time and your protections regarding trade secrets are discussed later.

OSHA does expect the compliance officer to sign a visitors' register or any other book or form used by the establishment to control the entry and movement of persons upon its premises. Such signature does not constitute any form of a release or waiver of prosecution of liability under the Occupational Safety and Health Act. In turn, the compliance officer will wear the normal pass that your company issues to all visitors. All parties should make sure that this pass is properly secured and does not dangle in a manner that could cause a safety hazard during the walkaround. It should not get pulled into moving parts, contact the surface of hazardous liquids, or get caught on a ladder being ascended or descended. Wouldn't that be embarrassing?

Inspecting Credentials

When the employer representatives greet the compliance officer, they will be courteous, serious, and very attentive. The compliance officer will probably display his or her credentials almost immediately. If not, do not hesitate to ask to see them. The compliance officer should not let those credentials out of his or her sight. Compare the photograph in the credentials to the person in front of you. You may wish to write down the serial number. If there is any doubt at all that the person in front of you is a compliance safety and health officer employed by the U.S. Department of Labor, OSHA, or the equivalent state agency, feel free to place a telephone call to the OSHA area office or equivalent state agency. Obtain the number on your own and make the call to verify the person's identity. It is very, very seldom that an imposter shows up. If there is an imposter posing as an OSHA compliance officer (a federal law enforcement officer, in effect), you and/or the OSHA area office should contact the Federal Bureau of Investigation and local authorities at once.

If the person attempts to sell you products or services, you know there is a problem. If the person indicates later in the day, after having escaped detection as an imposter, that he or she wishes to collect a fiscal penalty on the spot, once again there is a problem. The imposter might say that if the penalty is paid on the spot, a reduction will be built in. Sure, these cases are rare, but they have happened.

Another reason that a person could be motivated to pretend to be an OSHA compliance officer is to observe your trade secrets. That has happened, as well. He or she could be working for a competitor and wishes to see your new design, pattern, technical advancement, and so forth. A similar motivation would relate to secrets that threaten national security. There is also the threat of terrorism.

I have heard of at least one case where an imposter came and went. Only after the inspection did suspicion reign, and it was well founded. To my knowledge, the person was never caught, and none of the aforementioned reasons for the ploy seemed plausible. It appeared that the person was simply driven by a kick or a power trip. While I do not espouse paranoia, make sure that you are comfortable with the identity of your visitor.

The majority of the times when a management representative mistakenly thinks that an OSHA compliance officer was in his or her facility, it is simply a misunderstanding. Sometimes, the initial conversation is hurried, or interruptions throw things off track. Although the U.S. Department of Labor does not foot the bill for business cards, virtually all compliance officers carry them. You will probably be given one at the outset; if not, ask for one. Naturally, credentials have a greater degree of reliability, but you can take possession of the card. Frequently, employer representatives become nervous when they merely hear "OSHA" uttered by a live human being with whom they are face to face in the workplace. On many occasions, at a seminar or even in a social setting, individuals have told me that OSHA had visited their establishment when it simply was not true. It turns out that the visitor was a loss control representative (insurance), consultant, sales representative for a company selling safety supplies, local or state fire marshal or similar title, service representative for a crane or fire extinguisher company, local building official or inspector, or some other person whose job is related to occupational safety and health. In all of these cases, the visitor may have been carrying on legitimate business. Nevertheless, you have a strong need to know exactly with whom you are dealing, including what governmental agency, if any. The care with which OSHA inspections are handled should be exceptional. The stakes can be extremely high.

As a quick aside in regard to sales representatives, beware of attempts to sell you products that they claim are required by OSHA or OSHA-approved. Products are not OSHA-approved; they may be ANSI-approved, however. Products may meet OSHA specifications and may help you comply with OSHA standards if installed, used, and maintained according to the manufacturer's instructions. That is a distinct matter. In many cases, sales representatives have taken liberties with the truth in aggressively trying to convince an employer to purchase an item by insisting that without it the company could be subject to a large penalty. I know of numerous cases of such misrepresentation. In some instances, it appeared that there was a deliberate attempt to dupe the prospective buyer. In other instances, there was no plan to deceive; there was only ignorance on the part of the sales representative.

If there is a question or a doubt as to the relationship of a product or a service to OSHA standards, you can call the local OSHA area office and inquire. You might also ask the sales representative to present you with a copy of the standard(s) that he or she says apply. In fact, if I was in the business of selling products that were in compliance with OSHA standards (with all of the regular qualifiers), I would have handouts of reprints of the standards. Further, I would provide literature that explained exactly why and how the product fit in with those standards. I would be even more interested in offering this written information if there was a likelihood that the product was specifically required by OSHA (in other words, it did not just have to be of the particular type I sold, if the company *chose* to have one at all). Intelligent, honest sales representatives do generally provide the sort of documentation that I have described. If you are given a reprint of a standard, make certain that it is not taken out of context. You may have to go back to the original thick OSHA standards book and check.

Back to the compliance officer's arrival. You have verified that the compliance officer is who he or she claimed to be. Now he or she will ask to speak with someone in high authority. In technical terms, the compliance officer is to seek an owner, operator, or agent in charge of the workplace. As long as your employer representative selections are on an obvious level of significant authority, there should not be any question. Now it is time for practical hospitality. Go to the comfortable, quiet room that was already in your action plan. Brew up a batch or break out the Yoo-Hoo.

Cancellation of an Inspection

What if the company is in dire financial straits and has drastically cut back on its number of employees as well as functioning operations? That is not sufficient cause for an inspection to be halted. Even if the company has been formally adjudicated and determined to be bankrupt, the inspection will not cease. The relevant question is whether or not any employees might be exposed to occupational hazards. Therefore, if the company is still in operation, the inspection is to proceed. The company must comply with OSHA standards until all operations cease (i.e., until no employee exposure remains).

In the case of labor disputes, inspections may proceed regardless of work stoppages, strikes, or picketing. If the compliance officer identifies an unanticipated labor dispute at a proposed inspection site, he or she will consult with the assistant area director before any contact is made. Programmed inspections, usually called *"general schedule,"* may be deferred during a labor dispute either between a recognized union and the employer or between unions competing for bargaining rights in the establishment.

Unprogrammed inspections (e.g., complaints and fatalities) will be performed during labor disputes. However, the seriousness and reliability of any complaint will be thoroughly investigated by the assistant area director prior to scheduling an inspection. This is to provide reasonable assurance that the complaint reflects a sincere belief that a hazard exists. If there is a picket line at the establishment, the compliance officer will inform the appropriate union official of the reason for the inspection prior to initiating it. If no work is being performed, the discovery of alleged violations will be more difficult, but potential exposures may still be established.

Advance Notice

As soon as you hear the word *"inspection,"* you might want to know why you do not recall receiving advance notice of the visit. Wording in the Occupational Safety and Health Act states that the compliance officer may "enter without delay and at reasonable times any factory, plant, establishment, construction site or other areas, workplace, or environment where work is performed by an employee of an employer" [and that he or she may] "inspect and investigate during regular working hours, and at other reasonable times, and within reasonable limits and in a reasonable manner, any such place of employment and all pertinent conditions, structures, machines, apparatus, devices, equipment and materials therein, and to question privately any such employer, owner, operator, agent or employee."

The Occupational Safety and Health Act generally prohibits advance notice of inspections. There are very few exceptions to this statutory prohibition. Advance notice of inspections may be given only with the authorization of the area director and only in the following situations:

1. in cases of apparent imminent danger, to enable the employer to correct the danger as quickly as possible;

2. when the inspection can most effectively be conducted after regular business hours or when special preparations are necessary;

3. to ensure the presence of employer and employee representatives or other appropriate personnel who are needed to aid in the inspection;

4. when the giving of advance notice would enhance the probability of an effective and thorough inspection (a complex fatality investigation, for instance).

Occasionally employees claim that there has been advance notice of an OSHA inspection. Again, the actual cases of authorized advance notice are extremely rare. These employees are usually referring to what they believe, ostensibly or otherwise, was unauthorized notification. In my experience, the claims have invariably been faulty. I have already alluded to related misunderstandings that arose when a sales representative, loss control representative, building official, consultant, or other person tied to the field of occupational safety and health visited an establishment. Some employees knew that the individual, whom they thought was an OSHA compliance officer, was coming; in fact, an appointment may have been made. In other cases, word circulated through the facility in a general or very limited fashion that an OSHA compliance officer would be appearing. Shortly thereafter, a real OSHA compliance officer did arrive. Advance notice? No, not in the cases of which I have become aware.

In some cases, the prior knowledge (perfectly legal and not constituting any form of advance notice) was discerned because a compliance officer had been on the site one day unbeknownst to employees in a certain area of the facility. That compliance officer returned as planned on the following workday, after loose rumors that OSHA was coming had circulated. In some instances, the compliance officer was denied entry, and it was assumed that he or she would return, probably with a warrant in hand.

At times, complainants of their own volition tell fellow employees about their complaint. It is assumed that OSHA will be coming, even though no one knows when. Sometimes, an occupational fatality or catastrophe that must be reported to OSHA will lead to an inspection. Further, there are cases when severe injuries, fires, or explosions have been reported in the media. In all of these cases, management anticipates the likelihood of an OSHA visit and verbalizes that concern. This is another scenario where employees, not understanding the OSHA system and not having heard the whole story, may believe that unauthorized advance notice has been given.

Note that advance notice does not include nonspecific indications of potential future inspections. One example is when wording in a nonformal complaint letter from OSHA to the employer explains that an inspection may take place depending on the promptness, thoroughness, precision, and overall content of the employer's response. Another example is when a compliance officer tells the employer during a closing conference that an intraoffice referral may be made, usually safety to industrial hygiene or vice versa, and can lead to another inspection. Simple references to potential follow-up inspections are certainly not considered advance notice.

Criminal Penalties

When an employer has received true advance notice of an inspection and fails to notify the authorized employee representative, a citation (including proposed penalty) can be issued. Anyone, including OSHA compliance officers, who gives unauthorized advance notice of an inspection is subject to a criminal penalty. The Occupational Safety and Health Act and the U.S. Code provide for criminal penalties in such cases and in certain other cases unrelated to advance notice. Such penalties, including those for advance notice, are imposed by the courts after trials and not by OSHA or by the Occupational Safety and Health Review Commission, which hears contested OSHA cases. In each case, penalties can include fines, imprisonment, or both.

Criminal penalties can apply to giving false information to OSHA. You must be particularly careful when the information you are imparting is in written form. The Agency has no desire to seek legal action just because you make an honest (i.e., inadvertent) error. Nevertheless, you must strive to be totally accurate.

Be very cautious when composing a progress report. Be 100 percent certain that an item has been abated before declaring that it has been. Do not release to OSHA exposure records, sampling data, test results, inspection or maintenance logs, injury-illness information, accident investigation reports, miscellaneous statistics, manufacturer's operating instructions, material safety data sheets, or similar documents with which anyone has tampered.

You may have employees, which can include members of management, who want to alter or obfuscate a document or fact in order to put the company in a better light. More specifically and more damning, they may want to modify or remove certain information that would highlight the employer's state of noncompliance. Besides the criminal aspects of such actions (i.e., the giving of false information), this can bolster OSHA's documentation for the designation of "willful" on a citation (a civil matter). "Willful" violations are addressed elsewhere. There are several possible elements, other than giving false information, that can serve as a foundation for the citing of a noncriminal (again, civil) "willful" violation. However, when any "willful" violation of an OSHA standard, rule, or order causes the death of an employee, criminal penalties are also applicable.

It should come as no surprise that criminal penalties can be levied for killing or assaulting an OSHA compliance officer or for hampering the work of a compliance officer. This may seem evident, but most persons lack an adequate understanding of the definition of assault. Suffice to say that you need not physically harm a person to be fairly prosecuted for assault. Without trying to totally explain the legal definition along with the accompanying nuances, I stress that if you threaten the compliance officer and if it is your intent to put the compliance officer in a state of fear of physical attack, particularly if your goal has been achieved, you are in jeopardy of being charged with the crime of assault.

Denial of Entry

If you are considering what is often referred to as *"denial of entry"* to the compliance officer, I recommend that you at least sit down with the compliance officer and discuss the matter. Whether or not you have allowed a literal entry into your facility, insistence upon a warrant generally relates to not allowing the walkaround inspection. You might also want to place limits on the company paperwork that the compliance officer wishes to see. At any rate, I recommend that you at least proceed to the incipient stages of an opening conference. In my chapter on the opening conference, I address warrants and administrative subpoenas.

29

The Opening Conference

Warrants

I suggest that you be hospitable, even if biting your lip. Be practical in all respects. Do you want to deny the compliance officer the opportunity to perform an inspection? It is your legal right to do so unless he or she has a warrant. It is extremely rare for OSHA to obtain a warrant before an inspection. They might do so if your facility has insisted on warrants for several previous OSHA inspections or if your corporation has shown a clear pattern of insisting on warrants at other facilities. Another possibility (again, extremely rare) is when the Agency has clear reason to believe that a job will only last a short time or that job processes will be changing rapidly. Even then, the Agency will not generally seek a warrant unless there is also powerful evidence that a denial of entry is likely.

OSHA cannot conduct warrantless inspections without the consent of the employer. If you have any doubt about giving that consent, find out everything you can about why your establishment was selected for inspection, what the purpose of the inspection is, and what the intended scope of the inspection is. If the inspection was spawned by a complaint, ask for a copy of that complaint before settling into the full body of an opening conference.

You may insist that no inspection is to take place without a warrant (often referred to as "*compulsory process*"). Will such action make a large difference in how well you come out in an inspection? In other words, will the delay provide you with a solid opportunity to improve your compliance posture? For example, are you running a substantial amount of equipment for which guards are due to arrive and/or be installed within a few days? If not, could you effectively guard them, within a few days anyway, if you apply all available resources? Is the final, complete edition of your lockout/tagout program at the printing company? Is an electrical contractor scheduled to arrive the next day to correct several safety deficiencies? Are you running a particular unguarded machine as an experiment for an atypical reason, or for some other reason, only today and tomorrow? Scheduled or not, could you benefit and greatly ease your susceptibility to citation by hastening to perform maintenance on your cranes and ventilation systems?

These are the types of questions to consider, especially if pertinent to items on a complaint when the planned scope of the inspection is to be bounded (for the most part) by those items. Naturally, any delay in the start of a walkaround inspection and review of your required records can earn you more time to eliminate or reduce the severity of hazards. Although you may remove violative conditions, do not forget that OSHA can issue citations when there is documentation legally sufficient of recent noncompliance.

In the next chapter, I discuss the 6-month rule. Unless the inspection delay can significantly ameliorate what you perceive to be a major plight, you may not want to obligate OSHA to obtain a warrant.

If you deny entry or beyond without the presentation of a warrant, a well-seasoned compliance officer will attempt to tactfully persuade you to hold off on that decision. He or she will ask that the opening conference commence and that he or she at least be allowed to ask some questions so that proper paperwork can be filled out. This relates to the exact current name of the company, the number of employees, and so on. Although the compliance officer has researched such matters, the information scrutinized may not have been up-to-date. For instance, perhaps your company has gone through a major downsizing. Maybe your standard industrial classification (SIC code) has changed due to a shift in what you manufacture.

The compliance officer should point out that if you are kind enough to allow him or her to continue, you can certainly stop at any time and insist that he or she leave. The compliance officer is not about to run past you to sneak a look at an operation. However, if you demand that the compliance officer vacate the premises before you have even provided some basic company information, you run a substantial additional risk of receiving an administrative subpoena prior to being served with a warrant. In fact, such a subpoena will normally be issued if you do not permit the compliance officer to view your OSHA-required injury and illness logs or other such OSHA-required records as those spelled out in the hazard communication standard.

Subpoenas

Whenever there is a reasonable need for records, documents, testimony, and/or other supporting evidence necessary for completing an inspection scheduled in accordance with any current and approved inspection scheduling system, the regional administrator or authorized area director (who is, in fact, granted wide discretion by the regional administrator) may issue an administrative subpoena. The same applies to an investigation of any matter properly falling within the statutory authority of the Agency. This is not common, and it may occur at or near the end of an inspection if the employer was recalcitrant to the degree of hindering the gathering of evidence. It may also occur at the early stages of a planned inspection if the employer decides to forgo rudimentary cooperation, as previously described.

Even where the walkaround is limited by a warrant or an employer's consent to specific conditions or practices, a subpoena for production of records (subpoena duces tecum) will normally be served if the employer did not accede to a compliance officer's examination of OSHA-required records. The records specified in the subpoena will include, as appropriate, injury and illness records, employment data necessary for the calculation of lost workday injury rates, exposure records, the written hazard communication program, the written lockout/tagout program, reprimands to employees for violations of company safety and health rules, and records relevant to the employer's safety and health management program (e.g., safety and health manuals or minutes from safety meetings). Additional records specified in the subpoena might include those concerning employee task completion times, quotas, piecework, incentives, production schedules, and output. The idea is that the records sought can help determine if a causal link between injuries/illness and working conditions existed.

The area director may issue an administrative subpoena which seeks production of the foregoing categories of documents. The employer's ability to comply with the subpoena must not be unreasonably burdensome. Thus, the company should be able to easily retrieve and produce the records sought. The common practice is for the subpoena to require the employer to present the records at the OSHA area office. This can be an added burden if that office is not close to the plant. The subpoena may call for immediate production of the records, with the exception of the documents relevant to the safety and health management program, for which a period of 5 working days is normally allowed. Information gleaned from those records may reveal contradictions between your company policies and rules and your actual practice. Such a finding could certainly form the foundation of a "willful" violation. Similarly, safety meeting notes, which could include those of committee meetings, could set you up for a "willful" by showing a continual awareness of particular hazards and/or violations.

Since OSHA cannot cite for the absence of general safety and health program manuals and safety committee notes, for instance, the employer is not regularly obligated to share them with the compliance officer. Thus, by having essentially forced OSHA to issue the subpoena, you have placed yourself in a more precarious position. You have, by legal remedy, been involuntarily stripped of another layer of potentially protective insulation.

Yes, you may wish to contact your attorney, but unless the administrative subpoena is overly broad or burdensome, your chances of overturning it are small. The subpoena must be sufficiently limited in scope, relevant in purpose, and specific in directive. If the subpoena is challenged by the employer, in the case of a "programmed inspection," OSHA can be required to show that the selection of the company for inspection was the product of a systematic, nonarbitrary process. This would demonstrate that the selection was not based on the unbridled discretion of (for instance) the compliance officer. Seemingly, one area of the subpoena to attack would be the requirement for production of records voluntarily kept by the employer (i.e., those not specifically required to be maintained by the OSHA standards). Nevertheless, OSHA has historically been able to justify the inclusion of such records on their subpoenas.

The Occupational Safety and Health Act grants a broad subpoena power of books, records, and witnesses. Such power is legally considered to be customary and necessary for the proper administration and regulation of an occupational safety and health statute. In the exercise of their lawful investigatory powers, administrative agencies are conferred a wide discretion to require the disclosure of information concerning matters within their jurisdiction. On one hand, the Act contains no provision exempting self-audits or self-inspection reports from the subpoena authority. On another hand, the Act provides that OSHA regulations may include provisions requiring employers to conduct periodic inspections to determine their own state of compliance.

Thus, various records that are compiled and maintained by the employer on a voluntary basis are within OSHA's legitimate grasp through subpoena. Note that in exceptional circumstances the agency may seek to obtain records of third party (e.g., safety consultants, insurance inspectors) inspections. In my chapter on inspections I outline possible steps to prevent the employer from being forced to turn over such records. I caution that an attorney should be contacted for assistance in this matter. The subpoenas are often issued within 24 h of the compliance officer's notification to the area director. When the authority to compel testimony or evidence is defied, OSHA may apply to the federal courts for an order requiring compliance under threat of contempt.

Employee Records

OSHA standards obligate the employer, upon request by OSHA, to assure the prompt access of compliance officers to employee exposure and monitoring records. The standard also ensures that an employee or designated representative can have access to analyses that were developed using information from exposure or medical records about the employee's working conditions or workplaces. Personal identifiers (e.g., names, addresses, social security numbers, payroll numbers, age, race, and sex) must be removed from the data analyses prior to access.

When the employer fails to respond to the request, another scenario has been created whereby OSHA can issue an administrative subpoena for the production of these records, including voluntarily kept personal biological and environmental monitoring records. The argument of excessive burden to the employer is out the window because the compliance officers search through and copy records without requiring the time or expense of the employer.

When the Agency has a need to examine personal employee medical records, it can issue a medical access order through the OSHA national office. OSHA defines *"employee medical record"* as a record concerning the health status of an employee which is made or maintained by a physician, nurse, or other health care personnel or technician. This includes the following: medical and employment questionnaires or histories (including job description and occupational exposure); the results of medical examinations (preemployment, preassignment, periodic, or episodic) and laboratory tests (including X-ray examinations and all biological monitoring); medical opinions, diagnoses, progress notes, and recommendations; first-aid records; descriptions of treatments and prescriptions; and employee medical complaints.

The purpose is so that the Agency secures the proper tools with which to assess the effectiveness of control measures, the impact on employee health, and the need for further study and medical surveillance. The only OSHA area office personnel who are allowed to see these records are compliance officers (nearly always limited to one or two industrial hygienists) considered by OSHA to have a need to know and deemed by the Agency to be qualified to perform meaningful evaluations and make reasonable interpretations of those records. In the OSHA national office, there is a person (typically the medical officer) who is allowed to view these records, which may include the personal identifiers mentioned earlier. If the employer does not comply with the medical access order, the OSHA regional administrator can issue an administrative subpoena. The main reason that the authority is moved up to the regional administrator as opposed to the area director is due to the request for the release of the personally identifiable medical records for which the medical access order was obtained. In turn, refusal to honor it can result in a federal court finding of contempt.

The release of the personal identifiers to anyone other than those persons designated in the medical access order or to the subject employees themselves is strictly prohibited, even when a request has been filed under the standards or the Freedom of Information Act. The one exception—the only way that the personal identifiers may be released—is with a letter of authorization from the individual who is the subject of those medical records. That letter of consent must designate exactly to whom the employee is allowing the Agency to release his or her medical records.

Before returning to the subject of warrants, more explanation is needed regarding access to medical and exposure records. Although there are various provisions and nuances to the standard, I limit the discussion here to the chief sections. At the time of initial employment and at least annually thereafter, employees must be told of the existence, location, and availability of their medical and exposure records. The employer must also inform individual employees of their rights under the access standard and make copies of the standard available. Employees also must be told who is responsible for maintaining and providing access to the records.

When an employer ceases to do business, he or she is required to provide the successor employer with all employee medical and exposure records. When there is no successor to receive the records for the prescribed period, the employer must inform the current affected employees of their access rights at least 3 months prior to the cessation of business. The employer must also notify the director of the National Institute for Occupational Safety and Health (NIOSH) in writing at least 3 months prior to the disposal of the records.

Each employer must preserve and maintain accurate medical and exposure records for each employee. The access standard imposes no obligation to create records, but it does apply to any medical or exposure records created by the employer in compliance with other OSHA rules or at the employer's own volition. Exposure records and data analyses based on them are to be kept for 30 years. Material safety data sheets, required to be maintained under the hazard communication standard, are considered to be exposure records. Thus, they must also be kept for 30 years. There is a fine line alternative which basically allows the employer to maintain the key exposure data of the material safety data sheets. I recommend that you maintain the full material safety data sheets rather than attempt to glean from the standard the absolute minimum alternative requirement.

Medical records are to be kept for at least the duration of employment plus 30 years. Background data for exposure records (e.g., laboratory reports and work sheets) need to be kept only for 1 year. Records of employees who have worked for less than 1 year need not be retained after employment, but the employer must provide these records to the employee upon termination of employment. First-aid records of one-time treatment need not be retained for any specified period. OSHA does not mandate the form, manner, or process by which an employer preserves a record except that X-ray films must be preserved in their original state. Employers must notify the director of NIOSH 3 months before disposing of records.

Initiating and Securing Compulsory Process

The regional administrator may issue another type of subpoena. It is issued to require the testimony of any company official, employee, or other witness. This document is a subpoena ad testificandum. Further, the regional administrator can issue a subpoena requiring the production of physical evidence such as samples of material. He or she can also direct the area director to issue that type of subpoena. Warrants are not only sought upon refusal of entry. They are also sought for refusal to produce evidence required by subpoena.

The area director must consult with, and receive approval of, the regional solicitor prior to initiating the compulsory process. The Office of the Solicitor is not directly a part of OSHA. It is a separate agency of the United States Department of Labor. The warrant sought when the employer consent has been withheld will normally be limited to the specific working conditions or practices forming the basis of the unprogrammed inspection (complaints are always the best example). A broad scope warrant, however, may be sought when the information available indicates conditions that are pervasive in nature or if the establishment is on the current list of targeted inspections (including general schedule programmed inspections). The warrants are sought through written application to a United States magistrate or state equivalent. Information on the application includes a narrative of actions taken by the compliance officer and other involved OSHA personnel leading up to, during, and after the refusal.

Among the details on that narrative are the reasons stated for the denial by the person(s) refusing entry. The application also clarifies at what stage the denial was made (e.g., entry, the opening conference, the walkaround, etc.). Other information includes the following, as available and relevant: all previous inspection information, including copies of or references to previous citations; previous requests for warrants; description of the workplace and work process; known hazards and potential injuries and illnesses associated with the specific industry or process; investigative techniques which will be required during the proposed inspection (e.g., personal sampling, photographs, audiotape, videotape, examination of records, access to medical records, etc.); and the specific reasons for the selection of the establishment for the inspection, including proposed scope of the inspection and rationale. The application also refers to the legal enabling language and explains how all current OSHA procedures were followed.

In the case of a complaint or referral, the application will set forth the basis of reasonable grounds for believing that a violation that threatens physical harm or imminent danger exists and the standards (or section 5(a)(1) of the Occupational Safety and Health Act) that could be violated if the complaint or referral is true and accurate. A copy of the original complaint or referral, as well as a copy of the typed (usually "sanitized") complaint or referral, will be included. In the case of a follow-up inspection, copies of previous citations that formed the basis for initiating the follow-up will be included. For the follow-ups, any copies of settlement agreement stipulations and final orders will be included.

If OSHA truly adhered to its legal obligations and its own written policies, there should not be much of a problem obtaining the warrant. That *if* is a big one, and it is essential in a democracy. I do not mean to be overly didactic, but the point is that no warrant can be granted without just cause. The compliance officer who actually was refused entry or was refused continued entry should be the chief person to meet with the United States magistrate or state equivalent. He or she may be accompanied by the area director or assistant area director. These meetings are held ex parte (e.g., without the employer present). In most cases, the regional solicitor (really any attorney from that office) will not accompany the OSHA personnel, often due to geographical distance problems. In all cases, the OSHA personnel must be prepared to describe their communication with the regional solicitor or state equivalent and attest to his or her approval of the application.

The United States magistrate or state equivalent may challenge the OSHA personnel in the case of a complaint inspection if there is a labor dispute in progress. This can relate to strikes, negotiations, upcoming votes regarding the approval of a union, and so on. In such cases, OSHA might counter that the complaint did not relate to the dispute. However, even if the complainant was to some degree spurred on by the labor dispute or even if he or she simply declares a hatred for the employer, that is not sufficient reason for denying the issuance of a warrant. Motivation should not be a factor. The key question is whether or not OSHA believes that the hazardous conditions described by the complainant are likely to exist and constitute violations.

Frequently, employers gripe that the complainant must be a disgruntled employee. That argument is presented as if a disgruntled employee has no right to complain. He or she has every right. Why is that employee disgruntled? The answer may be a direct outcropping of the unsafe and/or unhealthful conditions at the workplace. Even if that employee is a chronic complainer about matters including those unrelated to safety and health, he or she may be offering well-conceived concerns about occupational hazards.

In order to obtain a warrant, OSHA need not show probable cause in the criminal law sense, but may produce specific evidence of an existing violation or a showing that a workplace has been selected for inspection based upon a reasonable administrative plan. That plan concerns programmed inspections. The Agency must demonstrate that the scheduling was based upon objective or neutral selection criteria. Workplaces are selected according to national scheduling plans for safety and for health (industrial hygiene) or special emphasis programs. These special emphasis programs may be national in scope or may be tied to local concerns, for instance, where particularly hazardous industries are concentrated. OSHA might concentrate on specific standard industrial classifications or the likelihood of encountering specific chemicals across an array of numerous, different SIC codes, for example.

At times, OSHA works through random number systems within emphasized SIC codes and potential hazards. What this means in the real world is that, for inspections that were not initiated by complaint, referral, fatality, catastrophe, imminent danger, or as a follow-up to an earlier inspection, there had better be a fair, logical, documented system. A compliance officer cannot just decide to walk into a building and attempt to initiate an inspection because he or she lives nearby. The OSHA area office cannot annually schedule a general schedule inspection of the same establishment, even though few violations have been found and the SIC code is not on what is often referred to as the high-hazard planning guide. The warrant-issuing official will need proof from the Agency that any programmed inspection was scheduled with adherence to the foregoing criteria.

Once a warrant is obtained, the inspection will usually begin on the next workday. The employer definitely has the right to limit the compliance officer's visit to areas and records spelled out in the warrant. If the compliance officer notes hazards that are in plain view, he or she may quickly attempt to gather enough information to document apparent violations. However, the employer may refuse to permit the compliance officer to linger in that area. In most cases of this sort, the employer consents to a reasonable expansion of the scope of the warrant. If that is done, no additional warrant is required. If the employer insists on a second warrant and if circumstances make it appropriate, a second warrant may be sought based on a review of the records or on those plain-view observations of other potential violations during a limited scope walkaround.

Whether you have insisted upon the original warrant alone or (and this is rare) that warrant plus subsequent one(s), consider that you may have set a regrettable tone. This is your right. Theoretically, an inspection by way of warrant should not yield harsher results than a warrantless inspection. There is the maxim that the compliance officer can only see what he or she can see. I suggest, with no disrespect intended toward OSHA, that reality and theory are not always the same. With a warrant in hand, perhaps the compliance officer's vision will miraculously become keener and perhaps he or she will bend or reach a little further when searching for hazards. Perhaps there will be less leeway in the interpretation of a standard. Sure, you can always challenge an interpretation of a standard. Not all standards are as simple as whether or not a machine was electrically grounded or whether there was a perimeter protection railing on a mezzanine. Maybe the particular penalty-reduction factors that are not cut and dried will turn out to be less generous than they could have been. Certainly, you can argue that you should have been given a greater penalty reduction relating to the assessment of your general safety and health program. Just do not lose sight of the fact that compliance officers and their superiors are human beings in most cases. They are not computers bereft of emotion. Enough said.

OSHA has been highly successful in its warrant litigation, and its administrative plans have been uniformly approved. A very small percentage of employers refuse entry. Going one step further, very, very seldom does an employer presented with a warrant still make the decision to refuse entry. Barring extraordinary circumstances, such as the warrant identifying the wrong location or equipment, you should plan to honor the warrant. The defiance in refusing to abide by the language of the warrant can get the employer in very serious trouble indeed. As with noncompliance with an administrative subpoena, spell that trouble c-o-n-t-e-m-p-t. The financial penalties levied by the courts (not by OSHA) can be assessed per each day of noncompliance with the warrant and accumulate until you allow entry. The final bill can be, and has been, staggering. Then there is still the matter of the inspection. There is not a choice of paying a set penalty for shunning the warrant or allowing a warranted inspection. Ultimately, you will do both.

Other Reasons for Postponing the Inspection

Some employers do not wish to insist upon a warrant, but prefer that the compliance officer come back another time. There is no provision in the system for this type of postponement. Will such a request ever be honored? It will not be honored simply because the employer is in a peak production period, in the midst of taking inventory, short some key staff members on that particular day, or inspecting the emergency feeds on the towel dispensers. There would have to be truly exceptional circumstances for the compliance officer to leave with the idea of returning without a warrant. In fact, my example may seem ludicrous, but it reveals just how remarkable a situation would have to be for the request to be honored.

In the fictitious situation, the inspection has been initiated off of the computer as a programmed, general schedule inspection. The establishment has never been inspected previously, and it is not a great driving distance from the OSHA area office. There are only 12 employees. The facility appears to be quite new and in good condition, albeit from the outside. The compliance officer enters the lobby and is greeted by a receptionist, who explains that the two brothers who own and run the company are not in; they are attending the funeral, in another state, of their father. The receptionist introduces the compliance officer to the person who was left in charge. This is the first time that the individual has been put in that position. He or she is clearly not a true management representative, but merely a veteran employee called upon in an emergency to keep the shop running for the three business days that the principals would be absent. In a sincere and nonthreatening manner, the individual asks the compliance officer to please hold off the inspection until the brothers have returned. The veteran employee says that the brothers have given specific instructions not to allow anyone but company employees into the plant. He or she adds that some of the key operations are not in service and will not be returned to service until the bosses, the only ones who run those operations, are back at work.

The compliance officer does not hear any excessive noise and does not smell anything that would give rise to a major concern. In this case, the compliance officer would probably call the assistant area director and explain the situation. As a result of that conversation, the compliance officer would extend sympathy and then depart after indicating that he or she would return but explaining clearly that no appointment could be made.

Joint and Separate Conferences

At one point or another, an opening conference will take place. It is time to start taking meticulous notes. If there is an authorized union representative (or where there is no union, another representative selected by the employees, e.g., the chairperson of the safety committee), that individual has the right to attend an opening conference. Further, an employee representative has the right to participate in all phases of the inspection. If either the employer representative or the employee representative prefers separate opening conferences, the compliance officer will abide by those wishes.

In deciding whether or not you want a joint conference, you should consider the advantages and disadvantages rather than just rashly uttering a demand. If you do decide on a joint conference, you will be in a good position to exercise more control over the scope of the conversations. Additionally, the compliance officer will appreciate and note the collective spirit that has been engendered. If you opt for separate conferences, the compliance officer would be open to a more candid discussion, at least in theory. If your safety and health program is in line with the elements that I detailed in the first part of this book, a joint conference would be fitting and beneficial.

Exemptions

If you believe that the facility has earned an exemption from general schedule inspections, you should state that early in the conference. You may have received this inspection exemption through approval into OSHA's Voluntary Protection Program (VPP). If your company is formally pursuing VPP approval status at a stage whereby you are already working with OSHA on the project, you may also be exempt. I address the Voluntary Protection Program later.

You might also be exempt if an OSHA-funded consultation is in progress—not just scheduled, but has actually begun—even if the consultant is not present at that very minute. That type of consultation is generally operated by the state.

Barring highly unusual circumstances, inspections initiated by complaints, referrals, fatalities, catastrophes, or concerns of imminent dangers are not likely to be exempt, no matter what pilot programs relating to exemptions are in effect at the time. However, the scope of those inspections may be limited as a reward of sorts for your participation in the VPP or OSHA-funded consultation program.

Reason for Visit

The compliance officer will explain the specific reason for the visit. If the inspection was scheduled due to a complaint received by OSHA, you will be presented with a "sanitized" copy of that complaint. This means that, unless specifically authorized by the complainant, the complainant's name and other personal identifiers will have been removed. In that vein, pet expressions or other wording that could point to an individual will also have been removed. The paper you are given will normally be typed. It will not be a copy of the original complaint with redactions.

Injury and Illness Logs

The compliance officer will want to examine your injury and illness logs. I discussed those logs in an earlier chapter. Depending on the number of employees at the inspected location, the compliance officer may request records as far back as 5 years. If you employ a large number of employees, the compliance officer may feel that the purpose of the examination will be served by a review of perhaps only 2 years.

Speak up if there are entries that you feel may not have been required. You can explain that, because there was some doubt, you chose to log the incident with the understanding that you could delete it if you subsequently learned that you were not required to record it. There are two types of situations that often fit this category. One, for instance, is when there is a close issue regarding whether or not the injury resulted in treatment beyond OSHA's definition of first-aid. The other is when there was lost time, for instance, but you are disputing whether the injury was work related. You can place an asterisk next to such entries, and be prepared to discuss them upon OSHA's arrival. That way, you cannot be accused of failing to record when required, and you leave yourself the opportunity to persuade the compliance officer not to include the incident if he or she works out a formula to quantify your injury and illness experience.

Surely, compliance officers will look for trends, as I have explained in the earlier chapters. Be prepared to provide the compliance officer with required supplementary reports on each of the entries for up to the previous 5 years. I remind you that injury and illness records, including the log and the individual reports, can serve as a basis for "willful" violations.

Documents Turned Against You

The compliance officer will ask to see any evidence of a general safety and health policy, plan, and rules. He or she will let you know that such a plan can help demonstrate your commitment to a safe and healthful workplace and your compliance with the Occupational Safety and Health Act. That is an accurate depiction on the part of the compliance officer, but the matter is not that simple. You may be eager to show off your grandiose documentation. But never lose sight of the fact that such paperwork can highlight your specific knowledge of standards and of particular or potential hazards in your facility. If you have followed through on complying with those standards and abating those particular and potential hazards, then you have much of which to be proud. The problem arises when your ego grabs hold and you rush to give the compliance officer documents that can be turned against you.

For example, you turn over notes from the last 6 months' worth of safety committee meetings to emphasize the diligence of the company in supporting such an oversight group. Perhaps the committee has performed admirably. Yet, those notes may reveal that the identical hazard that has been noted and earmarked for correction in every set of notes still remains a menace to employees. This is not just a matter of understanding that the employer could have detected the hazard and apparent violation with reasonable diligence. This is a case where the danger was definitely noted. As a result, the dreaded "willful" lurks nearby. Similarly, you may want to hand the compliance officer a copy of a consultant's inspection report. You reason that the voluntary hiring and related expense of the consultant demonstrates your efforts to uncover hazards. What if the consultant's report, which is dated 10 months prior to this visit by OSHA, details a high-priority item that has yet to be abated? Once again, the tables are turned.

Elements of a Safe Workplace

OSHA has always encouraged the development and implementation of comprehensive occupational safety and health programs. It has been standard practice in the opening conference portion of the plant inspection for the compliance officer to address this important issue. He or she asks several specific questions designed to help the Agency assess the value of the program. OSHA has embarked on a policy to place far greater emphasis in this area. Therefore, if you have a full throttle, well-steered program, avoiding the pitfalls illustrated earlier, the advantages to you can be many.

The Agency is still experimenting with means to fine tune the assessment. It is safe to say that the compliance officer's analysis will focus on four broad-based elements. OSHA has experimented with modifying the labeling of these elements and with splitting them into more distinct categories so as to form more headings. Regardless of how these elements are titled or broken down, the current idea is to rate them to determine the apparent effectiveness of the program. In addition to these four distinct elements, OSHA wants to find that there is a demonstrated system of regular program review.

It is no coincidence that the elements espoused by OSHA as major building blocks in the formation and maintenance of a safe and healthful workplace will look familiar. The first part of this book delves deeply into these requisite components. They are based upon sound tenets, and OSHA now gives the employer added incentive to apply them consistently. It is OSHA's intent to tangibly reward those employers who do so. The employer can now have more control over the level of the Agency's intervention. There would be a tally of numbers derived from the compliance officer's judgments of the existence and sufficiency of subelements. The resultant scores could form a foundation for the decision of whether or not the company would undergo a walkaround inspection, what its scope would be, and to what reductions in proposed penalty the company would be entitled.

The first element concerns management leadership and employee involvement. The first part of this area of the program should be characterized by visible management leadership serving as a motivating force and providing adequate implementation tools (e.g., resources, responsibility, authority, line accountability, program review). The second part recognizes that active employee participation is essential and assigns high merit to the philosophy and concept of inclusion. OSHA believes strongly that the employer must build a partnership with employees to promote and assure a safe and healthful workplace.

The second element is workplace analysis. That analysis should be integrated into the design, development, implementation, and changing of all process and work practices. There must be in-depth hazard recognition. Comprehensive inspections of conditions and behaviors must be conducted. Employees must feel comfortable in reporting or when asking questions about hazards and potential hazards; it has to be clear that their input is highly valued. All loss producing and near miss accidents must be investigated with the main objectives being to identify root causes and to preclude or greatly reduce a similar event from occurring. OSHA further expects the employer to search for trends in injuries and illnesses and to react in a suitable manner.

The third element addresses hazard prevention and control. It is always preferable to use engineering controls for this purpose. However, work practices must back up the controls. Then, administrative controls and personal protective equipment come into play. Facility and equipment maintenance, anchored by preventive maintenance, is extremely significant. OSHA also considers the availability and integration of an appropriate medical program to be a part of this overall element. Finally, an emergency preparedness and response plan is necessary.

The fourth element is safety and health training and education for managers, supervisors, and other employees. The sessions must be adequately scheduled, presented, monitored, assessed, updated, and documented. The format should consist of classes and hands-on sessions and be tied into performance and job practice training. The training and education are to cover the safety and health responsibilities of all personnel concerned with the site. Although there are several particular standards requiring hazard-specific training, it is understood that there are many other areas of training that are necessary.

At the end of the opening conference, or at the start of the walkaround, the compliance officer will want to examine those specific written programs required by the standards. However, the compliance officer will want to avoid delaying the start of the walkaround. If you have a large amount of documentation, he, she, or you might suggest that someone gather the information while the walkaround gets under way. The documentation can be checked later. I have chosen to discuss that documentation in the walkaround chapter.

Intraoffice Referrals

The compliance officer will explain the possibility of intraoffice referrals. They involve referrals to the OSHA office staff of the opposite discipline of the compliance officer. For example, a compliance officer who is an industrial hygienist may observe apparent hazards associated with cranes, electrical apparatus, or unguarded machinery. Since those categories are not normally within his or her field of high knowledge, he or she may put in a referral to the safety staff within the OSHA area office. The safety supervisor, who is also an assistant area director, may determine that the industrial hygiene compliance officer has gathered sufficient documentation to recommend a citation, with some in-office assistance by the safety staff. However, if the safety concerns are plentiful or involve complex matters, a safety compliance officer may visit the establishment.

The flip side of the coin is when a safety compliance officer discerns apparent industrial hygiene type hazards that are beyond his or her scope of expertise. For example, the safety compliance officer may become concerned about the use of certain chemicals. He or she may obtain copies of material safety data sheets and document questionable work practices involving the substances. That information will be brought to the industrial hygiene supervisor, who is also an assistant area director. An industrial hygiene compliance officer may then visit the establishment armed with an array of sampling devices. In one more example, the safety compliance officer takes sound-level meter readings, and it is clear that full-shift sampling (i.e., dosimetry) is needed. Again, an industrial hygiene compliance officer may visit.

Outside Contractors

You should inform the compliance officer if any outside contractors are performing work in your facility. Even if the compliance officer does not ask about this specifically, you should let him or her know of such work. OSHA can cite the contractor employer. This would usually be for cases where the contractor (and not your employees) are exposed to harm. More important to you, do not allow any misunderstanding whereby the compliance officer mistakenly believes that the contractor is your employee. It is not common for the host employer to be cited for work of the contractor and vice versa. Nevertheless, it does happen occasionally.

If your employees are endangered by the contractor and you could have known this with reasonable diligence, your company can be cited. Examples include the following actions by the contractor: smoking a cigarette near flammable vapors; welding near tanks of substances where phosgene gas can result; welding without barriers, thus exposing your employees to flash burns; working above your employees where there is no barrier or other means to protect your employees from being struck on the head by parts, tools, or materials; leaving unattended live, electrical parts in open boxes; and impeding access to your exits, extinguishers, electrical disconnects, or emergency eye-wash fountains. Similarly, the host employer has a responsibility to prohibit its employees from entering a contractor's truck trailer that has not been chocked or otherwise protected against inadvertent movement away from the loading dock. You must control, to a reasonable degree and within your range of understanding, the safe practices of contractors and other visitors on your premises.

The host employer also runs the risk of being cited if the contractor is exposed to hazards clearly in control of the host, whether or not your employees are exposed to harm. This can occur if the host provides an unsafe ladder for the contractor. It can also occur if the host provides an electrical extension cord that has split insulation and is missing the grounding prong. Your susceptibility to citation can arise if you fail to share your hazard communication information with the contractor.

Looking at the reverse side of these issues, does that contractor realize that he or she should not weld (without special precautions) in a particular area where vapor ignition is possible, and does he or she realize that welding in another area could, in effect, produce phosgene gas? Has the contractor been informed, where necessary, of permit-required confined space or lockout/tagout policies and concerns? Does he or she know what personal protective equipment should be worn in particular areas of your plant?

It is possible for the contractor employer to be cited even if only your employees are endangered. For example, the contractor is working overhead on a mezzanine that is equipped with proper perimeter protection. Unexpectedly, he or she throws debris over or through the railings to the plant floor below. Taking one more twist, the contractor is welding while wearing a suitable shaded face shield. However, he or she lights up a torch without warning, and your working personnel nearby are exposed to flash burns.

Other Topics

The compliance officer will explain the planned scope of the walk-around, conveying whether it will cover the entire establishment or to what specific departments, areas, and/or operations it will be limited. This can be influenced by the inspection initiating factor, the breadth of the complaint or referral (if one of those categories was the basis for the scheduling), the size of the facility, how recently a comprehensive wall-to-wall inspection was completed in the facility, inspection history (including follow-up inspections and/or other compliance-verification indicators), your standard industrial classification, the results of the evaluation of your safety and health program, and current OSHA emphasis and focus programs. The compliance officer should tell you that, if hazards are in plain view while passing through or near areas not originally within the planned scope of the walkaround, he or she is entitled to expand the scope and to document accordingly.

The compliance officer will tell you how the inspection will be conducted. He or she will work with you to decide upon a logical route. However, the final route plan and the duration of the inspection will be determined by the compliance officer. He or she will explain the use of instruments, sampling devices, a still camera, and possibly a video camera. Further, he or she will go on to relate other means of documentation, including interviews and the taking of dimensions. You will be encouraged to take copious notes, and you will do well to heed that sage advice. In the walkaround chapter, I detail the use of audiovisuals (particularly in relation to concerns about trade secrets) and the interview process, expounding upon the rights and the protections afforded the employer.

The compliance officer will inform you that there will be a closing conference after the walkaround. You will be notified that, at that conference, the compliance officer will discuss apparent violations, employer rights, procedures following an OSHA inspection, and so forth. That conference may be held immediately following the walkaround. However, the compliance officer might want to schedule it for the following workday or even a few days later. Situations that would instigate such a decision would include if the walkaround was finished late in the afternoon or if the compliance officer feels a need to organize lengthy notes or conduct research concerning several standards and/or other compliance questions.

If you prefer a gap in time before the closing conference is held, there is no reason not to express your views. You might want to gather purchase orders, work orders, documentation of safety program elements, updates that have recently been completed or are nearing completion, or similar paperwork. You might want to arrange for an in-house or contract stenographer. Further, you might want to bring in a consultant, a corporate safety director from out of town, and/or other key persons. It would make good sense to alert these persons well before the anticipated conclusion of the walkaround. In fact, why not do so as you prepare for the plant tour? You can then keep them posted as the walkaround progresses. In any case, do not waive your right to a closing conference. Even if you are as busy as a cat covering spit on a marble staircase, make it your business to be there.

The compliance officer will be emphatic when describing the protections afforded to employees under the antidiscrimination language in the Occupational Safety and Health Act. He or she will explain that, under section 11(c) of the Act, it is illegal to take any adverse action against an employee that would not have been taken but for that employee's protected activity. This has nothing to do with discrimination based on race, religion, color, sex, age, national origin, handicap, or sexual orientation. It has everything to do with protecting an employee who has, for instance, complained to OSHA or threatened to do so; been interviewed by an OSHA compliance officer, most notably when the interviewee provided information that could lead to the recognition of an apparent violation; participated on a safety committee; communicated verbally or in writing to management about safety or health concerns in the workplace; and so on.

These OSHA-related discrimination protections were addressed in greater detail in the introduction to this book. As you get ready to start the walkaround, I urge you to keep these protections in mind, as well as the exceptional value that OSHA ascribes to manifesting mature, cooperative employer-employee relationships.

30

The Walkaround

Here we go. The walkaround inspection is under way. You hope for the best, which is not to say that you are simply beckoning the forces of good luck. It is to say that you hope that all of the hard work on the part of your company (i.e., labor and management) has borne fruit. Achieving the goal of an inspection in compliance (i.e., one with no OSHA violations) can be quite elusive. However, it is possible. Even if you do not come out of the inspection 100 percent clean, the damage (mainly in the form of proposed penalties) need not be significant.

Whatever awaits you should be easier to bear if you start off with a good attitude. Granted there are more pleasant ways to spend one's personal or corporate time than accompanying an OSHA compliance officer. Yet, the experience need not be particularly painful. In fact, it will almost definitely prove to be educational and should result in the gathering of new insights and tools, which can serve to reduce your potential for risk, loss, and human suffering. In turn, and putting aside major OSHA penalties for now, if you heed the observations of the compliance officer by appropriately abating hazardous conditions and practices, your company will be in a much improved position to greatly reduce its future legal jeopardy and financial burden.

I explained many of the germane cost-saving factors in the chapter on accident investigation. Whether or not the company ends up disputing OSHA's findings, there is an exceptionally strong possibility that the compliance officer has noted unsafe or unhealthful situations, which even your safety director, loss control representative, and/or safety consultant overlooked. It is not uncommon for a company to fight OSHA citations and penalties but still take prosafety steps based upon what was noted by the compliance officer. This even happens for and when citation items have been deleted as a result of a settlement agreement or decision in a contested case.

Attitude Toward the Compliance officer

I suggest that you afford the compliance officer a good dose of professional and human respect, while not figuratively bowing down in awe. I also suggest that you avoid being smitten with the good nature of the compliance officer (yes, such affable personalities have been known to surface in these individuals), as this might lull you into too much of a comfort level. Be assured that while it is not generally his or her intent to trick you into lowering your guard, you must resist even self-initiated forays into OSHA-related off-the-record commentary.

Do not succumb to the flawed reasoning that there is much to be gained by seeking empathy through your exceptional candor. Always tell the truth, but do not reach to tell all of it if those extended portions can only get you into trouble or into additional trouble. Remember that telling the truth does not mean that you must be the initiator and that truth does not presuppose total disclosure or total frankness. If the compliance officer asks if your foot is in your mouth, and there are no two ways about it, admit your pedal insertion dilemma. If he or she does not ask, then why offer the information? If the compliance officer discerns this damning anatomical relationship, but incorrectly believes the insertion to be of only minor depth, you need not expound on the fact that you are choking on your kneecap.

In a similar vein, you should not make the OSHA inspection a time for catharsis regarding your possible sins against the working class. Be friendly, but not giddy. Try to promote a sincere air of cooperation, as well as a cautious rapport, and make every reasonable effort to avoid any decay in this productive atmosphere. Be businesslike, but not cold. Be mindful that the compliance officer (through the area director) can get you in a hell of a lot of trouble. However, he or she is engaged in gainful employment and holds no personal grudge against you. Feel free to disagree, but do not revile. You can vent without vilifying. More on this later in the chapter.

Type of Inspection to Anticipate

It is always wise to plan for a wall-to-wall inspection. OSHA usually refers to this full scope walkaround as being "comprehensive." Even if an inspection was initiated by a complaint regarding operations in a small portion of the facility, it can expand to a total facility survey. Some complaints, although only referencing a small number of alleged hazards, can relate to numerous areas of the establishment. In most cases, the compliance officer will want to evaluate your hazard communication program despite the lack of its inclusion on a complaint. In that evaluation, various sections of the facility may be visited to check for things like container labeling. Unless your establishment is huge or OSHA had conducted a comprehensive inspection not too long before this inspection, the compliance officer may plan to stroll through the facility for the purpose of determining whether there is a just reason to expand the scope.

Further, your SIC (standard industrial classification) code may be one that is being targeted at the time by OSHA due to a high frequency of citations in that SIC. In that case, once OSHA is present for any reason at the site, a "comprehensive" may be in the offing. If the compliance officer clearly indicates that the plan is only to inspect very specific, limited areas, it is best to escort him or her to those areas by way of the shortest or "safest" route. In other words, do not give a ready opportunity for the officer to spot hazards that are in plain view. Yes, they are citable, no matter what the initiating inspection factors are.

On the other hand, I submit that it would leave a very bad taste if, for instance, you insist on bringing the machine in question to the front lobby. This has actually been done! I would also recommend that you not make the compliance officer walk all around the perimeter of a building (let's say in the winter, to add insult) just to go in a rear door to see a particular room. In so doing, you may have avoided plain view hazards, but you will have set a terrible tone. Should that matter? Does the law indicate anything about setting a bad tone? The answer is: Use common sense. Find a route that is the best combination of being direct while not spotlighting major areas of concern. Be intelligent, but not ridiculous.

In all of these cases relating to expansion of the scope of the inspection, you may challenge that decision. You may insist upon halting the inspection until you speak with an area director or assistant area director. You may insist that the compliance officer leave and come back on another day, or with a warrant, or both. Maybe you will be successful. Maybe you will not be.

You might already be under warrant and be on firm ground to demand that there be adherence to the physical limits of the inspection, as spelled out in that warrant, and that the extent of the warrant not be exceeded. However, even with a warrant, if the compliance officer in the normal course of the inspection came upon apparent hazards in areas or operations not specified in the warrant, those hazards are open game. This again assumes that the apparent hazards were in plain view. A fair interpretation of that theory encompasses senses other than vision. The compliance officer may hear extremely loud noise emanating from a room adjacent to the one being inspected. He or she may smell an abiding and pervasive odor (e.g., that of solvents), with vapors that are likely to affect employees in another room. This may be the case even if the source is not located in that other room. So having completed the circle, be prepared for an inspection that can cover every nook and cranny of the building and grounds.

Follow the Process Flow

If there is to be a wall-to-wall inspection not initiated by a complaint, referral, fatality, or catastrophe, the compliance officer most often prefers to follow the process flow. For instance, if the wall-to-wall is to begin after the complaint areas or other "hot" areas have been inspected, the route would most likely then revert to the previously uninspected route of process flow. In all cases, there are plant areas that are common or peripheral and difficult to pinpoint within that flow. Once again, figure that OSHA will see it all.

If the building is multistoried, another approach by the compliance officer might be to start at the top and work his or her way down, with or without the same "hot" areas inspected first. Once the scope is set, although it can later be widened, the compliance officer is the one who can determine the route as well as the duration of the inspection. He or she is to proceed in a manner that minimizes work interruption, while still allowing for the purpose of the inspection to be met.

Main Purposes

The main purposes of the inspection are to determine if the employer is in violation of the Occupational Safety and Health Act and to determine if the employer is providing a safe and healthful workplace. The compliance officer is to identify potential hazards to employees and to ascertain whether or not the company is in compliance with the standards and with section 5(a)(1) of the Act (i.e., the general duty clause). The compliance officer will observe conditions and practices. He or she may use various instruments to collect samples of airborne toxic materials, measure noise levels, establish electrical voltage and the existence and adequacy of electrical grounding, quantify air movement for ventilation systems, provide pH readings of chemicals, reveal compressed air pressure, and so on. Some of these instruments are direct reading, but others are not, thus necessitating analysis subsequent to use. Whether by these methods or by other means, the compliance officer monitors employee exposure to vapors, fumes, gases, dusts, and similar substances and media.

Compliance officers also survey engineering and administrative controls. They will examine inspection, assessment, training, exposure (sometimes tying in medical records, either of a limited nature or of the sort requiring the presentation of a medical access order by the compliance officer), and maintenance records. In an earlier chapter, I addressed the standards and types of concerns that are most frequently the subject of these examinations. He or she will also examine material safety data sheets, labels, and documentation regarding numerous specific procedures and instructions. In some cases, these examinations may have taken place, to a large extent, during the opening conference or at least prior to the actual walkaround. That is when the injury/illness logs will have been scrutinized. The compliance officer will also check for required postings.

The compliance officer will also evaluate the effectiveness of the general safety and health program and the specific programs that were supposedly in place as required by particular OSHA standards. All too often, the compliance officer is shown such programs during the opening conference, but it becomes evident during the walkaround that these programs are dead on paper. Pertinent employees, including line supervision, do not know of their location or even of their existence, or do not possess the requisite working knowledge. This is generally determined by interviews. Sometimes, suspicion is aroused by the observation of activities that are contrary to what is called for in the program. This may mean that the program information was not properly and totally transferred to the floor personnel. If it was, but ignoring it was condoned by management, the results can be more dire to the employer. This can open the door to a "willful" violation.

When the compliance officer has reason to believe that a hazard exists, and certainly when there is reason to believe that a violation of the standards or the general duty clause exists, he or she will document the findings. He or she will note the identification of the exposed employee(s). In so doing, the compliance officer will, as it is usually played out, write down the name, address, and telephone number of at least one of the allegedly exposed employees. Then, the compliance officer will write down how many other employees are potentially or actually exposed, breaking down the information into how many per job title, as one method.

He or she will also note the methods and means used to arrive at any quantifiable device-indicated data, whether by a simple measuring tape or, for instance, by the use of an electrical circuit tester, a sampling pump, or a dosimeter. The compliance officer will note the procedures used surrounding the use of the more sophisticated devices as well as the calibration information, when relevant. In addition, the use of audiovisual equipment will be noted. For example, the compliance officer will make a written record of which frames of which roll of still photography film will be germane to the documentation of which particular items. A simple note, with a reference to the counter or actual time, can indicate that videotape was used also.

The compliance officer should exercise great care not to use flash systems where the hazards of vapor or similar ignition exist. Do not assume that he or she is aware of such hazards. Speak up clearly and firmly. In a similar vein, testing equipment that can result in spark production should not be done in such areas. The compliance officer should also make a strong point of not startling employees when taking photographs or when using testing instruments that emit light or sound.

Reasonable Diligence

The compliance officer must document how the employer had gained knowledge of the hazardous situation or how and why the employer could have recognized it with the exercise of reasonable diligence. As a general rule, if the compliance officer was able to discover a hazardous condition and the condition was not transitory in nature, it can be presumed that the employer could have discovered the same condition through the exercise of reasonable diligence. The hazard may be visually obvious. Attention may have been drawn to the hazard by an accident that had occurred. There are other elements of evidence that the employer had knowledge of the hazard, or could have with reasonable diligence. Had employees brought concerns about hazards to management's attention to no avail? It is not uncommon for that allegation to be within the text of an employee's complaint to OSHA.

Thus, you should think twice about foot-dragging when receiving complaints or inquiries from employees. The repercussions could be costly. Was the hazard addressed in notes of the safety and health committee? Was the hazard addressed in the rules that are a part of the safety and health program? Although you should always consider that hazards can relate to practices and are not limited to conditions, note that company programs are one area that generally specifies what unsafe practices are prohibited. Was the hazard the subject of a report of an inspection performed by a loss control carrier or consultant? In blatant cases, particularly when in combination, OSHA may view some of these situations as core documentation for "willful" violations.

Signs and Labels

Signs or labels on containers of hazardous substances can also serve to document that the employer should have had knowledge with reasonable diligence. Again, such indications of hazards may also be studied in considering whether or not a "willful" violation should be cited. These markings, as well as material safety data sheets, often refer to the need for ventilation, personal protective equipment, and eye and (sometimes) body flushing systems. If you feel that these printed recommendations constitute overkill, say so and explain why. The wording is often formulated with lawsuits in mind.

Sometimes, it seems that all labels on containers of industrial-use liquids, powders, and similar media call for a full suit of armor and the need to visit three surgeons if anyone even looks at the substance. There are those who would argue that such attorney-influenced, extra precautions are harmless. It is not that simple. There is the danger of the "boy who cried wolf" syndrome. When should the warning be taken seriously? Naturally, if there is any question (e.g., a close call or subjective decision) as to the need for ventilation, personal protective equipment, and/or emergency eye fountains or deluge showers, you should err on the side of safety. This is the cardinal rule. However, if the compliance officer explains that the warning itself might serve as the basis for the employer's knowledge, as well as for the basis of a citation, you may have a case to make. The compliance officer should be more concerned about the specific details, including data, on the material safety data sheet. He or she must also understand how your employees use or otherwise handle the material and with what volumes. What is the exposure?

Signs or labels that were originally affixed to a machine by the manufacturer and are still in place can certainly form a basis for employer knowledge. For example, the signs sometimes indicate that the machine is not to be operated without the guards in place. The compliance officer's attention, if not ire, is raised upon seeing that a machine has unfilled holes or securing points that are obviously intended to facilitate guard installation. Worse still, in many of these scenarios, the guard can be seen in close proximity to the machine (e.g., leaning against a nearby wall).

Similarly, the compliance officer may see a hinged or sliding guard set up where the machine is operated without the guard in its intended position. Another case involves when a company has two models of a particular machine—one that is quite new and one that is not. The compliance officer observes that one of the machines lacks guarding where there is a hazard associated with moving parts. The employer representative mentions that the newer model arrived with guarding at the same relative position. He or she even puts a hand on the guard. Therefore, the representative has aided the compliance officer in establishing the employer's knowledge that the machine manufacturer recognized the need for guarding.

There are situations where signs or labels on products other than machines or containers of chemicals can also get you in hot water.

Portable ladders usually have several warnings on them. For A-frame ladders, these warnings deal with not standing on the top level, fully opening the spreaders/supports, and other concerns. For extension ladders, printed warnings usually address the proper overlap of the two segments, adequate extension beyond the surface to be reached (against which the ladder is resting and where the climber might get off to walk), and the suitable angle at which the ladder is to be placed. General warnings on ladders use precautionary language about assuring stability, examining for defects, not placing in doorways, not using horizontally as scaffolds or other types of walking surfaces, and not using metal ladders where energized electrical parts can be contacted.

On many occasions, I have seen the incorrect respirator in use. As an example, a respirator is used by an employee who is spray painting a flammable and airborne toxic material. The original box for the respirator states *"not for organic vapors."* Other warnings on the box are also ignored at times. Examples include methods of storage and the necessity of a snug fit, including the proper use of all straps.

Some portable, hand-held tools are labeled with warnings. For one, grinders often have words explaining that abrasive wheel guards are necessary. In addition, various electrical tools have labels indicating that grounding is required. I have seen A-type fire extinguishers located in areas where a fire would involve flammable and/or combustible liquids. Commonly, this type of extinguisher is clearly labeled with words and symbols to warn that it should not be used to fight such fires.

Whether it be for the ladders, respirators, portable tools, extinguishers, machines, chemicals (substances), or other items, accompanying literature can be expected to include safety-related warnings commensurate with, or more extensive than, the affixed signs and labels. The compliance officer may ask to see that literature. He or she might also contact the manufacturer or distributor. In situations where there is no clear OSHA standard, but where there may be a violation of the general duty clause—section 5(a)(1) of the Occupational Safety and Health Act—this literature can become very important. It may support the contention that a particular product, operation, or process was not used or performed as recommended by the manufacturer or distributor.

Severity and Probability

The compliance officer must make a determination as to the types of accident or health hazard exposure that the violated standard (or general duty clause) is designed to prevent and the most serious injury or illness that could reasonably be expected to result from the type of accident or health hazard exposure identified. This comes under the heading of *"severity,"* which is one of the two chief factors to be addressed once the burden of proof for the employee exposure and employer knowledge issues has been guaranteed. The other chief factor involves the *"probability"* of injury or illness. The circumstances considered in determining at the degree of probability normally include the following: the frequency and duration of exposure, the proximity of employees to the hazard, the number of employees exposed, the use of appropriate personal protective equipment, medical surveillance programs, and mitigating and/or exacerbating circumstances.

Some of the situations that generally constitute exacerbating circumstances include the following: high noise levels, poor visibility, high stress including piecework, inadequate training, tripping or slipping hazards (e.g., near unguarded moving parts), crowded conditions, poor illumination, hazards that could surface unexpectedly (automatically starting/cycling equipment with unguarded moving parts, e.g., compressors) and any other elements of the scenario that can predictably increase the chances for the accident and then the injury or illness to occur, even if the symptoms of the illness will not be readily detectable for a long time.

Note, too, that some of these factors can also relate to severity. As an example, if someone sustains an injury that results in a significant loss of blood and if they are in a very noisy area that is poorly illuminated, their cries for help may not be heard and they may not be seen. Thus, the blood loss can continue, and further deterioration of other body systems is likely to occur. This may provide ample opportunity for shock to set in. In this type of situation as with many others (severe electrical shock being an excellent example), the lack of prompt emergency medical response, possibly including CPR, could even increase the chance of death.

Another factor has an interesting twist. When an employee is exposed for very short periods in a small, remote room (e.g., when working on or near an unguarded compressor), employers often argue low probability. Yet, adding in the foregoing factors, severity is actually increased. When will the victim be found?

Mitigating Factors and Equipment Not in Use

The compliance officer will also consider mitigating factors. There might be a tripping hazard, but the area around it is very well lit, and the hazard itself is painted with a bright color which contrasts with the adjacent floor. In some cases, exceptional training and experience of potentially exposed employees can be viewed as a mitigating factor.

A different type of twist to the mitigating concept can involve partial guarding or redundant protective measures, although neither one nor both in combination constitute total compliance. Warning signs, assuredly, do not take the place of machine guards, but they might be considered as mitigating factors.

Compliance officers should be extremely curious, posing a lot of questions and mentally predicting possible accidents. It is possible, though, that a situation may appear to be a hazard when it really isn't. It is also possible that, even if there is a hazard, the associated probability or severity may not be as bad as it looks to the compliance officer. Speak up and explain. I have run into situations where employers or employees have purposely placed their hand in what appeared to be the danger zone of a machine in my presence. They then removed their hand with no sign at all of pain or injury. The pressure simply was not enough to cause a problem. Similar demonstrations relating to moving parts of machinery have involved presence-sensing beams that were not immediately evident to the visitor, apparent pinch points that upon closer observation did not come close to a dangerous proximity, and rotating parts that could be stopped by hand pressure.

I urge you to use extreme caution in your decision to make such a demonstration. Not only might you help the compliance officer gather evidence (e.g., when your finger is plucked from the machine), but there might be other moving parts that present hazards. Maybe if the machine was not operated, the compliance officer would not have discerned those other hazards. At any rate, if you are not injured at all, a wise compliance officer will want to know if the machine can operate at greater speed, at higher pressure, in the other direction, or with a setup that will allow a closer and more dangerous pinch point.

At times, compliance officers may request that you demonstrate operations so that they can properly evaluate the situation. You are certainly not required to do so. There is no automatic exemption that allows for what would otherwise be violative conditions and behaviors resulting from demonstrations. This is not to say that compliance officers will attempt to entrap you. They should not, and such action is not condoned by the Agency. You may invoke that defense, however, if the only basis for a citation turns out to be that the demonstration was conducted at the request of the compliance officer. If you know there are obvious hazards, it is an easy decision to explain that you do not wish to start up the machine. However, if the hazards are so obvious and the use of the machine in its current state (e.g., without the addition of guarding) is a known fact, then there is another consideration. It is generally a good idea to try to avoid additional trips to your establishment by the compliance officer. If he or she does not see an operation run during the walkaround, particularly if that is the case with several operations, there may be a return engagement.

Other examples include if the compliance officer thought that there were exposed, energized electrical parts, but the circuit was dead, or if he or she thought that it carried 110 VAC, but it was only 6, 12, or 24 VAC. Further, what might appear to be a 110 VAC flexible electrical cord could merely be computer network wiring. It is possible that an exit door appeared to be locked so as to necessitate two actions to open it, but in reality one simple turn of the knob caused the door to open easily. A compliance officer may believe that a particular chemical presents a much greater hazard than it actually does. Explain about dilution or other mitigating factors.

Even though your substance containers are supposed to be labeled properly, and it is generally citable if they are not, I have seen containers marked with the identity of a troublesome chemical, while they contain a harmless one, sometimes even water. Do not be shy. Tell the compliance officer. Also, he or she may believe that a particular substance is flammable, although it is not. This could even involve water-based spray paint. The compliance officer may think that a 55-gallon drum, particularly if it is sealed and positioned so that it cannot be easily approached, is full. Let him or her know if it is empty. It may appear that there is no inspection tag on a fire extinguisher; however, the tag may have swiveled out of place and be tucked behind the extinguisher. The compliance officer may think that a particular pair of shoes do not have ANSI-approved steel toes or the equivalent. In reality, they are the ANSI-approved type in a new style unfamiliar to the compliance officer.

The compliance officer might push the horn on a forklift truck and, because there is no sound, decide that the horn is not operating. It might be that the vehicle key had to be turned or that a supplementary horn, located elsewhere, is always used by the driver. If the compliance officer thinks that a chain is a conventional power transmission unit, but it is only operated manually, he or she should be informed. A radial saw might not have an evident magnetic starter (manual reset), although it does in fact have one, which may even be in the form of a cord drop-out. Sometimes, soft grinder-mounted wheels at first appear to be hard abrasive wheels.

The compliance officer might at first think that lockout/tagout standards are relevant to a particular machine, the type of which is normally energized by conduit. To illustrate further, the energy source might be obscured from the compliance officer's view. Let him or her know if the power is actually fed by flexible cord and plug and that there are no other energy sources associated with the machine. The compliance officer might not realize that an employee exposed to high levels of noise is wearing earplugs because they are obscured by hair over the ears. A nozzle for a compressed air line (e.g., for blowing off dust or small parts in the open) may not appear to be the type that restricts dead-end pressure to 30 psi. This could be because it is not the kind with easily visible holes. However, it may be another kind that, nevertheless, properly vents excessive pressure.

The compliance officer might believe that a certain piece of equipment is in use, although it is not. Even if you make a powerful argument that the piece of equipment has not been used in many months, he or she might be in a firm position to contend that it could be used. This scenario concerns equipment that would obviously create a hazard if it was used. Examples include the following: a ladder with broken rungs, an extension cord with the third prong broken off, a portable electric drill with stripped insulation on the cord, a damaged material handling sling, a forklift truck with numerous obvious defects, a belt sander with an unguarded power transmission belt, and a discharged fire extinguisher.

To avoid this debate, and possibly a related citation, take definitive steps to assure that the equipment will not be used. The best way to accomplish this is to render it inoperable. Consider removing or negating power sources. Cut electrical cords after they have been unplugged, remove power transmission belts, and remove batteries. Destroy and discard equipment that will never be made safe and never brought back to service. Unambiguously put the equipment out of service. Also consider putting tarpaulins over equipment and using signs and tags. For areas that are unsafe, erect barriers to preclude entry and supplement them with warning signs.

All of these cases address instances in which the compliance officer did not initially realize that what appeared to be hazards and/or violations were not, that there were mitigating factors, and that equipment that was thought to be in use or available for use was not. Some employers reason that they will not correct the compliance officer's misunderstanding (or what the employer interprets to be a misunderstanding) during the inspection. The idea is that, when a citation is received, then the employer will come to the informal conference armed with these arguments. I urge you to turn against this supposed reasoning.

It is far easier to make the good arguments before a citation is issued. Oh, you might be victorious at the informal conference, but you need not accept that extra challenge. At the informal conference, your expostulation will be heard, but even OSHA (if not the Agency itself, then individuals in authority) has an ego. You may refute the documentation. They may realize that you are correct. Still, there is too often a reflex on their part to at least maintain the item, even if they offer a significant reduction in penalty. They are not too pleased with the idea of reducing the designation of the citation item (e.g., from "serious" to "other"), but may offer to do so as a way of admitting that you have presented good grounds for some modification of the citation.

This may frustrate the heck out of you. You know you are right and they know you are right, but they are not responding to your demand, request, or plea for a deletion. Besides ego, and maybe a little bit of stubbornness, the individuals with whom you are dealing are often concerned about perceptions of the Agency as inferred from their statistics. If, at informal conferences, proposed penalties are frequently and drastically reduced, designations are regularly downgraded, and deletions are common, a sour taste is left.

In these cases, the public, organized labor, politicians (including by way of congressional oversight committees), and others rush to brand OSHA as lacking in competence and/or as "giving away the store." There are many OSHA employees in authority who, when provided with information that truly alters the documentation, will act accordingly and fairly. You may get your deletion, but why go through the incredible hassle?

As the OSHA compliance officer or officers conduct the walkaround, they will perform a physical inspection. They will look up, down, under, over, and back, trying to catch virtually all angles. They will probably climb and may even get on the floor to gain a better view of something beneath or behind a machine. If you think that the compliance officers are about to endanger themselves, be sure to speak up immediately and state the reasons for your concern.

For instance, you may be aware of possibly unsafe conditions that may not be evident to the compliance officers. Such conditions could relate to mezzanine floors, ladders, live electrical parts that do not appear to be energized, doors that lead to an unexpected hazard (e.g., a sudden drop), material ejection areas of machines, or a welder who is about to strike an arc, and so on. Naturally, it is best if such situations do not exist at all. Moreover, their presence may point to an apparent violation. However, it is not worth permitting the snoopy visitor to be injured or sickened. In some of the cases, there might not be a violation, or the violation has been substantially mitigated. It could be that all of your employees know of the situations and employ great care, or even that none is actually exposed, despite the potential.

Members of the Tour Party

As per your OSHA visit action plan, you have selected the most appropriate employer representatives for the walkaround. At least one of your tour party knows what to call equipment, processes, and locations. He or she can provide designations for the compliance officer to note. There will be no miscommunication regarding what machine, electrical box, exit, or department is in question. Preferably, that person or another has a good grasp of how things work, to put it simply. Thus, it is a good idea to have the head of your maintenance department and/or the plant engineer on the tour. At least one of the employer representatives should be taking comprehensive notes. If he or she does not understand the compliance officer or if the compliance officer is going too fast, stop and ask questions. Make every effort to record all of the factors and elements that will comprise the compliance officer's documentation.

If you disagree in any way, including as to measurements, other facts, or opinions, be vocal but do not ramble. If you are rebuffed, be resilient without being sardonic. Aim to be succinct, but do not be afraid to take a few extra minutes to assure that your comments are understood, even if the compliance officer fails to see their merit. Emphasize that you would appreciate your comments being noted by the compliance officer. Then drop it.

Continuing to follow your OSHA visit action plan, you will be shooting photography and possibly videography along with the compliance officer within the framework and following the precautions described in the earlier chapter. Again reviewing that plan, you might have someone conducting side-by-side sampling as well. You might make a point of *not* discussing instrument calibration with the compliance officer. Perhaps, the compliance officer will fail to calibrate his or her testing-sampling equipment properly. This could set the basis for a later defense against citation item(s). The failure of the compliance officer to use properly calibrated instruments may well lead to the vacating (i.e., deletion) of the relevant item(s).

I do not recommend that too many persons stay with the inspection party throughout the establishment. If the group becomes a crowd, there may be unnecessary logistical problems. You may wish to have department forepersons or supervisors join in when the group is in their area. Management may permit the compliance officer to conduct the walkaround inspection without an employer representative. There is no sensible reason to do this. Period.

The compliance officer has the authority to deny the right of accompaniment to any person whose conduct interferes with a full and orderly inspection. The compliance officer will prefer to continue the walkaround, with the other remaining employer representatives. However, if the disruptive behavior resurfaces, the compliance officer may suspend the walkaround. This type of situation is barely nonexistent. For matters to deteriorate that far, management would probably have been afflicted with a case of severe cerebral-anal inversion.

Trade Secrets

The subject of the use of audiovisuals highlights the matter of trade secrets. A trade secret is, basically, any confidential pattern, formula, process, equipment, list, blueprint, device, or compilation of information used in the employer's business, which gives an advantage over competitors who do not know or use it. The employer may be able to bar the employee representative from entering a trade secret area without proper clearance. This can be a very sticky issue. For one thing, do not use this situation to arbitrarily expand the alleged trade secret area to far beyond its actual bounds. If the compliance officer intends to photograph or videotape what you fairly feel is a trade secret, make your feelings known at once. You may be able to cover critical sections of the machine, pattern, or process, for instance. It may simply be that the trade secret is adjacent to what the compliance officer really wants to capture in the audiovisual. That can be dealt with easily. If the compliance officer insists with good reason on taking an audiovisual that might reveal a trade secret, make it 100 percent clear that you desire the audiovisual to be handled in a trade secret manner.

Compliance officers' instructions from the Agency dictate that they are to use labels indicating "*Administratively Controlled Information*" or "*Restricted Trade Information.*" Of course, this also applies to the blueprints, lists, and other items that can fall under the trade secret heading. Compliance officers are not to label any of these items as "*Secret*", "*Confidential,*" or "*Top Secret*" unless they are also classified by some agency of the United States government as in the interest of national security. If compliance officers or, for instance, their supervisors release the trade secret information without authorization, they are subject to a fine of up to $1000, imprisonment for up to 1 year, and removal from office or employment. The labeled trade secret information, including formulas, audiovisuals, and all of the rest, will not be released without authorization even in the event of a request under the Freedom of Information Act.

Employee Representatives

Where employees are represented by a certified or recognized bargaining agent, the highest ranking union official or union employee representative on site can designate which employee representative will participate in the walkaround. If there is a dispute as to who is the representative authorized by the employees, the compliance officer has the authority to resolve the conflict. The representative must be an employee of the company. The employee members of an established plant safety and health committee or the employees at large may have designated an employee representative for OSHA inspection purposes or agreed to accept as their representative the employee designated by the committee to accompany the compliance officer.

If there is no authorized representative and no labor-management safety and health committee and if employees have not otherwise chosen a representative, the compliance officer may determine which other employee(s) would suitably represent the interests of employees on the walkaround. In practice, the chances of such a determination and resulting accompaniment occurring are very, very slim. In any case, management is not to select as an employee representative just any union member, safety and health committee member, or other individual(s).

Once again, if there is disruptive behavior, the compliance officer may deny the right of accompaniment to the offending individual(s). The employee representative (almost always in the singular) is to be advised that during the inspection matters unrelated to the inspection shall not be discussed with employees.

Interviews with Employees

Regardless of whether or not employee representative(s) participate in the walkaround, the compliance officer is likely to consult with several employees. In some cases, there will be private interviews. Those interviews are to be out of the hearing range as well as out of the intimidating visual range of employer representatives. OSHA considers such free and open exchanges essential to effective inspections.

Compliance officers are trained in interviewing techniques. Occasionally, they take written statements from employees. This is usually when the interview concerns an accident investigation. It can also be driven by the consideration of a "willful" violation. My chapter on accident investigation delves into methods employed for meaningful interviews and the gathering of meaningful written statements. Even if the interview and/or written statement does not involve an accident, those comments are relevant.

The compliance officer can conduct a one-on-one interview even if there is an authorized employee representative on site. The employee has the right to request the presence of an employee representative. I recommend that the employer not even bring that up to the employee. The compliance officer might mention that option to the employee without the employee representative present. In fact, the employee could insist on having an *employer* representative present. Nevertheless, I strongly recommend against the employer saying even one word about that to the employee. The compliance officer can interview off site, but that is very seldom the case unless there are exceptional circumstances.

If the interview of the particular employee will, at that time, cause the breakup of a work crew or in some other way be significantly disruptive to the work flow, the employer representative can ask the compliance officer to kindly wait a few minutes. Perhaps the selectee was involved in a critical task, was loading a truck that must leave the dock as soon as possible, or was an integral part of an assembly line and there was no immediate replacement available. Compliance officers are reasonable and respond well to this type of request. Just be certain to not try to influence the selectee on the company's behalf while the waiting time is lapsing. Some employer representatives suggest that other employees be interviewed. This is all right for the time being, but I recommend against even an implication that the selectee should be totally passed over.

This does not just relate to the selectee being engaged in important or team tasks. Sometimes, the employer representative tells the compliance officer that a selectee is a troublemaker, or is retarded, or was just hired. The employer representative then suggests that the individual would be a bad choice for an interview. Unless you are using reverse psychology, forget about that tactic. Those employees might well produce the most fruitful interviews. Simply let OSHA do the choosing. Not that you have a real choice, but do not even get involved in the decision.

If management absolutely insists that no employee interviews take place on site, it can be construed as an element of a refusal of entry, and a warrant may be sought. The common and prudent practice of OSHA area offices is that the compliance officer will notify his or her supervisor (who will begin the process of seeking a warrant), proceed with the inspection (with the employer's permission), and explain to the employer that a warrant may be sought.

Besides the interview question, this can relate to management's steadfast refusal to allow such activities as the taking of audiovisuals and/or the collection of industrial hygiene samples. The idea is to make the most of the inspection since the compliance officer is already on site. In most cases of disputes by management, such as those described here, the OSHA supervisor will attempt to speak with the employer representative on the telephone. Sometimes that allays fears and helps smooth out the wrinkles of doubt. Management typically feels more comfortable speaking with someone with more authority than the compliance officer.

The private employee interviews are most effective if conducted near the time that the compliance officer has begun to inspect the room or area in which the interviewee usually works. If a compliance officer inspects an entire room and then chooses an employee to interview, there can be problems. If the employee wants to tell the compliance officer about a safety concern in the room that the compliance officer admits having not noticed, he or she may be reluctant at that point. This is because, having supposedly finished the inspection of the room, the compliance officer's return to a specific piece of machinery signals that he or she was just tipped off by the employee. Even when a tip is given to the compliance officer after the room inspection, there are strategies that he or she can use to reduce the chances of implicating the interviewee. I choose not to divulge these strategies.

Protections Under Section 11(c)

At the beginning of the private interviews, the compliance officer is supposed to explain the protections under section 11(c) of the Occupational Safety and Health Act. I have explained those protections earlier. The compliance officer is to emphasize that any and all information and opinions given to him or her in confidence will, in fact, be held as confidential to the extent allowed by law. The confidential contents will not be divulged even under the Freedom of Information Act. The compliance officer should explain that, in theory, it is possible that the information can be tied to the interviewee in a court or hearing setting. In my experience, OSHA (along with the United States Department of Labor Office of the Solicitor) has refused to breach that confidence even in those situations.

Perhaps the employee was convinced to allow the disclosure. That is another story. Yet, in cases where the administrative law judge insisted on breaking the confidence (from what source did the information come?) and the employee maintained his or her desire for confidentiality, OSHA dropped that portion of the case rather than be involved in a corruption of the principle at stake. Although verbal and written statements can fall into these categories, so can an employee's voice on an audiotape or videotape or an employee's demonstrative actions on a videotape. This can relate to still photographs, but they are generally less revealing.

What if the compliance officer winds through the entire interview, including efforts to solicit any concerns or questions about workplace safety and health, and then at the very end of the interview explains about 11(c) and anonymity? The compliance officer will have erred in the interview process. The employee may have desired to speak up, but did not do so due to the fear of reprisals from management. The employee may have said, flat out, that there were no concerns or questions. In this scenario, after the compliance officer has finally explained the protections afforded the employee, most employees are too embarrassed to change their remarks. However, compliance officers do generally give out business cards to the employees they interview. They encourage employees to contact them, confidentially, if there are any comments or queries about working conditions that may be hazardous.

There should be little to worry about concerning employee interviews if you have an adequate safety and health program in place. As the first part of this book explained, such a program must be punctuated with employee awareness, employee acceptance, and employee participation. If you have been aboveboard with your employees, if your training has been thorough, and if your sincerity and commitment have shown through, you should fare well in the interviews. Compliance officers ask how management reacts when employees complain, inquire, or make suggestions about safety and health matters. If the interviewee responds that management displays a legitimate respect for the employee, is eager to listen, quick to act, and keeps the employees informed when called for, you will be cast in a very good light.

Interviews with Employer Representatives

What if the compliance officer plans to interview an employer representative? In this case, you might insist on the safety director, a company executive, or an attorney being present. I always suggest that it is preferable to avoid action that appears to heighten the perception of confrontation. However, if management has a significant fear that the employer representative's statement might needlessly place the employer in exceptional legal jeopardy, that is quite another matter. Such concerns are raised when the compliance officer is investigating a serious accident and when a "willful" violation may be under consideration.

Certainly, the specter of a "willful" may surface spontaneously. At times, even when there has not been an accident or a good reason to suspect that a possible "willful" was being examined, one short statement can cause the compliance officer's eyebrows to arch with interest. The interest might be toward a "willful," but it also might be simply toward a lesser designation of violation about which no suspicion had even existed earlier. So, if you do not feel confident that your employer representatives know what to say and what not to say, revert to the fail-safe position: Do not permit the representative to speak with the compliance officer one-on-one. Of course, you should have taken precautions to avoid this lack of confidence.

You should have prepared your employer representatives prior to any visit by OSHA (not implying that you know when there will be one) in careful, measured dialogue with OSHA personnel. In so doing, you should have elucidated the particular pitfalls that must be avoided in such conversations. Once again, this does not mean that you should condone, endorse, or passively accept the telling of falsehoods to compliance officers or other OSHA personnel. It does mean that there is no requirement for total candor and that the manner in which statements or responses are phrased can make a large difference. The interviewee must not feel obligated to fill in dead-air gaps in a conversation. He or she must comprehend that it is acceptable when accurate to state that he or she does not know the answer to a particular question. He or she is not expected to know every detail, about everything in the entire establishment. Do not guess.

On the other hand, where the admission of the lack of knowledge could point to a possible violation, with the most notable examples concerning the lack of required training, the individual must be very careful. If the training was up to snuff, the individual must be assertive in response. Why wait until later to explain that you know what you are supposed to know? If you forgo the timely opportunity to render the correct response and decide to provide the information only after checking with top management, then the compliance officer's perception of your eventual responses may be that they are tainted. Whether he or she can prove it or not is another matter. Speak up, while keeping within the scope of the question.

The Value of Ignorance

Do not flaunt knowledge. In one case, OSHA cited a company for several violations regarding electrical hazards. The employer representative (the owner, in fact) was quite vocal in revealing his extensive background in the field of electrical safety. Basically, he made certain that OSHA was fully aware that he had every opportunity, including by experience and education, to have discovered the hazards. He helped the compliance officer form the basis of a series of "willful" violations. Hold that ego in check. During the walkaround, do not eagerly point out or admit to probable violations or apparent hazards that you were aware of before the inspection, emphasizing that prior knowledge. You might feel that this will be in your favor because it highlights your safety and health knowledge. Forget it. Even if you quickly learn what the compliance officer is looking for, let him or her perform the inspection, not you.

When the compliance officer points out what he or she interprets as a hazard and apparent violation, you might be in a position to indicate that you are not sure what he or she means. After further explanation by the compliance officer, you might even ask a question or two so that your concern can be clarified. Then, perhaps, you will hesitate while integrating the input, study the relevant part of the machine, and then comment that you now see what the compliance officer means (i.e., you finally get it). You might add that you will look into it and that the compliance officer has an interesting point. It is better to appear somewhat enlightened by the expert abilities of the compliance officer.

To go one step further, it is all right and often advisable to appear a little ignorant. In many cases, companies have avoided the "willful" designation or had it downgraded after citation issuance by demonstrating that they had good intentions but screwed up. It may damage your pride to seek such negative recognition, but taking this tack is very often the intelligent decision. Avoid admitting that a certain situation constitutes a violation. However, if it is obvious that you should have known of the hazard and if you have issued a purchase order or work order, say so.

Abatement Steps

When compliance officers draw attention to an apparent violation or hazard, it is not their job to tell you precisely how to abate the situation. They should make clear what the point of the abatement should be (i.e., what the goal is). Some compliance officers are most reluctant to discuss abatement methods. You should kindly insist that you wish to learn at least what types of approaches can be successful. Compliance officers should bring up certain qualifiers to any abatement approaches they discuss. They should explain that there may be other methods, including those that are less expensive or more effective. They must make every effort to avoid leading you to a particular distributor, brand of product, or similar type of resource category, whereby there could be the appearance of a conflict of interest. Compliance officers should at least relate abatement means that they have observed, explaining the pros and cons. If they cannot offer constructive comments, they should refer the matter to their supervisor. This will be covered further in the chapter on the closing conference.

I recommend that you consider taking abatement steps as soon as you can. It is understood that, in many cases, matters are too complex or questionable to allow for a quick tackling of the problem. However, if practical corrections can be made quickly, particularly when there is only one sensible means of correction or you are comfortable with a certain means (and that means is within the expertise of your staff or of readily available resources in which you have full confidence), get to it. This does not mean that you are admitting that there was a violation.

You are merely responding to the watchful eye of a person (i.e., the compliance officer) trained in occupational safety and health matters.

Even if cited for the situation, you might still argue that it did not constitute a violation. At the same time, you are demonstrating good faith in seeking to remove or mitigate potential hazards, whether or not there are actual violations. OSHA has experimented with the concept of granting penalty reductions beyond the longstanding, general "good faith" reduction for hazards that are abated immediately and permanently. It must be underscored that this "break" would not apply to corrective measures that are temporary or superficial. Moreover, it would not apply to "willful," "repeated," "failure-to-abate" violations, or to fatality or catastrophe inspections. This type of system is controversial and, like other pilot programs, could be phased in or out.

The quick-fix strategy also applies to when you realize that the compliance officer is likely to uncover similar hazards in other parts of the plant. There is nothing wrong with you trying to abate these hazards before the compliance officer reaches those areas. If the compliance officer is coming back the following day or any other day, you might as well work overtime that night and clean up as much as possible. Do not fake corrective steps and do not clearly and purposely mislead compliance officers into thinking that "in compliance" machines observed on that subsequent day had been in that state for a long time. By the same token, do not even bring up the matter. If neither the compliance officer nor anyone else chooses to broach the question of when compliance was achieved, then so be it and be silently thankful.

Citations

Technically, OSHA can cite when employee exposure has occurred within the 6 months immediately preceding the discovery and/or establishment of the hazard. There are some nuances to the interpretation of this time period, but the 6-month period is the standard legal limit. Once in a long while, OSHA misses the 6-month deadline for issuing the citation after having observed the hazard during an inspection. In those cases, you should have little trouble having the citation or the particular items that were not cited in a timely fashion overturned.

The 6-month period also comes into play when OSHA uncovers a violation situation by means other than observation. As an example, the compliance officer may learn through interviews or a written complaint that a machine was in operation with an unguarded power transmission chain 4 months ago. Even if the chain was guarded 3 months ago, this unobserved hazard may be cited as long as the normal documentation criteria are legally sufficient. This happens most often in the case of accident investigations. Sometimes, the compliance officer discovers and searches out such conditions after reading the injury and illness logs. In such cases, a citation is likely. Yet, if the compliance officer learns during an inspection unrelated to an accident investigation or from a complaint that OSHA did not service very quickly about the chain that had been unguarded but was corrected several months ago, common practice is for no citation to be issued.

Returning to my earlier example, if you simply get ahead of the compliance officer and abate a hazard (true, complete abatement, not a halfhearted smoke screen or the shutting down of an operation or process), you still have to figure that it will be cited if detected. Whether or not a citation is actually issued for those particular items may well depend on the tone of the inspection, the general condition of the facility, and the spirit of cooperation that has been established among all parties. In this type of case, the authorized employee representative on the walkaround might take the side of the company in an effort to avoid a citation for these items. He or she might either intentionally avoid mentioning the recently abated situation or speak frankly to the compliance officer who has openly wondered about the guard that appears to be brand new. Certainly, there is no obligation on the part of the compliance officer to refrain from recommending a citation on the respectful recommendation of the authorized employee representative. It is fair to say, though, that the value of good labor-management relations should be evident.

Another type of situation involves very serious accidents, such as fatalities, catastrophes, and (even if there was no injury) fires, explosions, and major releases of toxic vapors. It is not uncommon in these cases for the accident to have occurred only a few hours prior to OSHA's directly related visit. When the compliance officer arrives, the entire operation is shut down or even no longer in existence. Certainly, there is the possibility that a hazard and violation can be established. If the proper violation documentation is in place, a citation should be expected.

In any case, it would be wise to assure that no employer representative provide the compliance officer with a written statement, including any statement written by the compliance officer in the representative's words and then signed by that representative, completely on his or her own. It is definitely prudent for top management and an attorney and/or consultant to get involved with such statements.

Positions of Authority

There is a significant question regarding the definition of an "employer representative." You can immediately figure that anyone with a title of supervisor or above (those titles above are generally evident) should be considered as an employer representative for OSHA purposes. At times, OSHA views the individual's position as that of an agent of the employer. Written and verbal statements given by such individuals could be interpreted by OSHA as emanating from a person speaking for the corporation. Therefore, when the individual spoke, the company spoke. In the simplest of terms, OSHA asserts that the supervisor represents the employer and that a supervisor's knowledge of the hazardous condition amounts to employer knowledge.

Any pronouncement from a "foreperson" (to be politically correct) will most likely be viewed by the compliance officer as coming from the company. You may gain some insulation by using alternative titles, such as "team leader," "group coordinator," or even "*working* foreperson." Whenever there is the claim by OSHA that a statement, action, or inaction pointed to a violation specifically because it came from a member of management, you should try to minimize the authority of the individual whenever possible.

Ideally, you will be able to counter the argument by explaining that the speaker had no actual authority to direct, supervise, order, control, schedule, hire, fire, promote, or discipline. Perhaps, there was a passive understanding that the speaker had a limited responsibility to coordinate and monitor activities, but it was not in his or her purview to make day-to-day decisions. Additionally, he or she served as a filter of sorts—simply the one to pass on information from management to the group or individuals. Even in that case, when arguing to OSHA that the person lacked actual management status, it is preferable if the person was not allowed to take any of the forms of personnel action that I have listed.

Your argument would be further supported if the speaker punches a time clock and is paid by the hour and if he or she is a member of a labor union. OSHA may also want to check hierarchal flow charts and rate of pay. Your case for the speaker being other than an employer representative will be enhanced when there are more layers of personnel (i.e., more titles) above the person in question. The case will also be enhanced when there is little difference in pay separating the person from the regular line employees.

I do not suggest that employees to whom significant responsibility and, more important, significant authority have been given be downgraded in any way at all. Front line supervision, no matter what their title, must have the tools with which to fulfill their duties. Yet, you do need to understand that there may be close issues upon which OSHA will base its view of whether or not the speaker represents the employer. The outcome of that view and the subsequent decision can be critical.

So, do not be too eager to proudly announce that the speaker is any sort of manager. I have emphasized, again and again, the value of employee participation and empowerment regarding safety and health. That is something to laud. Still, it is wise to avoid the temptation of making it seem that those with any form of fringe overseeing position possess real authority.

Isolated Events and Employee Misconduct

The foregoing issue becomes extremely important when the employer contends that an individual's unsafe and ostensibly violative conduct constituted an isolated event of employee misconduct. There are very few times when OSHA will accept your contention, meaning that it is rare for this reason or excuse to be sufficient to convince the compliance officer that a citation is not warranted. If the individual is not an employer representative, there is a larger window available for the contention. If the individual is, unarguably, a supervisor in the traditional role of line management, it will be quite difficult to convince OSHA of your assertion.

Even if the compliance officer at first rejects your argument that the supervisor's conduct was not foreseeable and that it was out of your control, that rejection is not immutable. Be prepared to offer an array of details to demonstrate that the supervisor was extensively trained in the avoidance of the unsafe activity, that he or she was supervised (in turn) so as to prevent that activity, and that the employer (in the form of upper management) uniformly enforces the germane, satisfactory work rules which had been effectively communicated to the supervisor. Provide dates and attendance sheets of training sessions and examples of disciplinary action that has been meted out. If the compliance officer cannot determine that even the supervisor had actual knowledge of the hazardous condition (sometimes resulting from or inextricably bound to unsafe acts), the knowledge requirement as obligated to sustain the alleging/citing of a violation will be met if the compliance officer is satisfied that the employer could have known through the exercise of reasonable diligence.

As a general rule, if the compliance officer was able to discover a hazardous condition that was not transitory in nature, it can be presumed that the employer could have discovered the same condition through the exercise of reasonable diligence. Key in on the word "transitory." For what period of time did the unsafe condition exist? What opportunity in that time frame was available for the employer (again, this could be a supervisor) to discover the situation? Although the duration may have been short, was the unsafe situation something that occurred frequently?

Looking more generally at isolated events of employee misconduct, sometimes referred to more compactly as *"isolated incidents,"* the fewer employees involved, the better. As in the cases mentioned earlier, the shorter the duration, the lower the frequency, and the smaller the chance that the supervisor or other employer representative had to discover the misconduct, the better for your argument against citation.

Many of these scenarios involve the personal protective equipment issue. I will present two examples that are a bit exaggerated to drive home the point. In both cases, an OSHA inspection is under way in a machine shop section of a factory. In both cases, the compliance officer is accompanied by one employer representative, and only one employee is found to be engaged in unsafe behavior. In both cases, the employer representative claims that the situation is an isolated incident. Keep in mind that the Occupational Safety and Health Act does not contain provisions for proposing penalties or other sanctions against employees.

In the first case, the compliance officer and employer representative have been in the room for several minutes. During all of that time, in clear view, an employee has been operating a large lathe. Long squiggle chips are flying from the machine in the direction of and very close to the employee's face. The employee is not wearing any type of eye protection; he squints as the projectiles come close to his eyes. The employer representative does not bring up this obvious hazard, although he or she is clearly aware of it. The compliance officer inquires as to why the employee is not wearing ANSI-approved industrial eye protection with side shields. The employer representative responds that the employee has been told "at least ten times" to wear the protection, but he simply refuses or just frequently fails to do so. The employer representative indicates that the employee is an adult, has been warned in a loose, nebulous sense, and that the employer representative cannot force the employee to protect himself. Further comments reveal that no one in the plant has ever been disciplined for failing to wear required personal protective equipment, that no formal assessment or training has been conducted, and that there is no written company rule regarding personal protective equipment. Without a doubt, get ready for the citing of the appropriate standard violation. It is bad enough that the employer representative's comments could not stop a citation. In fact, there probably was not much that he or she could have said that would have made a difference.

Unfortunately, the totality of the comments anchored by the "at least ten times" remark served to point up a likely "willful" violation. In my chapter on the closing conference, I explain the criteria for "willful" and the other specific designations of citations. Suffice it to say for now that my fictitious employer representative went out of the way to suggest the existence of a "willful." Add in a fairly recent related OSHA citation and/or a history of eye injuries at the facility (even worse if involving that operator) and that "willful" is sure to be guaranteed. Note that in no case where there is an apparent violation should management voluntarily mention that there have been related accidents, regardless of whether or not an injury resulted.

In the second case, the inspection in the room has just begun. The shift had just started. The regular foreperson was in a safety meeting. The only other employer representative in the room by any stretch of the imagination is the one accompanying the compliance officer. There are many machines in operation for which the operator would be required to wear the appropriate eye protection. Several conspicuous signs proclaim the requirement. The compliance officer and employer representative turn into a small area that is not easily visible from most of the main room. There are some chips flying in the direction of the machine operator who is not wearing any eye protection, but none of the chips are on a direct line to the face. The employer representative immediately confronts the operator of the small lathe, promptly reminds him of the eye protection requirement, and points to the confirming, large, unambiguously worded sign at that work station. The employer representative also notes the immediately available pair of appropriate goggles (generally to be worn over nonsafety prescription glasses, but approved for wearing over bare eyes) that remain at the work station. Also, the employee's appropriate plano safety glasses are on a table that is close by. The employer representative further reminds the employee of the written rule and the documented training sessions that were attended by that employee.

The employee appears to be sincerely embarrassed and remorseful—even shocked at his own neglect. He awkwardly reaches up to his face, seemingly surprised that he is not wearing the protection. This scene recalls how, on a few occasions, I have reminded car passengers that they did not buckle their seat belts, even though I knew that they had always worn them previously when in a car. In those cases, the passenger quickly put on the belt, with a movement and expression suggesting a feeling of a sudden awareness of one's nakedness. The employee apologizes and dons the eyewear, explaining that he has just taken an emergency telephone call from home and had, inadvertently, failed to put the glasses back on after rubbing his eyes. The employer representative sympathizes but explains that blindness can only make matters worse. He or she issues a stern but even admonition and states that it will be confirmed in writing. In fact, several months prior, another employee had been the recipient of a documented disciplinary action after having been warned twice about the failure to wear eye protection where it was required in this same room. There have been no eye injuries in the facility, and no personal protective equipment (eye or otherwise) related citation items issued to this company in many years. In this case, purposely skewed to an almost ridiculous degree in favor of the employer, the compliance officer would be unreasonable to recommend a citation.

Company Rules

I suggest that you do not smoke any form of tobacco when with the compliance officer, even if out of the plant or during lunch or break time. I suppose that the one exception could be if the compliance officer lights up first (hopefully after your permission has been solicited and granted). You might consider inquiring as to whether the compliance officer minds if you smoke although he or she is not smoking, but I would advise against even asking. For the most part, if you feel that you are in desperate need of a cigarette, excuse yourself for a short period.

In any case, do not be shy about assuring that the compliance officer obeys all of the company's rules for visitors. This should help protect the compliance officer, and it will reflect your commitment to safety. So, if your establishment is a totally nonsmoking environment and if the compliance officer indicates a desire to smoke, let there be no exception to your rule. Direct him or her to an area where smoking is permitted, even if that area is outside your gate.

As a sidelight to company rules and to make for a smoother flowing inspection, explain the regular lunch and break times to the compliance officer. Also explain the hours of operation, shift change times, and when certain operations may not be running. This will allow for the better use of everyone's valuable time. For any of numerous reasons, you may want to suggest to the compliance officer that breaks or lunch be taken at a certain time. Since the compliance officer is a visitor in your establishment, you might want to insist on certain times. In fact, you might have certain dietary needs or you might just be extremely hungry. However, it is very doubtful that you would feel the need to be demanding. If you explain the general benefits of certain times, things should work out. For instance, it might be beneficial to all to take a later lunch because a particular critical operation not yet observed will only be run for the next 45 min. For that same reason, you might suggest that the compliance officer put a hold on his or her inspection of a particular area or operation so that another one can be observed before it is down for the remainder of the day. Of course, you might prefer that the compliance officer not see the operation on that day. Then, your silence might be beneficial.

Be direct in explaining the personal protective equipment requirements in your facility. You have the right to require specific articles of personal protective equipment beyond OSHA's requirements for your employees or for visitors. You can expect the compliance officer to be wearing safety shoes and safety glasses. If, for any reason, the compliance officer is not already wearing the prescribed equipment, it should be in his or her vehicle. If pertinent, deal with the full realm of concerns that can reach beyond standard foot and eye protection. Does the compliance officer have long, unrestrained, or improperly restrained hair? Is there a hazard presented by any jewelry, clothes, or items that are being carried or are attached to his or her clothes? Must he or she wear a hard hat? Must he or she wear hearing protection? You might require metatarsal protectors, in which case you should have some interchangeable ones available for visitors. Do what has to be done.

Your diligence will be appreciated as long as you treat the compliance officer as you would other visitors who would be in the same plant areas. Some companies, essentially, try to make a fool out of the compliance officer. I know of no more accurate way to explain it. As an example, they will demand that the compliance officer wear a full, burdensome, *wholly unnecessary* complement of equipment, including a respirator, ostensibly because he or she will be a little closer to certain operations than other types of visitors or others in the inspection party. Don't do it. If there is a legitimate need, that is another matter.

Other Concerns

The compliance officer can refer an apparent non-OSHA problem to another agency through his or her supervisor. This is seldom done, but it can lead to another knock on your door. Once in a while, this involves employees who are apparently being paid under the table, who are illegal and undocumented aliens, are working excessive hours (related to state law), or who are being paid less than minimum wage and/or not being paid overtime as required. Another possibility is that, if you are in an industry that deals with food, beverages, or pharmaceuticals (maybe even cosmetics), the compliance officer might have professional concerns about health risks to consumers. A larger concern is that he or she would come across possible hazards that are in the realm of environmental protection. Whether in respect to state and/or federal laws, you do not want the compliance officer to believe that you are illegally disposing of chemicals into a waterway or drainage system, including a sink. It is unwise to bring up problems with roof top scrubbers, precipitators, or similar units and systems.

In this chapter, I have brought to light several types of comments and actions to avoid. You should revisit the chapter on accident investigation, wherein I detailed several excuses (mostly rationalizations) that should not be aired to explain why accidents, injuries, and illnesses occur or to defend a questionable practice or condition. I suggest that you refrain from uttering the following additional remarks to compliance officers, who hear them all too often:

"We're in a state of flux."

"If you had only come in next Monday."

"Anyone who puts his finger in there deserves to get hurt."

"It's only this way because we are in the middle of inventory."

"I know you have to find something—you have to justify your existence."

I recommend that you also dispense with "I'm going to have lunch with my good friend (fill in name of congressperson) on Friday."

General Light Conversation

As for general conversation, good taste and discretion should pervade all communications with OSHA. Although I have inserted these comments in this chapter about the walkaround, they apply to all aspects of the inspection process, as well as to all other dealings with the compliance officer and other Agency personnel. I recommend in the strongest terms that you absolutely eradicate any thought of injecting wit, humor, sarcasm, or innuendo that could be viewed as sexist or racist. Similarly, do not even entertain the idea of making remarks that could be loosely translated as deriding, belittling, or showing contempt for persons of particular religions, ethnic or national backgrounds, or those with disabilities. Tread lightly about politics. Yes, this is the United States of America. Yes, there is freedom of speech. Moreover, compliance officers can take a little kidding about the Agency or about some other agency or aspect of government. Do not get carried away.

I suggest that you stay away from any implication that you condone the militia or patriot movements or that you have a problem dealing with homosexuality. Do not bring up hot topics, such as gun control, abortion, or prayer in schools. I am on safe ground in recommending that you not use profanity. However, you should not be concerned if you or any employee slips and utters a mild epithet, as long as it is not used at or against person(s).

The bottom line, though, is to stay clear of needlessly offending the compliance officer. A reasonable degree and amount of verbal venting are acceptable. Compliance officers should expect and accept this in our democracy. It is one thing to concisely express displeasure with the Agency's policies, procedures, or standards, but do not direct anger at the compliance officer personally.

It is all right to engage in light conversation unrelated to OSHA or to occupational safety and health, as long as the inspection is not impeded. Most but not all of this conversation should be limited to lunch or break times or while additional members of the inspection party are being gathered. If you find the compliance officer to be resistant, drop back. Maybe he or she will eventually initiate some pleasant, gap-filling chitchat.

No one should lose sight of the purpose of the visit. Yet, it is not mandatory that the visit be burdened with an oppressive feeling of tension. You might try to alleviate the pressure by mentioning a movie or television program you recently viewed. Just remember to avoid controversy. How about chatting about restaurants? How about a recent sporting event or local team? If the compliance officer shows interest in a picture in your office (e.g., a scenic place, your family, or relating to a hobby of yours), perhaps you can engage in some minor, nonprobing dialogue. A little empathy or basic mutual recognition that the inspection party members are humans and not just faceless players brought into what can be a somewhat confrontational relationship can ease the process.

Bribes

What other types of behavior would be considered inappropriate? Do not even joke about anything that could be construed as a reference to a bribe. Even if it is fairly clear that it is a joke, an uncomfortable feeling can creep into the inspection. The compliance officer may well become uneasy and much tighter. There is no problem with offering a cup of coffee or a soft drink. Some compliance officers will not even accept such a libation without insisting on paying. This is a bit silly, but 50 cents is not worth arguing about. Anything beyond the simple nonalcoholic drink or perhaps something like a donut (brought into the room in a box of a dozen for everyone in the inspection party to share) should not be offered free of cost. In any case, if an employee representative is with you, be sure to offer him or her whatever you offer the compliance officer.

In some cases, compliance officers bring their own lunch. At other times, they prefer to eat in the company's cafeteria. All may wish to adjourn for a lunch break and agree to meet back in the lobby at a specific time. If you eat with the compliance officer, particularly if it is off site, be sure to invite the employee representative. No reasonable person should suspect that management attempted to bribe a compliance officer with a lunch. Nevertheless, I recommend letting compliance officers pay for their own. You do not want any hint of impropriety. I would definitely shun the idea of paying for the compliance officers' twin stuffed lobsters!

Generally, you should not offer to give the compliance officer samples of your parts or completed product. One exception would be if a small component could be used to aid the compliance officer's supervisor in understanding a process or in formulating a decision as to whether a hazard or violation exists. In some companies, very small samples are given to almost anyone who walks in the building. Further, a compliance officer may wish to purchase something that he or she sees. Although such activities would seldom constitute wrongdoing, they could be misinterpreted. A negative example would be when a person purchases a $50 item for a token single dollar. Receipts could be helpful, but are they worth the potential trouble? Someone could see a compliance officer carrying a bag from the building and believe that something illicit has occurred. Perhaps, you can direct the compliance officer to a company store as long as other members of the public (i.e., nonemployees) are allowed to shop there.

When I was an OSHA compliance officer, I inspected a few candy manufacturing facilities. In those plants, there were often pieces of product that had been put aside due to quality control problems. The pieces were edible, and everyone and anyone in the facility was allowed to eat them. If there were uneaten pieces, they were thrown away. Not wanting to stand idly aside while waste accumulated and not wanting to insult the company, I liquidated the evidence on the spot. It did not leave in a bag. It left in my stomach.

Chapter

31

The Closing Conference

Who Will Attend?

The compliance officer will hold a joint closing conference with the employer and employee representative(s) unless either side objects. If the proper spirit of teamwork is in place, none of the individuals involved will view the role of management or the role of the union (or employee group in general) as really being on different sides. Their goals, at least as related to a safe and healthful workplace, should be common. OSHA would prefer to see a joint conference because it serves as a manifestation of the proper spirit. Nevertheless, after weighing the pros and cons, you may feel that you can benefit from a separate closing conference. If, for any reason, the employer representative(s) or employee representative(s) do desire separate meetings, the request will be honored, and there will be separate, consecutive meetings. Generally, the conference with the employee representative(s) will be first. However, if the employee representative(s) prefer to meet with OSHA last, that will be done.

Meeting Before the Conference

The compliance officer may allow you a few minutes prior to the actual closing conference without the presence of the employee representative(s). Then, the full, joint meeting can begin. While being totally aboveboard, these few minutes can easily leave a residue of alienation. Don't forget that you might find it wiser in the long run simply to call the compliance officer on a subsequent day to convey whatever your special concerns were.

For example, if you had wanted to meet with the compliance officer prior to a joint conference to warn about something that an employee representative might theoretically misrepresent, it is probably better to forgo the private conversation and let the chips fall where they may. Again, you can attempt to straighten out the matter in a later telephone call to the compliance officer. You might also write to the compliance officer, but remember that the letter will be releasable under the Freedom of Information Act unless you have requested that it be handled in confidence and, in fact, you have clearly shown a very good reason for your request to be upheld. A realistic reason for communicating privately with the compliance officer, be it on site, by telephone, or by correspondence, could be that you want the Agency to know about a dire or rapidly changing fiscal status of the company. For instance, you might be considering layoffs, acquisitions, or geographical moves that could play a part in the time that you feel is appropriate for abatement dates.

Still, I cannot abrogate my continually espoused philosophy that, even in such special circumstances, employees and their representation as a unit should be kept informed. In all fairness, it is understood that you may not share my approach. It is also conceivable that your desire to talk privately with the compliance officer could relate to a truly personal issue, such as the (not generally known) extremely poor and failing health of one of your key personnel involved in your safety and health efforts. That is, most definitely, a fair reason for confidentiality.

The Setting

As with the opening conference, the setting should be comfortable and private. Make certain that there is plenty of room and that the furniture is arranged to facilitate the physical exchange and sorting of literature. There should be an adequate amount of writing surface for all present. If the closing conference will not directly follow the walkaround and if it appears that it will be lengthy, you might consider booking a super suite or a small conference room in a lodging facility for several hours. If you do anticipate a long conference and if there is any doubt that such a meeting held at the plant will not afford a suitable, practical setting, including the reasonable assurance that there will be no interruptions, barring emergencies, the off-site venue will be a better bet.

The Disclosure of Findings

There is a great deal of importance to the closing conference and the compliance officer will want to impart very specific information, but you should not feel intimidated. Try not to interrupt. Demonstrate a good measure of patience. Take copious notes. When there is a fair opening, keep in mind that this is a time for a free discussion of problems and needs. Air your logical arguments. If the compliance officer's reaction or lack of reaction causes you to become politely persistent, so be it. If he or she does not appear to be writing down your particular, legitimate points of contention, ask in a diplomatic but firm fashion that he or she do so. You want your well thought out, substantive, opposing viewpoints to become a part of the compliance officer's file. Do not vent for the sake of chest clearing alone. Vent with convincing sincerity and concrete meaning. Choose your words carefully and intelligently. You can maturely turn up your intensity without turning up your volume. Just do not permit that intensity to dissolve into invective.

Do not hold back when you have any doubt about the full meaning and implications of the compliance officer's comments. Further, if you think that he or she made omissions regarding your rights or responsibilities as employer, speak up. If you think that he or she has failed to mention an apparent violation during that dedicated portion of the closing conference, you might choose to ignore the issue. More than likely, the compliance officer will note the omission when studying the notes of the walkaround. In that case, he or she will call to inform you of the possible citing and to discuss the full scope of automatically attached, germane issues, such as possible abatement solutions, abatement dates, and mitigating and/or exacerbating factors. (Note that the same type of telephone call should be made when the compliance officer finds through research, or examination of a photograph, or videotape taken during the walkaround that he or she has discovered a possible apparent violation that was not identified during the inspection.)

At any rate, unless you are positive that the compliance officer will later realize that he or she neglected to bring up an apparent violation at the closing conference that was observed during the walkaround, why should you bring it up? In most cases, I suggest that you not be the compliance officer's helper. If you do not get cited for the apparent violation as viewed by the compliance officer, then that is fine. If you believe that there was a real hazard, you can still take steps to abate it. You need not be cited to feel motivated to remove an observed risk. Yet, if you believe that there might be a significant hazard, but you do not fully understand why and/or how to approach eliminating or minimizing it, you may feel obligated morally and in a good business sense to bring up the matter.

There is a middle ground with such scenarios. In taking such a route, do not jog the memory of the compliance officer. Maybe you will luck out, not be cited, and in turn, not be financially penalized. Later, you can seek the assistance of a consultant or similar person with high expertise regarding the area of concern.

Early in the conference, the compliance officer will begin to detail his or her preliminary findings. The compliance officer will discuss all situations that were noted as being unsafe or unhealthful and explain all apparent violations for which a citation may be recommended and issued. For some of the conditions or practices that the compliance officer believes to be hazardous, a citation may not be recommended. It may be that the matter was extremely minor, that there was no exposure, that there was no pertinent standard in place, and that a violation of section 5(a)(1) of the Occupational Safety and Health Act (the "general duty clause") was not applicable.

In slightly different circumstances or perhaps in subsequent inspections now that the situations have been brought to the attention of the employer, some of these matters may become citable. The compliance officer should be specific when addressing the apparent violations. He or she is probably about 95 percent certain as to what will be cited. However, somebody else does the actual citing. That job falls under the authority of the area director (or, as always, the state counterpart). Thus, the compliance officer should not declare that he or she will cite anything. Rather, the compliance officer should make it clear as to what conditions and actions he or she has thus far viewed as apparent violations and is likely to recommend for citation. Listen intently, as you should to all the words of the compliance officer throughout his or her stay at your facility.

As discussed in the chapter covering the walkaround inspection, the compliance officer should address abatement approaches. Along with the normal qualifiers and disclaimers, he or she should be able to put you on the right track while not steering the train. If he or she cannot offer any ideas for abatement, including after a few days of discussion with other OSHA personnel or through research, it is of prime importance that you make notes of all related conversations with the compliance officer. I further address the issue of impossibility of abatement in the next chapter.

This brings to mind another salient point. If the compliance officer at any time during the inspection agrees to send you information or to call you with answers to your questions, be sure to follow through if he or she fails to fulfill the promise. For instance, your questions could deal not only with abatement methods but with interpretations of standards, technical details concerning chemicals, information regarding current OSHA training courses, the status of proposed standards (including comment periods and public hearings), the availability of OSHA personnel for speeches, and so on. If the compliance officer does not provide the information despite your concerted efforts to obtain the literature or verbal responses, you should contact his or her supervisor (usually an assistant area director). In a real-world vein, I do not recommend that you lose your patience too quickly and get on the wrong side of the compliance officer. However, if you are put in a position whereby there is a need to contact the supervisor, that individual will probably be earnestly appreciative of your diligence, as well as dismayed by the lack of professionalism on the part of the compliance officer.

It is understandable if the compliance officer calls you to say that he or she is still seeking the correct answers or to explain that certain requested pamphlets are not in stock. You should be kept posted. This is not simply a lesson in courtesy, though. If you are not receiving the necessary assistance from OSHA (without implying that the Agency is required to give you unlimited hours of telephone consultation to the disadvantage of other employers or seekers of information), keep a clear record of all efforts to do so. For one thing, it is owed to you. Of more specific importance, the lack of suitable and timely response from OSHA may directly relate to the slowing of your abatement efforts. If this occurs, employees are in danger for a longer period than should be the case, and you may be in jeopardy of entering into a "failure-to-abate" status. I address means to achieve extended abatement dates a little bit later. Do not lose sight of the fact that, in the final analysis, it is still the employer's responsibility to comply with the standards.

Mitigating Circumstances

When the apparent violations are discussed, be sure to help yourself out at every opportunity. In other words, whenever and wherever you can spell out clearly mitigating circumstances, endeavor to do so in an effective, no-nonsense manner. Explain, where relevant, the general lack of exposure. This may be due to the remote location of the operation, the operation being seldom run, and/or the fact that very few employees are ever in close proximity to the operation.

Other mitigating circumstances were discussed earlier. Some such circumstances to describe could include the following: extensive training and experience of potentially exposed personnel, partial guarding or other protection, the use of personal protective equipment where it can reduce the probability of injury/illness or reduce the severity of any injury or illness, good housekeeping in the immediate area (relating to a lack of slip/trip hazards and an excess of working/walking space), good lighting, warning signs, the visibly obvious nature of the hazard (not necessarily a factor in your favor, but could be helpful), and slow speeds or little force regarding moving parts of machinery.

Another type of mitigating factor could involve an impediment to immediate access to an exit. Ideally, you can point to other totally available exits in near proximity to the problem one or even to fully opened overhead doors, while understanding that they do not constitute official exits. Similarly, if there is impeded access to a fire extinguisher, you might be able to show that there are several other extinguisher locations that are well positioned and well maintained in the area. These factors will not and should not put a halt to the proper citing. Even if you have 10 more exits and 10 more extinguishers than the standards require, each and every one must be immediately available at all times. Again, the purpose here is to show mitigation. In fact, you may be in a position to add your sprinkler system, extra wide aisles, and a significant dearth of ignition hazards (e.g., no smoking or welding in the area) and particularly troublesome fuel (e.g., no flammable liquids and very little wood) to your argument.

One more example would be if a forklift truck or crane did not have a posted capacity. Perhaps you can show that in no case are weights that even remotely approach the actual capacities ever lifted.

If you really keep these concepts in mind, you may surprise yourself with the spate of defenses that you will offer the compliance officer. Such defenses can result in lower proposed penalties or in citation classifications of "other" instead of "serious." Also, as mentioned in the previous chapter, there are many times when you can reveal or demonstrate that what at first appeared to be a hazard or violation was not one at all. I have already examined several examples. This type of case is usually made by the employer at the time the concern is noted by the compliance officer during the walkaround. If you failed to assert yourself adequately during the walkaround, you might propose another walk out into the shop. The compliance officer may wish to wait until the completion of his or her list of apparent violations before going out to see where loose ends could be tied up. There may be a few locations that, with your insistence, qualify for another look.

Abatement Dates

The compliance officer should be prepared to tell you which particular standards are under consideration for appearance on a citation. He or she should at least be able to reference a fairly tight section of the standards if not the exact subparagraph. If the compliance officer is not on the money, which is understandable, he or she can call you within a few days following the inspection to explain exactly which standard is involved. At other times, the compliance officer might reference section 5(a)(1) of the Occupational Safety and Health Act. I remind you that, technically speaking, this is not a standard. In all cases, the compliance officer must be able to tell you precisely what the hazard is. It is not sufficient to tell you that there is a standard (i.e., that it's in the book). You want to know what can happen that can cause an injury or illness, how it can happen, and what the resultant injury or illness can be.

The compliance officer will want to work with you to determine what abatement dates would be reasonable. Compliance officers do not set the abatement date. They develop information to assist in arriving at recommendations to give to the area director. Unless there is an imminent danger, which is extremely rare, nothing has to be abated until a citation or citations arrive.

For numerous reasons, it behooves you to start working on the elimination and control of hazards as soon as possible. This is regardless of whether or not the company will receive a citation for the item. Those arguments have been made in an earlier chapter. Suffice to say, for now, that the compliance officer will solicit your input to help in arriving at equitable abatement dates. Each item of alleged (once it appears on a citation, the word "*alleged*" is appropriate) violation can carry its own abatement date. In fact, once in a while, standard items are split so that some instances within the item carry a different abatement date than other instances within the item. For example, if access to three distinct exits is impeded, it could be that two are blocked by minor storage and one is blocked by several near-ceiling high stacks of palletized goods. In such a case, barring exceptional circumstances deeming the significantly blocked exit to be of current, critical importance, the employer would probably be given 1 day after receipt of citation to clear the first two exits and a period of a few to several days to clear the other one. So that there is no misunderstanding here, the exact dates would be given on the citation.

Although I have given an example of abatement 1 day after receipt of citation, you can put the Agency in a posture whereby it can indicate on the citation that there was immediate abatement. This is frequently very doable in the case of blocked exits. In order for a citation to indicate that there was immediate abatement, it is not enough for the employer to state in the closing conference that the item has already been corrected. The compliance officer must actually witness the in-compliance status that has been achieved. In some cases, the abatement may have taken place right before the compliance officer's eyes during the walkaround. If items have been fixed since the compliance officer continued on from the location of the apparent violation, it is a good idea to keep a running list of such actions and then to go over it at the closing conference. Then you can ask the compliance officer to accompany you to the relevant areas so that it can be verified that quick abatement was accomplished.

If, for instance, you continually request 2 years for abating violations, you will not be doing the company a favor. You will be throwing away an opportunity to work with the compliance officer in good faith. You will be leaving it up to the compliance officer to recommend your abatement date fate without what would most likely be your leveling input. You will be given shorter abatement dates than if you offered sensible reasons for longer ones.

On the other hand, if you continue to indicate that each item will be corrected before the citation arrives, this will appear unrealistic in the great majority of cases. Indeed, you will not even know which items will definitely be cited. Once more, I am not suggesting that you postpone your efforts to address hazards, cited or not, until they actually appear on a citation. However, there are times when abatement could be very costly (and unnecessarily so, particularly if you do not fully comprehend all of the reasonable abatement method options), could relate to more of a technical violation than a significant hazard, and might not be required once the compliance officer researches the matter or discusses it with his or her supervisor. In those cases, you should not be hasty in attacking the situation. If you are simply being sarcastic when promising to abate the violations quickly, you are again losing a valuable opportunity. If you convince the compliance officer that you really only need extremely short abatement periods, your vision may be shortsighted and may not be considering the large number of corrections that lie ahead of you. You may be setting yourself up for "failures-to-abate."

When discussing abatement dates, take your time. The compliance officer will want to consider how easily the item can be addressed. Within this category, questions arise as to the availability of parts and labor. Is the machine very old? Will parts have to be special ordered? From a great distance? Will guards (always a good example) have to be custom made or fitted? Do you have in-house expertise on the matter? Will you need to hire outside contractors? Will you need to shut down operations? Is a shut down period scheduled for the plant (this does not imply that OSHA is always willing to wait several months for operation shut down, even if that is the only time or way effective, complete abatement can be achieved)?

Other elements that can be considered by the compliance officer include the probability of an injury or illness occurring in the near future and the most likely severity. Also, how many persons are subject to being exposed? Do not insult the compliance officer by rationalizing that only one or two employees can be severely injured. This is not the time to inject that type of supposed levity (or stupidity) into your conversation with the compliance officer. Do not diminish the value of any human life.

The compliance officer will want to gather input and then use his or her professional judgment to recommend to the area director the shortest interval within which the employer can reasonably be expected to correct the violation. This takes into consideration, to some extent, the number of corrections that will have to be made, barring deletions by way of settlement agreement or contest. Thus, the compliance officer will understand that priorities will have to be determined. While the compliance officer with the area director's approval will attempt to establish the shortest practicable abatement dates, periods exceeding 30 calendar days will not normally be given for safety items. Situations may arise, however, especially for health items, where extensive structural changes (e.g., for ventilation or noise reduction) are necessary or where new equipment or parts cannot be delivered within 30 calendar days. When an initial abatement date is granted that is in excess of 30 calendar days, the reason (if not self-evident) shall be documented in the compliance officer's file. This does not mean that final abatement must absolutely be achieved within the initially granted (i.e., printed on the citation) abatement dates. Employers have a right to seek extensions of abatement dates, even when there is no contest and no settlement agreement (settlement agreements and contests are discussed in the next chapter). The other mechanism available for requesting and obtaining extensions is called a *petition for modification of abatement date.*

Extensions

If there is any reasonable concern that an abatement date set forth on a citation will not be met, it is essential that the employer seek an extension. Note that no extension is needed if all exposure and potential exposure are completely removed. Therefore, if you decide to cease an operation (with all guarantees in place) and do not allow it to resume until full abatement has been achieved, you have actually entered into a state of compliance.

As for excuses for why exposure remains beyond an existing abatement date, forget it. What if OSHA discovers that you are not in compliance after the abatement period has elapsed, and you explain that this is because of difficulty in obtaining necessary parts? Maybe they were difficult to locate or secure. Maybe they were generally available, but the incorrect ones were sent to you. What if you explain that the electrician or other contractor who was scheduled to come in and correct the problem broke his arm? In these cases, you still could have and should have requested more time. If the employer requests additional abatement time after the 15 working day contest period has passed (to be discussed), that employer should file the petition for modification of abatement date. In all cases, it should be in writing.

The petition must be filed with the area director no later than the close of the next working day following the date on which abatement was originally required or required by previous extension. Petitions can be filed even after a date had already been extended by settlement agreement or contest. The written petition should specify all steps taken to achieve compliance, the additional time needed to achieve complete compliance, the reasons such additional time is needed, all temporary steps being taken to safeguard employees against the cited hazard during the intervening period, that a copy of the petition was in fact posted in a conspicuous place at or near each location where a violation occurred, and that the employee representative received a copy of the petition (as with all references to the employee representative, it is understood that there is not always such an individual or individuals in place).

After 15 days following the posting, the area director shall determine the area office position, agreeing with or objecting to the request. This shall normally be done within 10 working days following the 15 working days. In some cases, a compliance officer will be dispatched to the workplace to conduct a monitoring inspection, the results of which will help the area director reach a decision. When the petition is the second one or beyond (i.e., at least one previous petition had been granted for the specific item), the likelihood of a monitoring visit is heightened. The area director will notify the employer in writing as to the decision.

If the area director denies the request after consultation with the regional solicitor, he or she will notify the Occupational Safety and Health Review Commission (to be discussed later) of that decision. Once the commission notifies the area director that the matter has been docketed, the director will file a response within 10 working days setting forth the reasons for opposing the granting of the petition. Regardless of whether or not the area director objects to the petition request, affected employees or their representatives may file a written objection with that area director within 10 working days of the date of the posting of the petition by the employer or its service upon an authorized employee representative. This will also set into action a docketing before the Occupational Safety and Health Review Commission.

At the closing conference, the compliance officer will inform the employee representative(s) of that right to contest abatement dates, including original abatement dates. For original abatement dates, do not forget that they have been set forth by OSHA; this does not involve a request for extension. Such contests are nearly nonexistent. In those cases, the original abatement date will often pass before the case is heard. Therefore, the contest by an employee representative will have, in effect, given the employer more time to abate. This means that risks continued longer than they would have without the abatement date contest, constituting a counterproductive result. The employee representative(s) are also told by the compliance officer that, if the employer enters a notice of contest, the employer is obligated to inform them. In turn, employees may elect "party status" before the Occupational Safety and Health Review Commission.

If the petition was clearly not filed in good faith (i.e., no meaningful abatement action had been made), the abatement date has passed, and exposure remains, a "failure-to-abate" citation can be issued. When considering the posting period, the original abatement date often passes while the process is in midstream. This is within the legal boundaries of the system and does not, in itself, cause the determination of a "failure-to-abate" citation issuance. The question, once more, is whether or not the petition was filed properly, fully, and in good faith. In reality, if you are sincere and meticulous, there should not be a problem.

The Agency is fairly liberal in granting extensions of abatement dates as long as all of the pertinent facts were set forth. The employer will certainly not be granted an extension if he or she merely indicates in vague, generalized terms that the company has been too busy to cause abatement. The petition must be detailed. If a company had been given a 30-day abatement period and did not even seek a contractor until the 29th day, the Agency will view this as a breach of good faith and give serious consideration to denying the request for additional time.

On the other hand, the inclusion of purchase orders, letters from suppliers explaining delays, and so on can be very persuasive. If the employer files the petition far in advance of an abatement date (in the rare situations where several weeks had been granted originally or by a previous extension), OSHA will probably ask the employer to hold off until a clearer picture of abatement-compliance status has been drawn (i.e., until the employer can make a more realistic prediction of how much more time will be needed).

The Agency does not want to go through the paperwork process of granting extensions every couple of weeks for the same item. It can request monthly updates in exceptional circumstances that deal with major structural renovations. It may tie in those monitoring visits in such circumstances. Further, OSHA may respond to the petition by working with the employer to establish mileposts. For example, if the company truly needs 8 months to install a great deal of complex, expensive guarding on 80 machines, OSHA may require that a certain number of machines be completed per month. In so doing, hazards are removed in project phases within definite slots of time as opposed to the possibility of all hazards remaining until the last few weeks.

Only the regional administrator can approve employer extension requests that would bring the abatement date to more than 2 years from the issuance date of the citation. This is seldom an issue. Where it is obvious that a great deal of time is needed for final abatement (possibly related to an expenditure of several hundred thousand dollars for noise abatement), the area director will try to work with the employer so that an extension of several months is given at one time. Additional petition(s) could be filed if adequate progress is demonstrated.

Citation Posting, Penalty, Informal Conference and Contest

The compliance officer should explain that, if citation(s) are to be issued, the package will arrive by certified mail along with a return receipt. Occasionally, citations are hand delivered. The compliance officer should explain the format of a citation, including the reference to the relevant standard or to section 5(a)(1) of the Occupational Safety and Health Act; the synopsis of the standard or section 5(a)(1); the specific charging language describing, for instance, the specific machine, tool, chemical, practice, etc.; the abatement date; the proposed penalty; and the citation classification (I discuss the definitions of those classifications, as well as related penalties, in the next chapter). The compliance officer should briefly explain the penalty system and criteria, including possible reductions, and make it clear that the proposed penalties are owed, unless modified or eliminated by settlement agreement or contest. There are still a few naive employers who believe that the penalties are negated upon timely abatement. This is not the case.

The compliance officer will tell you that, as indicated on the citation, each page must be posted at or near the apparent violations for 3 working days or until the pertinent items have been abated, whichever period is longer. Because a particular page may address alleged (that word is now applicable) violations in various parts of the facility, photocopies generally need to be produced. If the shop is extremely small, one posting of all pages at a bulletin board or time clock can suffice. If there is any doubt, fall back to the safe position.

The compliance officer should explain that, upon receipt of the citation(s), if you have any questions at all, you should call the area office. The employer has 15 working days (that expression always translates to weekdays not counting federal holidays) in which to file a notice of contest. There is certainly no obligation to contest, but your right to do so will expire after that time period. Unless your entire upper management staff has been in a coma, do not figure that you or your attorney can get around the time limitation.

If you contest, the file will be sent to the U.S. Department of Labor Office of the Solicitor. The employer also enjoys the right to request an informal conference in the area office, with the chief goal being to obtain some degree of relief from proposed penalties, abatement dates, citation classifications, or the items themselves. This would be accomplished by signing a settlement agreement. That conference must be held, not just scheduled, within the 15 working days following receipt of the citation(s). Once a notice of contest has been filed, the area office will not hold an informal conference because the case has moved beyond OSHA to the regional solicitor. If you have contested but still wish an informal conference within the 15-day period, you will have to rescind the contest. You can reinstate it, in writing, at the close of the informal conference.

One more comment on this for now because it should be explained during the closing conference. Once you sign a settlement agreement, your right to contest has been waived. In fact, words in the settlement agreement specifically clarify this. In the next chapter, I examine the contest and the informal conference in depth.

Progress Reports

You will be told about the employer's requirement to send progress reports (also referred to as letters of corrective action) to the area office. These reports should address each instance of alleged violation that appears on the citation(s). You must be specific in explaining, in an orderly fashion and with reference to the citation, item, and instance numbers, exactly what has been accomplished and what steps are planned. It is not satisfactory to indicate that everything has been fixed. You must clarify, while providing the unambiguous citation references. Examples could include descriptions of railings installed, presence-sensing devices installed, material safety data sheets of specific chemicals obtained (with copies attached), ladders destroyed and replaced, exit accesses cleared, grounding and bonding wires (or equivalent) put in place for dispensing of flammable liquids, the purchasing and assured use of particular items of personal protective equipment (there may be a need to attach a copy of the certifications for assessment and training, as well as a memo sent to employees), forklift brakes repaired, grinder tongue guards added, electrical knockouts plugged, and so on. The list can be exhaustive.

When addressing abatement methods that are not totally clear in a few words, you should expand your comments. For example, there may be a need to explain how and with what material a certain danger point or series of danger points were guarded on a machine. Photographs can be beneficial in verifying your claims and in removing gray areas. However, a note of caution is necessary here. Make certain that no other apparent violations are evident in the photographs. For example, if you are sending a photograph of a guarded machine, you would not want the compliance officer to see a blocked exit and uncapped spare oxygen cylinder in the background. Copies of purchase orders and work orders can also bolster your claim of compliance.

If you do not send timely progress reports and/or if any reports that you do send fail to address the abatement status and means thoroughly, the chances of a follow-up inspection will increase dramatically. During follow-up inspections, you can always be in jeopardy of having the compliance officer uncover "repeated" violations or "failure-to-abate" violations. The penalties can be staggering. More on that in the next chapter.

Remember that ordering a part, hiring an electrician, putting out bids, or forming a committee to complete a required written program does not constitute compliance. If you indicate in a progress report that a certain item will be abated in 3 more weeks, an important question remains. Will you still be within the abatement date? If not, you must make it indelibly clear that you desire to file a petition for modification of abatement dates. If you do so, the area office will send you a sample form or a list of the required information that you must submit.

Do not reason that, if you do not specifically mention the need for an extension, OSHA will figure out that one is required. If you are not specific, it may be inferred that the referenced 3 more weeks are actually within the abatement date. In all cases, keep in mind (as the compliance officer should inform you) that heavy penalties can be levied for lying to OSHA, particularly on progress reports. I have addressed this issue earlier. I recommend that progress reports, petitions for modification of abatement dates, contests, and other important letters sent to OSHA be sent by certified or registered return receipt mail or by a reliable delivery service. In all cases, you should consider telephone calls and/or facsimile transmissions as backups. When speaking with any OSHA personnel, when discussing the form, content, or receipt of letters bearing on verification of abatement, as well as such subjects as interpretations of standards or suitable means of abatement, obtain the name of the individual(s) and document the conversation in your OSHA file.

The compliance officer should tell you if an OSHA intraoffice referral is likely. This usually concerns an industrial hygienist referring to the safety staff, or vice versa. Another inspection can result. You may be able to head off that visit by suitably addressing the issue in a progress report or an earlier letter to the area office.

Variances

The compliance officer will explain that the employer's requirement to comply with a standard may be modified through the granting of a variance. Employers may ask OSHA for a variance if they cannot fully comply due to shortages of materials, equipment, or professional or technical personnel or if the employer can prove that the facilities or methods of operation provide employee protection at least as effective as that required by the Agency. An employer will not be subject to a citation if the observed condition is in compliance with either the variance or the standard. In the event that the employer is not in compliance with the requirements of the variance, a violation of the standard shall be cited with a reference in the citation to the variance provision that has not been met. The detected instances of employers in violation of the requirements of variances granted to them is extremely uncommon. The granting of variances, in general, is uncommon.

Sometimes, letters of interpretation from the OSHA national office in direct response to inquiries from employers, associations, manufacturers, or distributors serve a similar though less formalized purpose. In many cases, employers request variances only to learn that they are already in compliance with the germane standard, at least by way of OSHA's interpretation. For example, there are cases where the employer has used a state-of-the-art material not addressed in the standards. In some of those cases, the material has been set up in a manner where dimensions or unsecured spans do not meet with the letter of the law. OSHA has seen fit, if the employer can prove with documentation (often including specification sheets and/or engineering studies), that the state-of-the-art method was actually at least as effective as that described in the standard and considered compliance to have been achieved. The burden of proof is on the employer. When the request for interpretation is sent to the OSHA national office, the documentation should be included. If the documentation leads OSHA to consider the employer in compliance, no variance is needed. Whenever there is any doubt, be certain to maintain available copies of the response from OSHA. You may have to provide a copy to a compliance officer.

Employers located in states with their own occupational safety and health programs should apply to the state for a variance. However, if an employer operates facilities in states under federal OSHA jurisdiction and also in state-plan states, the employer may apply directly to federal OSHA for a single variance applicable to all the establishments in question. OSHA will then work with the state-plan states involved to determine if a variance can be granted that will satisfy state as well as federal OSHA requirements.

Variances will not be granted simply because exposed employees have had many years of experience and education. They will not be granted just because proper compliance will be costly. Further, they will not be granted merely due to the fact that frequency and duration of exposure are very low. The variance system calls for requesting and granting prior to OSHA inspections. In fact, the point is to request a variance before employees are exposed to the related hazards. Thus, if you wait for a compliance officer to visit for the purpose of an inspection and then ask about variance procedures because you are not in technical compliance, you will have erred. You might be able to secure a variance later, but if you are in violation at the time of the inspection, you can expect a citation. When a citation is outstanding, you are not prevented from filing a variance application. Additionally, variances are not retroactive. An employer who has been cited for a standards violation may not seek relief from that citation by applying for a variance. Generally, the compliance officer will not offer further details concerning the variance system at the closing conference unless you request that he or she do so. However, I will address the issue further, at this point.

A temporary variance may be granted to an employer who cannot comply with a standard by its effective date due to the unavailability of personnel, material, and so on, as described earlier. A different reason would be that the necessary construction or alteration of facilities cannot be completed in time. Employers must apply for the variance within a reasonable amount of time prior to the effective date and must demonstrate to OSHA that they are taking all available steps to safeguard employees in the meantime. The employer must also put in force an effective program for coming into compliance with the standard as quickly as possible.

A temporary variance may be granted for the period needed to achieve compliance or for 1 year, whichever period is shorter. It is renewable twice, each time for a maximum of 6 months. An application for temporary variance must identify the standard or portion of a standard from which the variance is requested and the reasons the employer cannot comply with the standard. The employer must document those measures already taken and the measures to be taken, including dates, to comply with the standard. The employer must certify that employees have been informed of the variance application, that a copy has been given to the employees' authorized representative, and that a summary of the application has been posted wherever notices are normally posted. Employees must also be informed that they have the right to request a hearing on the application.

A permanent variance (i.e., an alternative to a particular requirement or standard) may be granted to employers who prove that their condition, practices, means, methods, operation, or processes provide a safe and health workplace as effectively as would compliance with the standard. In making a determination, OSHA weighs the employer's evidence and arranges a variance inspection and hearing if appropriate. If OSHA finds the request to be valid, it prescribes a permanent variance detailing the employer's specific exceptions and responsibilities under the ruling.

When applying for a permanent variance, the employer must inform employees of the application and of their right to request a hearing. Anytime after 6 months from the issuance of a permanent variance, the employer or employees may petition OSHA to modify or revoke it. OSHA may also do this of its own accord.

So that employers may continue to operate under existing conditions until a variance decision is made, they may apply to OSHA for an interim order. Application for an interim order may be made either at the same time as, or after, application for a variance. Reasons why the order should be granted may be included in the interim order application. The concept of an interim order appears to be contrary to what should be the continuing obligation of the employer to assure compliance by standard or variance. Nevertheless, OSHA does maintain the interim order system. It is rare, indeed, for such orders to be granted. I cannot encourage you to seek this route. If OSHA denies the request, the employer is notified of the reason for denial.

If the interim order is granted, the employer and other concerned parties are informed of the order and the terms of the order are published in the *Federal Register*. The employer must inform employees of the order by giving a copy to the authorized employee representative and by posting a copy wherever notices are normally posted.

If an employer is participating in an experiment to demonstrate or validate new job safety and health techniques and if that experiment has been approved by either the Secretary of Labor or the Secretary of Health and Human Services (of which the National Institute for Occupational Safety and Health is an agency), a variance may be granted to permit the experiment. One more type of variance can be granted if the Secretary of Labor finds that it can be justified when national defense is impaired.

OSHA as a Source of Information

The compliance officer will furnish you with numerous free handouts. Be sure to let him or her know of your company's needs for other written information, usually in the form of pamphlets, or additional copies of the literature given to you. All written material printed by OSHA is in the public domain, and you may photocopy, or otherwise reproduce it. You need not be concerned about copyright laws or even attribution. Some OSHA literature must be purchased from the U.S. Superintendent of Documents, Government Printing Office. This includes the actual standards.

The compliance officer will apprise you of the area office's desire to serve as a major resource center for occupational safety and health information and assistance. Anyone can call with questions. You need not identify yourself or the company with which you are associated. You may visit the full service office to discuss standards, including those in various stages of proposal, or abatement methods with compliance officers. You may avail yourself of resources in the office library, although OSHA generally restricts anyone from borrowing any of the books. A good supply of free handouts should be available. The area office may have audiovisuals available for free loan. At any rate, the regional office will have a large selection of audiovisuals for borrowing. Upon request, OSHA personnel should be able to direct you to other agencies or to professional organizations whose primary functions include the promotion of safety and health interests.

The compliance officer should inform you of the availability of OSHA personnel for speaking engagements at no charge. You would not want OSHA personnel to speak on the floor of your plant, however. The problem would be that apparent hazards may be visible. Although the OSHA person does not desire to observe hazards in such situations, an uncomfortable and potentially conflictive scenario could exist. Indeed, the speaker could be put in a position whereby he or she is obligated to consider a citation.

OSHA personnel would not be used to provide detailed training sessions. While they should be willing to speak with your employees or to a corporate gathering, for instance, they are most often asked to speak at meetings of associations and similar organizations. Of course, they will gladly speak to union groups and other aggregates. In a similar light, the compliance officer should apprise you of the OSHA Training Institute in Des Plaines, Illinois (greater Chicago area), and its various branches throughout the country. For most available courses, the public is allowed to attend. You can request a listing of courses, dates, and locations.

Voluntary Protection Programs

The compliance officer will explain OSHA's Voluntary Protection Programs (VPPs), which represent one part of the Agency's efforts to extend employee protection beyond the minimum required by the standards. These programs endorse, and essentially require, a vigorous three-way partnership among employers, employees, and OSHA.

The idea is for this partnership to be grounded in a spirit of cooperation as opposed to an atmosphere of coercion. The programs are designed to recognize outstanding achievement of those employers who have successfully incorporated comprehensive safety and health programs into their total management systems. Further, the programs are designed to motivate others to achieve excellent safety and health results in the same outstanding way.

OSHA views the benefits of participation in a Voluntary Protection Program as including the following: improved employee desire to work safely, leading to better quality and productivity; lost workday case rates that are generally much less than half of industry averages; reduced workers' compensation and other injury/illness-related costs; positive community recognition and interaction; further improvement and reenergization of already good safety and health programs; that all important partnership with the Agency; and exemption from certain types of OSHA inspections.

There have been numerous names for programs within the VPP system. The individual programs are separated, for the most part, by the stringency of the floor requirements. There are usually about three programs available at once. Employers apply for one at a time, although the programs are not all stepped. The employer can apply directly for the one with the most difficult criteria for acceptance. Being that these programs are changed periodically, along with their names (e.g., STAR and MERIT), I will not fully detail each of them.

The most demanding and most prestigious of the programs are open to companies who clearly show evidence of a comprehensive safety and health program that is exemplary and injury rates that are below the national average for the industry. Illness rates are definitely a concern but are considered a bit more difficult to factor in, at least when analyzing those that developed over several years. The work required to attain acceptance (i.e., to be approved) for a VPP is highly labor intensive. To be successful, the employer must go far above and beyond the regular criteria that could earn the company "good faith" penalty reduction or scope reduction relating to a compliance inspection.

The first part of this book will give you plenty of directions to take. Making an effort or being in compliance with OSHA standards is not nearly sufficient to qualify a company for VPP acceptance. OSHA will desire to see policies that are not even addressed by standards but can be very beneficial to all. For example, they will be impressed if you can show them specific, detailed evidence of your policies to preclude or avert workplace violence, to assure that all motor vehicle occupants are protected by belts, and to prohibit smoking anywhere in or near the facility. The total safety program and allied culture has to be of a truly superior quality that sets an outstanding example to be admired. A tremendous commitment is required.

An employer may make application at the nearest OSHA regional office. OSHA will thoroughly evaluate the Voluntary Protection Program application, which includes several specific elements that must be addressed. In putting together your application, you should make it easy for OSHA personnel to trace your references to documents and required program elements. Do not just wheel in several cartons of loose-leaf books and tell them that "it's all in there." Once OSHA is satisfied that, on paper, the employer qualifies for the program, an on-site review will be conducted to verify that the safety and health programs described are in operation and are being suitably implemented at the workplace. This is usually performed by a team. That OSHA team, or a part of it, will conduct many private interviews with employees to confirm that the employer's program is really being followed, that it is working, and that it is reaching the employees to be served by it.

An overview type of facility audit will also be performed. It will not be as detailed as a regular compliance inspection. Yet, if there are many evident hazards observed, particularly those of a serious nature, the existence of even an outstanding safety and health program will not be enough to save the application. Therefore, do not just have all of the paperwork in order. It would be embarrassing at best for this special OSHA team to quickly note unguarded power transmission equipment, a plethora of tripping hazards, mezzanines without perimeter protection, and employees standing on bare raised pallets of forklift trucks.

The OSHA review team presents its findings in a written report for the company's review prior to submission to the Assistant Secretary of Labor, who heads OSHA. If approved, the employer receives a letter from the Assistant Secretary informing the site of its participation in the Voluntary Protection Program. A certificate of approval along with a flag that displays a special logo are presented at a ceremony held at or near the approved worksite. Press releases herald the acceptance of the employer into the program. That acceptance is considered to be a great honor. As with variances, I have gone into more detail about VPPs than the compliance officer would at the closing conference.

Evaluations are conducted on a regular basis—annually for most VPP programs, whereas others have been scheduled for evaluation every 3 years. All participants must send their injury/illness information to their OSHA regional office annually. Sites participating in the VPP are not scheduled for programmed inspections. However, and it is critical that this is understood, any employee complaints, serious accidents, or significant chemical releases that may occur are handled according to routine enforcement procedures.

Consultation Services

The compliance officer will inform the employer about on-site consultation assistance. These programs are not run by OSHA enforcement personnel. The distinction is paramount. The programs are largely funded by OSHA, and the services are provided at no cost to the employer. Primarily targeted for smaller employers with more hazardous types of operations, the consultation service is delivered by state government agencies employing professional safety consultants and industrial hygiene consultants. All consultation services are provided at the request of the employer.

The process begins with the employer's request for consultation and the commitment to correct any serious job safety and health hazards identified by the consultant. When delivered at the worksite, consultation assistance includes an opening conference with the employer to explain the ground rules for consultation, a walkthrough to identify any specific hazards and to examine those aspects of the employer's safety and health program that relate to the scope of the visit, and a closing conference, which will be followed by a written report to the employer listing the consultant's findings and recommendations. No penalties are proposed or citations issued for hazards identified by the consultant.

The service is provided to the employer with the assurance that the company and any information about the workplace will not be routinely channeled to OSHA enforcement staff. Possible violations of OSHA standards will not be reported to OSHA enforcement staff unless the employer fails or refuses to eliminate or control employee exposure to any identified serious hazard, including imminent danger situations. In such highly unusual circumstances, OSHA may investigate and begin enforcement action. If an employer undergoes an OSHA enforcement inspection, the compliance officer should not request a copy of any state consultation visit of the type described herein.

Employers who receive a comprehensive consultation visit and demonstrate highly commendable achievements in workplace safety and health through the abatement of all identified hazards, and who develop and implement an excellent safety and health program, may request participation in OSHA's SHARP program. SHARP is an acronym for Safety and Health Achievement Recognition Program. There are certain similarities between SHARP and the Voluntary Protection Programs. However, the VPP review and approval system is generally a much more intense process. Employers who are accepted into SHARP may receive an exemption from programmed inspections, but not from complaint or accident investigation inspections, for a period of 1 year.

A Second Closing Conference

In some cases, the employer and employee representative(s) will be told that a second closing conference may be held. This is most frequently the case when industrial hygiene hazards must be evaluated following the initial closing conference or when the compliance officer will have to await laboratory reports. At the end of the initial closing conference (and the second one, if relevant), the compliance officer should remind all parties of the protections afforded by section 11(c) of the Occupational Safety and Health Act. Also, with all closing conferences, if you have any questions before the compliance officer departs, pose them. You may want to caucus with your key staff to determine if any issues left unresolved would be worth bringing up while the compliance officer is still on site. With hearty handshakes and cordial farewells all around, the inspection is history, and that's all I have to say about that.

32

Citation Remedies: The Informal Conference and the Contest

Within just a few days after the compliance officer's departure, you should assemble the key staff to formulate a strategy for dealing with the apparent violations and other hazards noted. Of course, if you are positive that no citation will be issued to your company, you may choose to put down this book and pay a visit to the refrigerator. After reflecting upon the inspection, I suspect that you will forgo the confectioneries and read on.

Postinspection Meeting with Staff

You may not receive the citation(s) for several weeks, but as I have explained earlier, you should consider attacking the hazards regardless of legalities. You should have a pretty good idea of what will appear on the citation(s), anyway. So, start to assign responsibilities and map out a division of labor. Hold a brainstorming session, with the purpose being to bring to the surface any thoughts about the apparent violations that might be useful to bring to the compliance officer's attention. You will want to include your safety director, maintenance chief, plant engineer, and any management representative who accompanied the compliance officer during any aspect of the inspection.

Although it may seem radical, you are (hopefully) comfortable enough to include the employee representative(s) as well. If you are, you can start with your commitment to abate all true hazards regardless of the legal outcome of the whole affair. You can explain that it will be beneficial to all if the proposed penalties are minimal and if the citation(s), including proposed penalties, do not put the company in a bad light with its customers or the community at large. Any handling of occupational safety and health matters deserves teamwork. Any clear-thinking employee representative(s) will appreciate your being up front. This approach has an idealistic glare to it, but if your relationship with the employee representative(s) in matters related to safety and health has been built from mutual respect and trust, then your efforts should come back to reward you in even better labor-management relations and put you in a better position to demonstrate your safety consciousness to OSHA. Whether prior to citation issuance or subsequent to the receipt of the citation(s), the Agency will place great stock in what employee representatives have to say.

At the meeting, you can discuss mitigating factors and other concepts discussed earlier. It is not too late to do this. There is certainly no prohibition against communicating with the compliance officer between the completion of the inspection and the receipt of the citation(s). If you have something worthwhile to pass on—perhaps some facts that the compliance officer did not realize—structure your thoughts and deliver them to the compliance officer in words devoid of extraneous and verbose content. Be particularly cautious when committing those words to print. Do not put your foot in your mouth. Enlist those that you trust to critique your words, particularly if they will be memorialized on paper. I have already addressed pitfalls that can push a "serious" alleged violation to a "willful" alleged violation.

Contact with OSHA

During the inspection, you made every reasonable case that could cast the company in a better light or that could diminish the significance of an apparent hazard. Now study the notes again to determine if you have additional cases to offer. The correct notion is that, despite the fact that you will have opportunities to offer your arguments after the receipt of the citation(s), it is easier and more effective to persuade the compliance officer prior to the issuance. In fact, once the citation(s) have been issued, you will have to deal with an OSHA area director or assistant area director. At that point, regardless of the merits of the arguments that you would make during an informal conference, ego comes to rest at the OSHA area office. With all due respect and without sarcasm, that very human quality, consciously or otherwise, can cloud the logic of the OSHA authority and impede his or her eagerness to modify a citation to a measure commensurate with the merits of your arguments.

Once the alleged violations are on the official form, it seems that it hurts OSHA individuals and the fictional embodiment of the Agency to alter them. Sure, it is possible to convince OSHA to modify citation items. This can include an outright deletion or a change of classification. However, even if you enlighten the OSHA personnel with very important facts (e.g., facts previously unknown to them but now revealing in indisputable fashion why an item should not have been cited), you may be pushed into a negotiation. In other words, you may well have to give something back in a settlement agreement. In such cases, they can rebut accusations by auditors or others that they gave away the store. If they deleted an item or changed its classification from "willful" to "serious" with no give back by the employer, it looks more like the correction of an acknowledged mistake than the product of negotiation. I am not suggesting that this is fair; only that it is reality. Later, I address settlements in detail. For now, I cannot urge you enough to make every ethical and wise effort to eliminate items from being included on a citation and, where that is not a realistic goal, push for the least damaging classification.

The Arrival of Citations

The letter carrier arrives with certified mail. It is a package from OSHA. The authorized employee representative will also receive a copy of the citation(s). First, be sure that it relates to your company. Do not even think about ignoring it. Remember that as soon as the receipt is signed your contest period starts to run. If for any reason the area office does not receive a return receipt, OSHA personnel will trace the package or simply hand carry another one to the establishment.

You cannot hide from a citation. However, there is a statute of limitations with respect to the reasonable promptness of citation issuance. No citation is to be issued after the expiration of 6 months following the occurrence of any alleged violation. I delve more deeply into this loophole in the next chapter. The relevance here is that, if you know that nearly 6 months have elapsed since the completion of the inspection, a tactful, subtle thwarting of the citation delivery may be worth the effort. Maybe that final week of the 6-month period is the time for a plant shutdown. I am not rendering legal advice in this or any other case. I am merely casting light on theory.

In fact, note that the 6-month limitation language indicates issuance as opposed to explicitly addressing the delivery. The further fact is that citations have actually been issued and hand delivered on the last day of the 6-month period. In those cases, OSHA just made it under the wire. I have also heard of cases where the statute of limitations expired prior to issuance. As a separate matter, you should check the abatement dates right away to see if the citation(s) were issued within the period and also to see if there are any abatement periods that must be met within the next few days.

After receiving citations, you could fire off an angry letter to your U.S. representative. Yeah, right. He or she will get back to you after the health-care crisis, drug blight, and selection of a permanent baseball commissioner are all solved. Maintain your composure. Get focused. Convene a meeting of your previously selected team, identifying which individuals will spearhead your efforts to comply with the citation(s) if you feel that such a course of action is proper. You should also identify those persons who will be able to contribute to your efforts to fight aspects of the citation(s), if appropriate. There is no implication here that you should turn your head away from items that you deem to be unworthy of a place on a citation. I maintain that you should abate all hazards. The point here is that, if you believe that no realistic hazard existed and if the cost of abatement (in any of the various ways that cost can be measured) is substantial, you may plan to seek a deletion as opposed to only a penalty reduction or change of classification.

Be sure to keep a close eye on the abatement dates for those items, however. Note that the formal process covering petitions for modification of abatement dates discussed during the closing conference addresses actions to be taken after the contest period has passed or after a settlement agreement has been signed at the area office level. Even if you attend an informal conference or file a notice of contest, you might not do so until some of the abatement dates have passed. There may be abatement dates allowing just a few days for your corrective actions. If you truly cannot meet those dates and if you can make a strong, no-nonsense case, you should notify the area office at once.

This is where one more means of abatement extension exists. OSHA can issue a simple amendment to the citation(s). There have also been situations where the employer has apprised the area director of this type of dilemma, and the area director has responded that no "failure-to-abate" would be considered until the employer has had the opportunity, at the informal conference, to seek an extension. The promise has been honored. The area director meant well, but that abatement date was still hanging out there. I encourage you to seek an amendment, at least as a stopgap measure. There is a school of legal thought (or interpretation) that would submit that no abatement date within the contest period could be tied to a "failure-to-abate". There should be no great reason to test that theory.

Back to the team meeting. Give each person a copy of each page of the citation(s). In addition, make certain that copies of the citation have been prominently posted in the appropriate places as explained at the closing conference. Plan to monitor those locations daily to assure that they remain posted and are not defaced or visually impeded. You might prefer to plasticize each page.

Next, you should take pains to identify each location, machine, process, and document that is referenced in the allegations. Be certain that there is no misunderstanding between the company and OSHA as to what is the subject of the alleged violations. Check the abatement dates one by one. While you already scrutinized them for fast-approaching deadlines, you may find that there are several different dates. All too often, employers have quickly scanned the citation(s) and misread an abatement deadline by looking at the wrong portion of the abatement date column. When you provided input to the abatement date discussion at the closing conference, there was no guarantee that your suggested timetable or that which was seemingly favored by the compliance officer would be reflected on the citation. The citation(s) probably display abatement dates that are roughly what you anticipated, but do not miss a date and then claim that you had been promised a longer period. Due to your input at that conference, the abatement dates should be grouped to facilitate logical project planning and an economical use of abatement resources.

Attorneys and Safety Consultants

If you feel that the proposed penalty is out of line, allegations are unfounded (perhaps there was no real hazard, or no potential exposure, or the cited standard does not apply), practical abatement methods are not clear to you, or you question the legal protocol of the inspection, it might serve you well to contact an attorney, a safety consultant, or both. In the case of the attorney, he or she must be thoroughly familiar with the OSHA process. I recommend that you do not seek counsel from an attorney simply because he or she is a personal friend or relative. If he or she is a friend in tune with OSHA, that is fine; but if the attorney's only field of expertise is tax law and you had planned to avail yourself of his or her services just because of the relationship and a low (or no) fee, it would be an exercise in poor judgment. Further, you may be aware of a high profile attorney, who enjoys a wonderful reputation for criminal defense. Again, if he or she does not know OSHA, make another choice. There are attorneys who specialize in labor law and are well-versed in how to mount an effective defense for the employer in an OSHA case.

However, in the majority of situations, the citation(s) should be straightforward and, if you have a solid understanding of the paperwork, you should not require the professional assistance of an attorney at this stage. My purpose is not to dissuade you from hiring an attorney. In fact, you may have one on retainer. I simply do not wish to imply that legal counsel is a must for every company that receives citation(s). Nevertheless, if there are extremely high proposed penalties involved or if you feel overwhelmed or significantly confused, then that is another matter.

As for the safety consultant, I recommend that you only hire one that has extensive experience with the technical issues and OSHA procedures and policy. You should know right off the bat that the chances of winning any aspect of your case due to an OSHA blunder on the legal protocol issue are very slim indeed. If your only gripes relate to a need for extensions of abatement dates, you should not need an attorney or a safety consultant, at least not unless you are experiencing resistance from OSHA when attempting to secure those extensions.

Informal Conferences with OSHA

Proceed to damage control. Decide if you want to request an informal conference at the pertinent OSHA area office. If so, do not call the office near the end of the 15-working day period for an appointment. Leave some breathing room. That 15-day period for contest or for settlement agreement through informal conference will not be extended simply because the parties involved could not work out an appointment. Further, the holding of an informal conference will not extend the contest period.

There are few good reasons to pass up the opportunity to participate in an informal conference. Even if your hopes of reaching a satisfactory settlement agreement are slim, there is still good reason to attend. If you are at all contemplating filing a notice of contest, by all means attend. Reestablish the lines of communication. You should be able to gain a better understanding of where OSHA seems susceptible to movement or seems to have less than pat documentation. Find out what cards the Agency is holding. Such information can afford you an improved position from which to mount a defense.

I will paint a scenario wherein you may not wish to participate. Let's say that the proposed penalties on the citation(s) were relatively low in respect to your particular company, you did not foresee any major problems in achieving compliance (e.g., by cost of abatement, complexity of abatement, or by production hindrance due to the abatement methods), you are not concerned about a major lawsuit relating a severe injury or illness to violation(s) of OSHA standards, and you believe that you have no substantive arguments with the allegations. Say, too, this is accurate and you were sent an expedited settlement agreement offering to reduce the penalty by 50 percent if you sign, commit to total abatement and seek no other modifications other than perhaps reasonable abatement date extensions. Your attendance at an informal conference may not have a strong point to it under such circumstances. You can still always ask questions by telephone, fax, or correspondence. OSHA's proposal of such a simplified settlement agreement is, in the great majority of cases, only offered when there are no alleged "willful," "repeated," "failure-to-abate," or high-gravity (e.g., high severity and greater probability) "serious" violations, and the employer has an excellent (or no) history with the Agency.

Choose wisely when deciding who will attend the informal conference. It is all right to show a little emotion, but leave the short-fused individuals back at the plant. As the employer is obligated to post a notice announcing the time, date, and place of the informal conference, affected employees and/or employee representative(s) may wish to attend. They have the right to participate fully.

Regardless, affected employees or the employee representative(s) could have requested the informal conference on their own. If that is the case, but they choose not to attend, the OSHA authority holding the conference should make a reasonable attempt to contact the party or parties to solicit their input prior to the signing of a settlement agreement, unless the only modification is to adjust the penalties. If one of those parties objects to the proposed modification, a separate conference is to be held prior to the OSHA authority making a final decision as to the settlement agreement offer. In the case of authorized employee representative(s), even if they did not request a conference, their views are to be solicited as just outlined.

I have addressed the presence of an attorney and/or a safety consultant. A safety consultant who can speak the language of OSHA and of occupational safety and health (there is not a total overlap) and one that is respected by the Agency can be very beneficial. The consultant can supply OSHA with a level of comfort that the company is now in good hands. The right consultant can also assist with procedural issues and can spot holes in the compliance officer's case, including documentation. In a slightly different scenario, you could be called in by the IRS to discuss the disposition of their audit of your firm. Wouldn't you hire a tax consultant to accompany you?

If you think that you can simply stroll into an informal conference, smile, and be rewarded with a major penalty reduction and the deletion of a few items, you are mistaken. The area director (and, in effect, the assistant area director) is authorized to enter into settlement agreements that revise citations in order to avoid prolonged legal disputes and so that speedier hazard abatement can be attained. However, in the real world of OSHA, do not consider this to be automatic. There are cases where this is done without many other concrete reasons, and it may be more common when the percentage of contests has risen significantly. One reason for OSHA to avoid granting automatic or expected penalty reductions for the majority of employers is that the word would spread and the floodgates would open.

Recall that the mailed expedited settlement agreement system is at the disposal of the Agency but is only employed when special criteria have been met. For the most part, the OSHA authority must document the specific reasons for any and all citation modifications. He or she had better be particularly ready to explain item deletions and classification changes which negate or downgrade "willful" and "repeated" alleged violations. This documentation and preparation must be even keener if a fatality or serious injury or illness was involved. The importance of a good explanation increases further if there was a major media event.

Most of the old, broad, employer excuses do not work anymore. If you are one of the principlals of a small company, do not claim to be impoverished if you are going to wear a $10,000 watch and park your $60,000 car in a lot that can be seen from the window of the OSHA office where the meeting will be held. If the company is truly in dire financial straits, be prepared to hand over your fiscal records for OSHA to review confidentially. Short form summaries will usually suffice, particularly if notarized or similarly documented and certified in a convincing, formal manner. If you feel that this is none of their business, then do not cry that the company is poor. What if you can prove that your company lost several thousand dollars in each of the last 3 years? Be ready to answer questions about your personal salary or that of family members employed at your company. Point made.

You may choose to gripe that the inspection stemmed from a complaint from a disgruntled employee, that inflation and foreign competition are destroying your business, and that you aren't too crazy about the conduct of the IRS. None of this will help your case, but it might make you feel better. OSHA ears have become immune. Make it quick and move on to the real issues. The fact is that they came, they videotaped, and they cited. A little groveling may have some positive effect on your cause, but do not yell, "BS," filibuster, or verbally attack individuals. This will probably be your last chance, short of a contest, to obtain some relief from the original citation(s).

When you are in serious pursuit of a just and binding settlement agreement (actually referred to as an "*informal settlement agreement,*" when consummated at the Area Office level within the contest date and without a contest in effect), you can be respectfully assertive as long as you are not boorishly aggressive. I remind you to heed the warnings that I brought to light earlier concerning how the flaunting of knowledge and other indiscretions and demonstrations too frequently slip out during the inspection. These attempts at defense or some degree of justification can backfire with powerfully adverse results. Of greatest concern, they can strengthen OSHA's contention that a "willful" existed. In fact, it is conceivable that OSHA's documentation for the "willful" nature of the alleged violation was on thin ground as to legal sufficiency. Now, at the informal conference, you have provided a better foundation for their classification. Further, you may have entered the informal conference to discuss a "serious," but unwittingly talked yourself into a "willful." The Agency seldom upgrades a classification after a citation has been issued, but they are not prohibited from doing so when new information has been introduced. Later in this chapter, I address arguments to offer against the designation of particular citation classifications.

Bridle any inclination toward arrogance and express your thoughts and convictions in a coherent and determined manner. It is your right to make your case. It is your right to attempt to rebut the Agency's position to this point. Be a good active listener. You may even learn something.

Now that you are in the right frame of mind, the OSHA authority will explain all rights and procedures, emphasizing the time restraints for settlement agreements and contests. You should have heard virtually all of it before at the closing conference. Be sincere. Bring something to the table. There are many valuable and ethical negotiating tools and concepts available to you. Be thoroughly prepared to make specific and logical arguments. If need be, bring prepared notes with you. These notes are not for giving to OSHA personnel, but to jog your memory and assure that you do not forget your key points.

You should, however, bring a letter to present for the OSHA file. It can serve as a combination of a well-organized progress report and a list of citation and item-referenced respectful arguments for modifications. It will be impressive if, after a short period following receipt of the citation(s), you can show that substantial progress has been made toward abatement. Of course, in the majority of cases, this will only be the first of a series of progress reports. The believability of your compliance claims can be enhanced by the inclusion of photographs. Exercise care to assure that apparent violations, even those not cited by OSHA, do not appear in the background of the photos.

Citation Modifications

Regarding your arguments for citation modifications (which may include claims that OSHA did not know of or did not believe, at least prior to the citation issuance), there are advantages that the letter has over a verbal recitation alone. You will not have to worry about whether or not the OSHA authority is writing down all of the good points that you make. As a corollary, you will have relieved the OSHA authority of the tedious and tiring task of taking all of those notes; he or she can refer to the included letter in his or her informal conference notes. The letter will become an integral part of the official file, and while I am not endorsing brevity for brevity's sake, the meeting can proceed much faster. The signature at the bottom of the letter, made by a high management official of the company, adds credibility to your claims. Even if the letter is not notarized, it is a certification of a sort. You can point out that you are aware of the severe penalties that may be imposed against an individual who lies to OSHA. Thus, there is no misunderstanding. Your company stands behind the contents of the letter.

The modifications that you can request include deletions, reductions in proposed penalties, change of classification, and abatement date extensions. You might also request different wording in the charging language if you feel that the original wording misrepresented the actual conditions, even if there was a true violation. For instance, this can relate to the referencing of an incorrect machine number, an overstated volume of chemicals, an "unguarded" moving part of a machine where "inadequately guarded" would be more accurate, a mezzanine with "no" perimeter protection where "partial" or "insufficient" would be more accurate, or an overstated electrical voltage. For another example, the charging language in the citation might indicate that "all employees were not wearing" industrial eye protection with side shields in a certain area or when operating certain machinery. That can be misleading and, as with the prior examples, anyone reading the citation, including by press release excerpt, can get the impression that the company had placed employees at greater risks than were actually the case. More than likely, the eye protection language should have indicated that "not all employees were wearing" the protection. The difference can be significant.

You could request that one cited standard be replaced with a different standard. This could be because you believe that the incorrect standard was used. The change would be particularly beneficial if the replacement standard would have the net result of requiring less burdensome abatement and/or specify a type of hazard that carries a less negative aura. For example, you would rather be cited for an unguarded "floor hole" than for an unguarded "floor opening." You would rather be cited under one of the general machine guarding standards than under the standard for unguarded mechanical power presses. That mistake is made occasionally when OSHA incorrectly characterizes a machine as a mechanical power press. In that case, not only can the Agency hold you to a greater burden of abatement, but several other standards that are only citable for mechanical power presses can kick in. As one more example, you may have been cited for "exposed live electrical parts," although you should have been cited for "locations where electrical equipment would be exposed to physical damage."

Settlement Agreements

Attempting to arrive at an informal though binding settlement agreement that is mutually acceptable to all parties—OSHA, the employer, and where relevant, employee representatives—is not like purchasing an automobile. However, the art of negotiation does come into play. Although the OSHA authority can simply refuse to bend any further (whether or not he or she would have been justified to do so), you should not merely present your case and await an offer. Do not refrain from actively participating in the determination of the specific elements of the potential settlement agreement. Ask for more than you think you will be given (e.g., more penalty reductions, more classification reductions, and so on). Without being totally ridiculous, do not worry too much about insulting OSHA authorities. You might be surprised at their willingness to accede to your request or to find your request only a bit out of line. If you make requests for modifications and agreement is reached too rapidly, you may rightly feel that you started with needlessly conservative, preconceived ideas. It is preferable to persuade the OSHA authority to make the initial settlement proposal. Barring exceptional circumstances, do not accept OSHA's first offer, even if it is better than you anticipated. Do not rush to a decision. Pause. Slow down. Consider improvement of the deal. By all means caucus with your group, even when you are pleasantly surprised by the OSHA authority's generous proposal. It is very possible that he or she expects you to seek a better deal and that OSHA's bottom line has not yet been reached. You are involved in a parley, not an execution or an inquisition.

The negotiating process can be uncomfortable, frustrating, and downright unpleasant. However, you can minimize these negative feelings if you are well prepared and have practiced how you would present your case. Try to anticipate the reactions of the OSHA authority and how you would counter that individual's rejections of your arguments or the introduction of new arguments against your position. Rehearse with a person who will take the side of OSHA. Take all necessary paperwork, including not only documentation that you may wish to provide or refer to but also your notes. Be certain that the paperwork is properly organized and that you will not have to fumble through it when searching for material.

If your company is to be represented by more than one person at the informal conference, assure that a chief spokesperson is assigned well in advance of the session. I do recommend that more than one company representative be present, but it is of paramount importance that you form a cohesive team with each member having the same good attitude and the same understanding of the facts. This is not the place for quick-tempered individuals. Know your team and be sure that they are in harmony. You must present a united front. One of the group members may have more technical knowledge than the others, but there must be no conflict in your stories. Avoid bringing someone who has been away on a vacation for an extended period and who is to meet you at the meeting without having been thoroughly brought up to speed. Assure that all members of your team are well-rested, focused, sharp, and not in a rush. Do not complete a negotiation if one of your group is tiring, has become too tense or is otherwise not at peak performance. You may try to iron out a few minor details, but ask to return another day, or after lunch, for the continuation of the meeting.

Even if you are accompanied by a professional safety consultant, assure that the individual has a clear, comprehensive grasp of all relevant facts. Do not allow the consultant to simply join you at the informal conference and receive a briefing ten minutes before the meeting. The consultant might have many questions that you cannot answer at that time. Further, with adequate time, the consultant might have been able to bolster your case with information gained from research of standards, interpretations, case histories, technical information, and so on. You should make every effort to thoroughly acquaint him or her with the case several days before the informal conference. If you have properly acquainted the accompanying professional safety consultant—one who has extensive first-hand knowledge of the OSHA settlement process—it is advisable to permit the individual to take the lead in presenting your case. You will not abrogate your right or authority to be the ultimate decision maker. You are the one who will sign the settlement agreement for the company. I do not recommend that you provide OSHA with legal paperwork authorizing a consultant to sign a settlement agreement in the company's behalf.

Be prepared to display a healthy dose of optimism, born of a positive outlook rather than a cocky disposition. You should leave no doubt that you view the citations and the protection of employees as extremely serious matters. At the same time, you must come across as the personable and honorable individual that you are. Show an interest in all aspects of the citations and not just in the dollar amount of the proposed penalties. Give arguments that are based in easy to follow logic, rather than vague intangibles or visceral attacks on the enforcement system. Lay out those arguments in methodical procession. However, this is not to suggest that you set yourself up to be perceived as an analytical robot. Do not be afraid to reveal your feelings, as well as intellect. Attempt to humanize yourself in the eyes of the OSHA authority, and you may be able to summon some personal empathy or sympathy from him or her. It is helpful, in most cases, if you are seen as an individual and not only a body filling a job slot in a company. Realize that a little sincere charm can help see you through some of the rough spots in the conversation. Other traits you will want to demonstrate are honesty, persistence, endurance, patience, and respect. You can be strong-willed while gracious, firm but fair. Do not interrupt. Be attentive. Do not permit yourself any distractions. There is nothing wrong with a little small talk to break the ice, but then stay on track, only returning to the weather, ball game, or best pizza in town when there is a stretch and coffee break.

Know the value of silence and do not feel compelled to fill gaps in the conversation. This is particularly important when you are weighing your next comments. Do not be hasty. Allow yourself the necessary time to sort out and frame your thoughts. There is no reason to be embarrassed by the period of silence. If you have asked a question, and the OSHA authority is slow to respond, so be it. When there are periods of silence, you might reach for a mint. Avoid the temptation to jump in. Perhaps the OSHA authority was just about to tender a better offer. If you speak too soon, offering a concession, a golden opportunity may have been lost. Additionally, although no deal is final until a signed agreement is executed, it is rather difficult to go backwards in discussions—to withdraw a concession. Do not argue just to filibuster. Resist the all too common instinct of railing against every bit of the citations when you know that there is no *reasonable* argument to make. This type of mistake can destroy your credibility and cause the OSHA authority to raise his or her guard. The barrier will act as an impediment to negotiations. Be ingenious, not ingenuous.

Try to dispense with the more trivial elements of the citations first—the easy stuff. Clear away as much smoke as possible and get to the heavy items last. Show that you can accept reasonable elements of citations and that you understand the Agency's position. Make it clear that you can be somewhat flexible, but that you expect the same respect from OSHA. Nevertheless, if you can offer irrefutable arguments to show that a portion of a citation was plainly in error, then do not decide that you must be flexible on that issue just to demonstrate a spirit of cooperation. If OSHA has made a factual mistake, hold your ground and, in a level tone, explain that there is no reason to compromise on that issue. For example, even if the OSHA authority deletes an item that undoubtedly should never have been cited, no trade-off should be expected from you in return.

Prior to entering an informal conference, you should have established a target. This will consist of the minimum that you will accept in the negotiations and what it is you are willing to give up. This is not something you mention at the meeting—at least not unless you have reached the final stage of consummating the settlement agreement. If you are very close to attaining your goals, communicate your appreciation for the OSHA authority's understanding and the specific modifications that he or she has offered. Then, in a conciliatory manner, explain two things. First, indicate that you are resigned to the fact that you will not attain all that you had hoped for. Second, reason that much time has been invested by OSHA and the company at this meeting, and that it would be a shame to have an impasse when you would be willing to sign the agreement if a little more ground was given by OSHA.

Refrain from buckling and offering large concessions when you are close to an agreement. Do not throw in the towel simply so you can get back to the office. In fact, over the course of the meeting, your offers of concessions should steadily decrease. Each of those concessions should be presented slowly and with an air of reluctance. Attempt to place a value on every one of your real concessions (when not disputing a minor item for which you had no logical defense whatsoever, that is not viewed herein as a concession), including those of the virtually inconsequential variety, offered as part of a final trade solely to close the deal. No concession is offered for free. Ask what will be given in return. On the other hand, if you are faced with an unreasonable demand, do not feel the need to give a counteroffer. If the OSHA authority does not reduce the demand, then suggest an equally absurd concession on the part of OSHA. Ideally, this type of awkward game-playing will not surface. Yet, you must refuse to be bullied. This does not mean that you should be a bully. Although you may engage in a fair amount of nibbling as the negotiations near a successful conclusion, I suggest that you show some sensitivity and do not absolutely insist on one more concession by OSHA, simply so you can gloat. Although not likely, the whole deal could collapse at that point. More importantly, the settlement agreement and the informal conference may well be only a part of your continuing relationship with OSHA. There may be further interaction with the Agency and it is wise to have developed a cooperative relationship and a climate of trust.

At any given point, the OSHA authority can refuse to give another inch, in which case your basic decision will be whether to sign the on-the-table agreement or to file a notice of contest. You always have the third option of doing neither and letting the original citation(s) become a final order of the court. In almost all cases, that would be pointless. After all, unless your refusal to sign an agreement stems from some burning need to make a political statement, why would you neglect either option? Even if a settlement agreement falls short of your goal, it would probably be advantageous to the company when compared to the original citation(s). The OSHA authority may assert that he or she lacks the authority to grant certain requested deletions, penalty deducting percentages, or other modifications. This may well be accurate. If the decision can be made only by the regional administrator, then ask that that person be contacted. Explain that you can wait or be contacted later, as long as the contest period is still in effect.

If you are not going to contest, how is it that there could be any time when signing a settlement agreement would be a poor choice?

As an example, the settlement might grant minor reductions in penalties, which would no longer be viewed as proposed, but at the expense of your contractual agreement to guarantee costly above-the-standards commitments. Companies have been too eager to sign settlement agreements, which would have contained "give backs" including even significant penalty reductions, without fully comprehending the long-range fiscal impact of the full package. The problem was that the tradeoffs would have carried very high nondirect costs associated with the company's new obligation to file numerous, detailed reports with the OSHA area office, provide a great deal of formalized training not required by the standards, and assure redundant and/or particularly production-cramping means of abatement that were not necessary to bring the company into compliance. There is a special consideration regarding settlement guarantees, that the employer will provide OSHA with copies of inspection reports (self-inspection or third party), safety and health committee notes, or other documents that may indicate possible violations or hazards. Even if all such deficiencies have been abated prior to sending the documentation to OSHA, there is the chance of an FOIA request. I again suggest contacting an attorney to help you frame leading words to preclude the release of the information via FOIA. This goal should be achievable.

Remember that, if you sign a settlement agreement, you will be waiving your right to contest. You may wish to take a draft of the Agency's last-offered proposed agreement and return within the 15-day limitation with your final decision. One reason to schedule the initial informal conference early in the period is that there will be adequate time to study any proposed agreement and discuss it with your key staff, if necessary. Remember that despite the last-offer concept, you may be able to win further small concessions from the Agency just to get the matter over with. Prior to the drafting of the settlement agreement, be certain that all loose ends are tied up and that there are no misunderstandings. Carefully read and reread every word of the draft document. If to your advantage, suggest slight modifications for the recrafting of the final version. Details do matter.

The OSHA authority (as a reminder, this generally means the area director, assistant area director, or state equivalent) may point out that he or she does not mind receiving a notice of contest and then passing it on to the U.S. Department of Labor Office of the Solicitor. I mean this in the context of the OSHA authorities trying to impress upon you that they do not care if you reach an agreement and that the paperwork will be out of their hands and moved to the lawyers. It is not that simple, though. When there is a contest, even though the file will be out of OSHA's hair for a while, the area office is far from through with the case. There may be extra compliance officer time spent in scrutinizing and mounting photographs and referencing videotapes, extra clerical time spent in transmittal obligations, and time-consuming interactions with the solicitor.

If the case is not settled by the Office of the Solicitor, the actual hearing before the Occupational Safety and Health Review Commission (OSHRC) can require much preparation time. Invariably, it will require special meetings with the particular solicitor(s) involved. More than likely, the compliance officer and assistant area director will have to answer long written interrogatories presented by the employer's counsel. Further, there may have to be a search for expert witness(es). The testimony time can tie up OSHA personnel for several days.

From a different angle, although a high contest rate is not necessarily a negative indicator for an area office, it can certainly prove to be when federal auditors become involved. They might find that the rate was inflated not due to a fundamental pattern of firm but fair enforcement but due to a trend of unreasonably high proposed penalties, nitpick citation items, poor documentation, questionable procedural methods, and/or the lack of any realistic efforts toward settlement.

Thus, for the most part, the area office would prefer to reach settlements as long as exposed employees will be afforded (in a timely manner often thwarted by the contest process itself) the protection to which they are legally entitled, and the employer will receive the proper message and be keenly motivated to avoid continued or repeated violative behavior. Therefore, enter any settlement discussion with the knowledge that the Agency itself has a stake in the case and that the OSHA authority does not want the file to go up to the solicitors.

One more twist is that the OSHA authority might realize that he or she would be taking a chance at an OSHRC hearing. Not only could a substantial portion of OSHA's case be lost, but the resultant court record documenting the decision and the reasons for it could be called upon in subsequent cases to support claims that similar OSHA citation(s) were faulty. Such precedents do get overturned rather frequently, but any establishment of a written record favoring employers can only be viewed as something to be avoided.

On the other hand, an employer may agree to comply with the citation(s) if given a fair shake by way of an informal settlement agreement. Settlement language can also call for the employer to agree to abate a particular situation, even though the related item(s) are deleted. In that case or even if the item(s) are deleted without special settlement wording to guarantee abatement, there is no judge's decision that can be later held up to the judicial light by a defense attorney in a later OSHRC matter. It may be a case of bluff, or who blinks first, or when OSHA will choose to fight a particular battle over a major controversial issue.

To what types of special prosafety and health steps can the employer commit in an informal settlement agreement as part of a trade for reduced penalties, deleted items, reclassifications, extended abatement dates (again, not usually too much of a stumbling block), or other breaks? In most cases, timetables, including starting dates, for the various elements to be implemented will be incorporated into the settlement language. The promises should not be accepted by OSHA if they are vague. Do not overextend yourself and do not sign open-ended commitments (e.g., ones that are too broad and/or do not give a termination date to the particular constituent of your commitment when one would clearly be appropriate). Do not accept vague defined-as-you-go "arguments in principle." Do not make promises that you (i.e., the company) cannot keep. The commitments you can offer include the following:

employee training of specific types, durations, and frequencies;

the hiring of a qualified safety and/or industrial hygiene consultant and/or ergonomist (generally including how soon the services will be engaged and at what intervals for what period of time those services will be utilized);

the hiring of college co-op student(s) to assist in the development, implementation, and/or monitoring of your program;

the hiring of a full-time safety director, industrial hygienist, or similarly titled person (do not get cute by hiring and separating within a short span of time);

specific improvements to be made in the comprehensive safety and health program;

the formation of, or the addition of specific desirable conditions to, a safety and health committee;

the addition of a dedicated occupational safety and health block or column to the company's written evaluations for management personnel (so that managers' safety and health performance can influence their pay, position, and benefits);

enrollment and successful completion of occupational safety and health courses (possibly general in nature, but depending on the results of the OSHA inspection, the courses may have to deal with particular subject matter) presented by accredited colleges, any location of the OSHA Training Institute, or by individuals certified by the OSHA Training Institute to instruct in the particular course (assure the inclusion of wording to protect the company if, despite your early applications, there are no available openings; in the case of the OSHA Training Institute, the OSHA area office can assist you in being accepted);

the completion of certain occupational safety and health certifications;

memberships into societies, associations, and similar groups that are largely concerned with promoting the improvement of occupational safety and health;

the written translation of the overall safety and health program, or specific portions of it, into particular languages other than English;

the hiring of interpreters (not necessarily full time) to explain occupational safety and health rules, practices, and so on in languages other than spoken English (i.e., verbally explain in foreign languages and/or use American Sign Language where appropriate);

the purchasing of videotapes, magazine subscriptions, or other training aids (subject matter limited to occupational safety and health in general or locked into particular subjects);

an invitation to OSHA personnel to speak at the corporate meeting.

In big money cases, OSHA might be willing to negotiate quite a bit if a corporate-wide settlement agreement is to be signed. Such proposals, once committed to writing, must be studied in great detail. They may require, and it would be advisable to secure, the signature(s) of individual(s) of greater authority in your facility or in the corporation than those who might normally sign a settlement agreement. If a corporate-wide settlement agreement is signed, the Agency would notify all relevant OSHA offices (i.e., all offices where your corporation does business). Beyond that, there would be a clear record of the agreement on the OSHA CD-ROM. All of your facilities, anywhere in OSHA's jurisdiction (including the American territories that are not part of the 50 states and the District of Columbia), should be notified. Do not be hasty in entering into such an agreement. If you go for it, leave plenty of lead time to assure that all of your facilities understand exactly what is required of them and that they will be in full compliance with each and every element of the settlement agreement language.

You may be able to make other proposals. Be creative, but do not put all of your ideas on the table right away. In many cases, you can ask "what if" questions, in an effort to formulate alternative propositions that will be acceptable to both sides. Maybe you can work toward an excellent agreement and then nibble away to obtain a little more relief by throwing in one more commitment. In a few cases, innovative approaches that have led to significant penalty reductions involved the company or corporation providing major funding for occupational safety and health courses presented by technical schools, colleges, unions (a bit of prudent thought is needed here), and similar groups or organizations. In a comparable vein, the company can offer to pay grants for studies or research or to financially back laboratory services, the production of videotapes, the publication of books or pamphlets, a seminar series for the respective industry, and so on. The OSHA authority may have some other suggestions.

Do not bite off more than you can chew. Stress how the fulfillment of your legally binding commitment (that could not have been required by the standards themselves) will enhance the safety and health environment and culture at your company. Be patient in your negotiations. Keep in mind that everyone, even the Agency as an entity, wants to come away feeling something like a winner. Always be sure to emphasize that the most important victors will be the employees that must be provided with safe and healthful workplaces. Be crystal clear about that. Show that you recognize that OSHA and your company are partners in the process of assuring a safer and more healthful workplace. It is all right to explain that straight business considerations are involved in the settlement talks, but keep highlighting your sincere dedication to lowering risks at your workplace. You might even be permitted to guarantee the outlay of a specific amount of money to be poured directly (keep that word *"directly"* in clear sight) into your occupational safety and health program within a certain period of time. No matter what, urge that you be allowed to apply a large amount of the original proposed penalty to your abatement efforts and related activities rather than simply give it to the government as a form of punishment.

There are two more ideas that you can offer for consideration. They are on a totally different track than the others that have been addressed. You can suggest that the abatement periods be even shorter than those that the OSHA authority had already expressed a willingness to grant. Be careful. While understanding that you may later petition for the modification of abatement dates, OSHA will obviously take a harder line in response to such a request if you are now trying to use shorter abatement periods as a negotiating tool. Figure that you will not have to request extensions unless there is a completely unforeseen occurrence that causes unpredictable delays. Drive home the point that removing employees from exposure earlier than would normally be allowed must surely be worth a further reduction in penalties.

As an offshoot to this idea, how about offering (again, as a part of the settlement language) to forgo the easily granted installment payment plan (over many months, where relatively large penalties are involved) to render full payment within a certain, uncommonly short period? You may have to expound upon your logic. Tell OSHA that the area office will be able to avoid all of the computer tracking, file tickling, and tedious check recording and transmittal. To boot, the government will have the full amount longer, even though government profit is not the point, and the file can be closed at an earlier date.

Citation Reclassifications

I have discussed attempts to gain citation reclassifications. Most commonly, employers seek to have "willful" reduced to "serious." There is also a desire to modify a "willful" to a "repeated." That can result in a significant lowering of the penalty, but it is not necessarily the case, particularly if the standard or a substantially similar one has been violated repeatedly. The -*ly* suffix can be critical in the upholding of the classification by a court decision and in the introduction of a greater coefficient to be applied to the gravity-based penalty. Therefore, a "repeated" (or "repeat") may carry an even higher proposed penalty than a "willful."

The excess stigma attached to a "willful" is another story. Many companies have opted to gain a reduction in classification, generally down from "willful" to "repeat," while agreeing to accept a larger penalty. Their main objective is to be purged of the adverse public perception attached to a "willful," regardless of the associated penalty or even which standard was involved.

The Agency generally offers a curious alternative to a "willful." Upon acceptance of the alternative, the company must agree to pay the entire original penalty or a sum that has only been reduced by a small percentage. Further, the company will be required to make significant added concessions of the more burdensome type that I have outlined. If OSHA is satisfied and the employer is amenable, the classification will be changed to a violation of section 17(a) of the

Occupational Safety and Health Act. That paragraph states, "Any employer who willfully or repeatedly violates the requirements...may be assessed a civil penalty of not more than $70,000 for each violation, but not less than $5,000 for each willful violation." If section 17(a) is invoked, there is no record of a "willful" violation per se, even on OSHA's management information system. Any possible adverse perception is at least clouded. In this case, if a company is asked by any group, branch of government, or other company that is putting out bids, the subject company can fairly state that it has not been cited for a "willful" violation.

At the employer's request, in most cases, exculpatory language will be added to the settlement agreement. This is essentially a nonadmission clause. It has little legal force, but the employer can point to that paragraph when defending a third party suit or when communicating with the media. The exculpatory language will not prevent the citation items from being used as a foundation for later OSHA actions and classifications, most notably the citing of alleged "repeat," "willful," or "failure-to-abate" violations. In all cases, the settlement agreement will include the employer's agreement to post a copy of the document, as was required with the original document, and a statement that if the amended penalties are not remitted as agreed they will revert to the original amount plus any accrued interest, delinquent, and administrative fees.

Contests

OSHA does not enter into informal settlement agreements that only address portions of the citation(s). You cannot obtain a partial, precontest settlement agreement and then contest the remaining issues. Of course, if a contest reaches the formal hearing stage, the parties may stipulate at the outset to specific agreements concerning modifications and/or the lack of challenge by the employer to certain elements of the citation(s).

If the last day of the contest period has arrived and you cannot decide whether or not to sign the settlement agreement, it is advisable to file a written notice of contest. The notice (i.e., the letter) need not be lengthy. Occasionally, due to some emergency that crops up, or some logistical problems, the germane company officials and/or OSHA authority become unavailable to nail down all of the i-dotting and t-crossing on time. Again, you should contest. The OSHA area director, while accepting the notice of contest, can provide details of the circumstances to the regional solicitor. That office can then release OSHA to complete negotiations and arrive at terms of settlement that will be acceptable to all parties. Subsequently, those terms will be forwarded to the Office of the Solicitor, where a formal, postcontest settlement agreement will be drafted. Once the 15-day contest period has elapsed, even in the case of such good faith and near-settlement status, OSHA cannot consummate a settlement agreement on its own. Although the terms have been worked out by the OSHA area office, it is in the purview of the Office of the Solicitor to complete the documents and sign off on them. On occasion, with special permission, some regional solicitors have allowed the area director to sign a postcontest settlement agreement.

These formal postcontest settlement agreements must also be posted. In virtually all instances of this type of settlement, an attorney should not be needed. There is no formal hearing. No testimony is heard in any court or by deposition. The format of the settlement is basically the contents of an informal settlement agreement set forth in a more formalized document.

The regional solicitor will inform the Occupational Safety and Health Review Commission (OSHRC, or the Review Commission) of the settlement. OSHRC has the authority to review all settlements reached after the filing of a notice of contest. It can disapprove settlements that it determines are inconsistent with the purposes of the Act. That power is very rarely exercised. The type of formal settlement agreement worked out by an OSHA area office will generally occur before the complaint submitted by OSHA (more precisely, by the Secretary of Labor, who is the complainant of record) was filed with the Review Commission, in which case no administrative law judge has yet to be assigned to the case. In such circumstances, it is often the chief administrative law judge who will review the settlement. If the complaint has been filed and the answer (submitted by the employer, who is the respondent) has been filed in turn, a specific administrative law judge (ALJ) will be assigned to preside over the case. Thus, if the settlement reaches OSHRC after the receipt of the complaint and the answer, the presiding ALJ will be the one to approve or disapprove.

OSHRC is an independent agency of the United States government and is not connected in any way with the Occupational Safety and Health Administration or the U.S. Department of Labor. It is composed of three commission members who are appointed by the President of the United States for 6-year terms, and it employs administrative law judges to hear cases. Those judges are assigned to cases upon docketing. OSHRC is a fact finder, and the ALJ serves as an arm of OSHRC for that purpose. OSHRC's only function is to resolve disputes that arise under the Occupational Safety and Health Act.

If you do not file a written (verbal will not suffice) notice of contest within the 15 working days following receipt of citation(s), then those citation(s) become a final order of OSHRC, and the employer has no further recourse for review by any court or agency. That is the basic language in the Occupational Safety and Health Act. Nevertheless, there have been a handful of exceptional cases in which OSHRC has granted relief to employers from the 15-day limitation and permitted late contests. OSHRC deemed that this course of action is proper where the delay in filing was caused by OSHA's deception or failure to follow proper procedures. Some examples of situations that can result in the extension of the 15-day contest period include the compliance officer telling the employer that an informal conference serves as a notice of contest and silently participating in a miscalculation of the date by which a notice of contest is due and informing the employer of an incorrect (i.e., miscalculated) date by which the contest is due. Note that these situations are remarkable, and as long as OSHA personnel have not misled or improperly confused an employer as to how to file a notice of contest, OSHRC will not extend the contest period. A compliance officer should document in his or her case file the explanations of contest rights that were given to the employer. This is of paramount importance particularly when dealing with an apparently unsophisticated employer of a very small company, where that employer lacks legal counsel and is involved in his or her first interaction with OSHA.

The notice of contest need not be on any particular form or follow any particular format. It must clearly identify the employer's basis (although briefly—you need not get into the details of your defenses) for contesting the citation(s) themselves, notice of proposed penalty, abatement period, or notification of failure to correct a violation (still widely identified as a citation with a "failure-to-abate" classification). For instance, it should be specific as to which items, penalties, and dates are being contested. I have already addressed various lines of defense. In the final chapter, I place these in perspective and introduce several more strategies and legal arguments to explore.

For uncontested portions of the citation(s), the abatement requirements must be fully met, and the employer must pay whatever penalties were proposed for the related items. The abatement requirement is suspended for contested portions of citation(s) until the Review Commission issues a final decision, unless the contest was filed solely for delay or avoidance of penalties. The interpretation of the motivation is not always an easy task. Legislation has been proposed to require abatement under certain circumstances, even while a contest is in effect. This type of legislation could be passed for situations of imminent danger (where a temporary restraining order may have been issued by a court, anyway) or for circumstances in which clear OSHA documentation points to extremely dire hazards that can be abated at relatively small financial costs and in a relatively short period of time. On the other hand, there is nary a chance of the enactment of legislation obligating across-the-board abatement for contested items still in contest, where the abatement could be removed if the items are deleted by the Review Commission. The unrecoverable effort and cost put forth by the employer in such cases could be staggering. There are also a plethora of legal challenges available.

A copy of the notice of contest must be given to the employees' authorized representative. If any affected employees are not represented by a recognized bargaining agent, a copy of the notice must be posted in a prominent location in the workplace or else served personally upon each unrepresented employee (a most rare practice). Within 15 days of the receipt of the notice of contest, the OSHA area director forwards the case to OSHRC, which (as explained earlier) assigns the case to an ALJ. The judge may disallow the contest if it is found to be invalid (very seldom the case), or may schedule a hearing in a public place that is usually near the employer's workplace (may be a few hours' drive, however).

After OSHRC has been notified of the contest, the employer will receive a notice from the Review Commission that the case has been filed. Along with that, the employer will receive forms to use for notifying affected employees and their union, if any, that a Notice of Contest has been filed. The employer and the employees have the right to participate in the hearing. There is no requirement that they be represented by attorneys.

I suggest that, at this stage, it is prudent (certainly for the employer) to engage the services of an attorney. This is but one factor to keep in mind when considering whether or not you will contest. There will be less need for an attorney if you can arrive at a settlement with the Office of the Solicitor without having to appear before an administrative law judge, but the stakes are still very high and each line of an offered settlement must be studied and understood. In fact, the significant majority of cases are settled prior to the hearing stage. You may try to settle at that level and, if the case will proceed to a formal hearing, then hire an attorney. In any event, consider the costs of an attorney and of the time to be expended by whomever you wish to have testify. You may even call in expert witnesses. If you contest, particularly if a formal hearing does take place, it will be a very time-consuming process.

For the most part, informal settlement agreement offers are made by OSHA at the informal conference. In some cases, the proposed modifications are distinctly minor or, in the opinion of the employer, scant. If you contest, when OSHA sends the file to the Office of the Solicitor, it will include a transmittal which details the last offer proposed by OSHA (i.e., the best offer rejected by the employer) and the major stumbling blocks to a mutually agreeable settlement. The transmittal may also describe any weaknesses in the Agency's case. This information is usually sent within a format branding the memo as an internal document between client (i.e., the OSHA area director) and attorney (i.e., the regional solicitor), so that it is confidential in theory and not releasable through the Freedom of Information Act or by way of discovery. Your attorney might be able to argue successfully against the efficacy of OSHA's position on this matter.

At this juncture, employers often believe that, when the case reaches the Office of the Solicitor, the review and evaluation will officially begin with the last proposed settlement. In other words, they think that they have already, in effect, had the severity or scope of the citation(s) mollified. This is incorrect, although in the circumstances whereby OSHA and the employer were very close to signing a settlement, simply ran out of time, and then the settlement was effectuated by the regional solicitor, it is a depiction that is fair even if not legally accurate. Even if you had engaged in several hours of tedious negotiations with the area office and despite the fact that the OSHA authority explained the last offer to the employer, the whole matter now reverts to the original citation(s). Note my earlier comments about OSHA only entering into full informal settlement agreements. It is back to square one.

OSHA's complaint must be filed within 20 calendar days of the date on which the Review Commission received the employer's notice of contest. A copy must be sent to the employer and all other parties in the case. The complaint sets forth, in detail, the alleged violation(s) for which the employer received the citation(s). It must state the basis for jurisdiction, the time, place, and circumstances of each contested violation, the basis for the proposed penalties, and the factors upon which the proposed abatement period and proposed penalties are based. It must show that those dates and penalties are reasonable. The employer's answer must be filed with the Review Commission within 20 calendar days after receipt of the complaint. A copy must be sent to the Secretary of Labor and any other parties in the case.

There may be amendments to citations and to pleadings. Amendments are frequently made to OSHA's complaint and may occur before, during, or after the hearing, or during review by the OSHRC provided that the employer is not prejudiced by the amendment. In addition, informal prehearing conferences may be conducted at any time prior to the hearing if ordered by the ALJ or upon motion of any of the parties. Generally, the purpose of the prehearing conference is to define and narrow unresolved issues in preparation for the hearing. After the prehearing conference, the ALJ will issue a prehearing order. This order details agreements that were reached between (or among) the parties and any prehearing rulings governing the conduct of the formal hearing to follow. Prehearing orders may require the parties to submit the names and resumes of witnesses and to list documents or audiovisuals to be introduced into evidence at the hearing.

Simplified Proceedings

If all of the parties are in concurrence, they may collectively elect to have the contest heard under the Review Commission's simplified proceedings rules. Otherwise, conventional proceedings will be in effect. Simplified proceedings exclude the use of formal pleadings, and they eliminate discovery, including requests for admissions (except upon special order of the ALJ) and the application of the federal rules of evidence. In the first stage of the procedure, there is a conference attended by all parties before the ALJ. This is where (ideally) some issues are resolved, and others are at least clarified.

During the second (last) stage of the simplified proceedings, the unresolved issues are tried before the ALJ. The parties may present testimony for the record and documentary evidence. Oral argument is allowed at the close of the hearing, generally in lieu of posthearing briefs. Simplified proceedings are frequently processed more quickly, and they tend to encourage settlement because of the informal manner in which they are conducted.

One reason for choosing simplified proceedings is because the time and expense of going to court will probably cost more than the penalty. Another reason is that, even if the penalty is not too high, the available abatement means will require a significant sum of money to complete. This can relate to the direct costs of planning and design, parts and materials, labor, and possibly the associated costs of complex local or state permitting processes. You should also figure in downtime, when equipment must be removed from service for long periods. In that case, if your orders cannot be held up that long or for other reasons, you may have to contract out work temporarily, accruing another expense. One more concept is that, even where the penalty and abatement means, including indirect costs, are not extreme, you may feel that once compliance has been achieved the operation will be slowed enough to severely cut into your profit and/or cause you to lose a competitive edge. While that is very seldom the case, it is conceivable. All of these considerations regarding abatement can, in fact, be the very reason(s) why you contested in the first place.

Simplified proceedings are not always an obvious choice. At any rate, not all cases are eligible for simplified proceedings. They are designed for cases that are not of a complex nature either legally or factually. If you have contested the citation(s) only because you believe the penalty is excessive, you should consider simplified proceedings. There are other circumstances in which very few issues are involved, such as if OSHA alleges that the guard in place on a relatively simple and well-known type of machine does not adequately protect the exposed employees within the meaning of the standard. Again, the simplified route may be your best choice.

There are certain types of cases, however, that are not generally allowed to be litigated by this route. They include complicated and highly sophisticated technical issues. Common examples include noise exposure, ventilation, radiation, and various matters of exposures to toxics. Citation(s) alleging violations of section 5(a)(1) of the Act are also not suitable. If you believe that you will require the testimony of expert witness(es) to establish your case or that you will need to obtain information in the possession of another party in order to present your case in a suitable fashion, then simplified proceedings are probably not the way to go.

Whether simplified proceedings or conventional proceedings are in effect, the Review Commission encourages the settlement of cases. Cases can be settled at any stage of the contest proceedings. OSHA and the employer must agree to the settlement terms. Additionally, the affected employees (or their union or other embodiment of employee representation) must be supplied with a copy of the proposed settlement for 10 days before that settlement will be approved. They may only object to the reasonableness of abatement periods.

Conventional Proceedings

The administrative law judge presiding over the hearing will enable the parties to present evidence on the issues that have been framed by the complaint and the answer. OSHA has the burden of proving that the citation(s), including penalties, are correct. This proof must be by a preponderance of the evidence. It need not be by absolute certainty. The contesting employer answers the allegations. Each party to the proceedings may call witnesses, introduce documentary evidence, and cross-examine opposing witnesses. The ALJ may only consider properly introduced evidence in reaching his or her decision. Although the hearings are essentially governed by the federal rules of civil procedure and the federal rules of evidence (the state review system should have similar rules), the ALJ (through the OSHRC) is far more liberal in admitting evidence. Nevertheless, attorneys on both sides make objections, and those objections are either sustained or overruled. So, despite the less restrictive approach to the admission of evidence, you will know that you are in a court of law.

As prescribed by the code of federal regulations, the duties of a presiding ALJ are to:

administer oaths and affirmations;

issue subpoenas and rule on petitions to revoke subpoenas;

hold conferences to simplify issues or to effect settlement;

regulate the orderly conduct of hearings and strike testimony for witnesses' failure to answer proper questions;

rule on motions to dismiss complaints;

make rulings on evidence, including the use of depositions;

rule on procedural motions, including those referred to him or her by the OSHRC;

rule on motions to amend pleadings;

order hearings reopened;

order consolidation of hearings;

make decisions pursuant to the Administrative Procedure Act;

call and examine witnesses;

introduce evidence, including documents, into the record;

request statements of parties on their position on any issue at any time during the proceeding;

adjourn a hearing as justice and good administration dictate;

take all other actions necessary to the exercise of the specific powers as authorized by the rules of the Occupational Safety and Health Review Commission.

A transcript of the hearing will be made by a court reporter. A copy may be purchased from the reporter.

After the hearing is over and before the ALJ reaches a decision, each party is given an opportunity to submit to that judge, for study and consideration, proposed findings of fact and conclusions of law with reasons why that judge should decide in its favor. There is never a jury. The decision will be made by the ALJ. That decision must be rendered in writing, and must include the findings of fact, the conclusions of law, and the reasons or basis for those particular decisions.

Once the decision becomes final, do not neglect to watch the new abatement dates carefully. If for some overwhelmingly convincing reason further extensions are needed, be sure to send in the proper, detailed paperwork. The general practice is for the OSHA area office to be back in the decision-making role. In any case, even if the whole matter has been in limbo for several months, resume the filing of your detailed progress reports.

Appeals

Once the ALJ has ruled, any party to the case may request a further review by OSHRC. Any of the three OSHRC commissioners also may, at his or her own motion, bring a case before the commission for review. Commission rulings may be appealed to the appropriate (circuit) U.S. Court of Appeals. Further, a small number of OSHA cases have reached the U.S. Supreme Court. States with their own occupational safety and health enforcement programs have a state system for review and appeal of citation(s), penalties, and abatement periods. The procedures are generally similar to federal OSHA's, but cases are heard by a state review board or equivalent authority.

There are many, many intricacies incorporated into the different types or levels of appeal. These include, for instance, restrictions on the scope of judicial or OSHRC final orders. As only one example, appellate review of penalties assessed by OSHRC is limited to whether the penalty represents an abuse of discretion or is arbitrary and capricious. As it is not my intention to present a complete in-depth explanation of every facet of the contest process, including the total workings of the Review Commission, I do not delve more deeply into this matter.

An exhaustive study of this overall legal process would require an extremely lengthy and detailed account. If you choose to become a student of this subject, you may want to conduct research into procedures, regulations, and the history of cases and court decisions including the following: provisions for the withdrawal of employer contests, the discovery process, interrogatories, requests for admissions, requests for production of documents and entry upon land for inspection, subpoenas, protective orders for trade secrets, discovery of the names of informants (generally referring to complainants), rules of evidence, and the enforcement of discovery orders.

Enforcement

OSHRC does not have the authority to enforce its orders. Those orders are to be enforced by the U.S. Court of Appeals. OSHA may request and obtain enforcement of OSHRC orders by petitioning the court in the circuit where the violation occurred or in which the employer has its principal offices. If the employer did not file a proper and timely petition for judicial review (certainly not required in itself), the OSHRC order will be determined to be final and binding, and the order will be enforced. Unless the employer does file for judicial review, I advise in the strongest terms that the intervention of the U.S. Court of Appeals be avoided. If, through some grievous and foolhardy error on the part of the employer, the U.S. Court of Appeals has entered into the matter, the swift reversal of that blunder will be essential.

33

Defenses to Citations

Interpretation of the Occupational Safety and Health Act and its enforcement by OSHA are steadily evolving processes. Some areas have been clarified by the judicial process, whereas other questions have not been raised or have been decided on both sides of an issue. Any of the defenses raised in this chapter may be profitably mounted up during an inspection and/or at an informal conference, as well as at a formal hearing. There are many grounds upon which to mount legal defenses to OSHA citations. Broadly defined, this reference to citations includes penalties and abatement dates. The validity of the Agency's allegations can be challenged from various strategic angles.

However, numerous arguments have been used unsuccessfully by employers. Those arguments do not fall within the acceptable scope of those challenges. OSHRC has routinely rejected various inconvenience defenses. These include such things as difficulty of compliance, difficulty of maintaining the physical abatement means, interference with production or other work, and impracticality.

Low Win Percentage Challenges

You may offer constitutional defenses. One that is most commonly attempted involves search warrants. Now that OSHA and the issuing U.S. magistrates have gained significant experience in this area, it is highly unlikely that a defense in this category will be successful. Nevertheless, as with other low win percentage challenges that I will list, it is important that you understand their availability. You may raise the issue that the inspection was performed without consent of the employer. The courts have found that such consent did not have to be expressed. In other words, a failure to object to the inspection generally constitutes consent, albeit passive. Maybe the individual who welcomed the compliance officer into the facility did not hold a true management position. If so, you might claim that this individual did not understand the implications of an inspection and that the compliance officer should have waited to meet with a person in real authority, who had a major stake in the inspection and would not feel so intimidated as to allow entry.

Another defense alleges that your facility was selected for inspection by way of a discriminatory method. I have addressed inspection scheduling, particularly in relationship to general schedule inspections. Unless the OSHA area office has inexcusably erred in following its internal guidelines, you should not have much of an argument on this one either.

Another constitutional argument alleges the violation of due process. This can relate to the vagueness of a standard, a subject that also falls under another category of defenses. As concerning due process, the employer may contend that a standard that is somewhat general in nature was sufficiently vague so as to fail to provide adequate notice of the conduct required. OSHA must prove that the employer had actual or constructive notice of just what the standard required. The language must be explicit and reasonable. If the language was not adequately explicit, then constructive notice may be established by considering the following factors: industry custom and practice, the injury rate for that particular type of work, the obviousness of the hazard, and the interpretation of the regulation by OSHRC as evidenced by its history of decisions.

OSHA's Burden of Proof

Moving on to different types of defenses, OSHA must not fail to meet its burden of proof on elements of violation. From a general viewpoint, when alleging violations of standards, OSHA must prove the following:

1. The standard applies to the cited condition.
2. The terms of the standard were violated.
3. At least one of the employer's employees had access and potential exposure to the cited conditions.
4. There was an actual hazard.
5. There is a feasible and effective abatement method.
6. The abatement date is reasonable.
7. The employer knew, or with the exercise of reasonable diligence could have known, of the violative conditions.

OSHRC will vacate the citation if OSHA has not met its burden of proof on any one of these elements. In earlier chapters, I provided several examples of where OSHA's documentation might be faulty or might simply lack sufficiency.

The Particularity Requirement

Under constitutional defenses, I explained the concept of the unenforceably vague standard. Next is the defense of lack of particularity of a citation. Citations must contain a description of the alleged violation and a reference to the standard allegedly violated. The citation must be reasonably specific in apprising the employer of the alleged violation so that corrective action may be taken and so that the employer may be able to decide whether to file a notice of contest and, if so, how to proceed with the contest. The employer need not be informed by OSHA as to exactly how the hazard should be abated. A citation will only be dismissed when it does not contain a description of the alleged violation or a reference to the standard allegedly violated.

The reasonable particularity requirement applies to descriptions of the locations, measurements, and nature of alleged violations. The citation must identify which of its provisions have been allegedly violated. Employers, for the most part, will only win on the grounds of lack of particularity when they can demonstrate prejudice to their ability to defend themselves. You should not count on fabled loopholes, such as the incorrect (or no) inspection date or a misspelled street name, to give cause for the vacating of a citation.

Reasonable Diligence

For a "serious" violation to be upheld, OSHA must prove that the employer "knew, or with the exercise of reasonable diligence, should have known of the presence of the hazardous condition." OSHRC has held that OSHA has the burden of proving employer knowledge in both "serious" and "other" violations. OSHA's own literature at times refers to "could have known" rather than "should have known." You can definitely formulate a stronger defense if you reference the "should." That definition would be much less burdensome to an employer. At any rate, "reasonable diligence" has been defined as including terms such as "watchfulness," "caution," "foresight," and "prudent." Recall that a supervisor generally represents the employer, and a supervisor's knowledge of the hazardous condition amounts to employer knowledge. I explained the nuances to this concept when detailing isolated incidents in the chapters dealing with the inspection itself.

As an example that illustrates why it is much easier to defend against the "should" position than against the "could" position, I point to an electrical hazard in a receptacle. Before detailing the hypothetical circumstances, I emphasize once more that I am not an attorney. However, I can fairly say that, through my experience with OSHA (including as an employee of the Agency), I would find the following to be a good argument on behalf of the employer.

The compliance officer might use a circuit tester device of the type whose light patterns when plugged into the receptacle will indicate if there is, for instance, an open ground or reverse polarity. He or she will test several receptacles, even if there is no reason to believe that there is any problem with the wiring. In some cases, the compliance officer has an extra motivation to perform the test. He or she might learn that employees feel a tingle when plugging equipment into or unplugging equipment from that receptacle. He or she might learn that sparks have been emitted from the receptacle. Particular receptacle(s) may not outwardly appear to be in a violative condition, but the compliance officer chooses to test them because a pattern of many open grounds or reverse polarities has been established with other receptacles in the area. If the receptacle looks physically damaged, the compliance officer is likely to test it, although with added caution.

In the added motivation cases (particularly the last one), OSHA is probably in a good position to assert that you should have and could have known of the hazard. However, if it is not visibly obvious that there was at least outward damage, there was no pattern of receptacle hazards, and there was no evidence of shocks or flying sparks, then the employer should be in a position to assert that there was no reason to determine that he or she should have known. You can support this argument with the fact that the electrical receptacles were installed by a licensed electrician and that you have periodically used a circuit tester to spot-check receptacles in general. After all, you are not automatically obligated to test all of the receptacles every week.

There are examples of when you might be able to claim that the requirement for reasonable diligence does not apply. This can relate to the previously discussed issue of isolated incident of employee misconduct, most notably when the unsafe behavior only existed for a short period and no member of management was in the vicinity. If you are successful in establishing the whole isolated incident case, then you can argue that there was no violation at all; you are not arguing for a lesser classification.

There are situations observed by a compliance officer that may have entered the violative stage within minutes of the observation and in a gap of time during which no management was in the area. A forklift truck may have just sprung a hydraulic leak or just struck an electrical receptacle, a guard, or a railing. An employee or a trucking contractor may have placed a pallet of goods in front of an extinguisher, exit, or electrical disconnect only a few minutes before the inspection party reached the scene. The handle actuator on a deluge shower may have broken off earlier that morning and the report was about to be filed.

In that vein, it could be something that was not visibly obvious, at least not without positive action. An example is that you test the emergency eye fountains once a day, but a plumbing problem occurring since yesterday's eye fountain test is now found to have resulted in inadequate streams. It is not as if the eye fountains should be run constantly so that the streams can be visually monitored all through the shifts. You might also come upon a broken guard for the radial saw blade. Did it get broken just that morning? Besides, had it been used since it was broken? That is often a key question. However, in the case of the eye fountain and the deluge shower, your contention that there was no hazard since the unit was not used may be rather weak. The question would be whether or not there was face splash exposure to the relevant chemicals while the unit was not in a state of suitable readiness. However, with the foregoing example, you would still be left with the defense that the "should have" could not be applied.

Another way to help defend against OSHA's contention that the employer should have known of the hazard concerns purchase orders and work orders. In the case of general standards or when hazards were not evident, a vigorous reliance on this defense can be sustainable. One of the best examples concerns the standards for general (i.e., nonspecific) machine guarding. These standards are frequently assailable, at any rate. Although the compliance officer will point out the possibility of entry into a danger area, you may be in the position to explain that, due to the manner in which the machine functions and the particular way the machine is operated, the operator and other employees are nowhere near that area. The compliance officer will look at potential exposure, and you should counter with the argument of reasonableness, also explaining the vagueness of the standards. You could argue that exposure is so minimal that, even with reasonable diligence, there was still no actual knowledge. Your points may have validity, although I would recommend the guarding in any case.

Still, it is definitely preferable to produce the purchase order or work order upon which your specifications and associated language were accompanied by words obligating the seller, installer, or similar party to assure that the machine would be in compliance with all OSHA standards, including those under 29 CFR 1910.212. This is not to say that such language would always get you off the hook. It will not, but when it is buttressed by the other argument, the combination creates a more solid defense. The compliance officer must substantiate why and how the employer should have known.

As a related matter, OSHA has an extra burden to show employer recognition when citing violations of section 5(a)(1). In all cases, this general duty clause can only be cited for "serious" violations. The hazard must be recognized. Recognition of the hazard can be established on the basis of industry recognition, employer recognition, or common sense recognition. OSHA can add strength to the industry recognition idea if you belong to a trade association or receive trade publications that have noted the hazard. So, do not be so hasty in your efforts to demonstrate safety and health consciousness by bringing up those situations, activities, or memberships. Recognition of the hazard must be supported by satisfactory evidence and adequate documentation in the compliance officer's file.

For industry recognition, OSHA's burden of proof can be challenged if the Agency can only demonstrate that there was recognition by an industry other than the industry to which the employer belongs. As for actual employer recognition, this brings us back to warnings that I offered earlier relating to what to say and what not to say during an inspection or informal conference, and what documentation to provide and what not to provide to OSHA personnel. This can parallel cautionary notes regarding the avoidance of "willful" violations.

There is a special asterisk in the 5(a)(1) case, however. During a previous OSHA inspection, a compliance officer may have informed you of a hazard actually occurring in your establishment and/or that should be kept in mind because it is often in existence in your industry, even though there is no standard which covers it. He or she may not have been in a legal position to recommend a citation under section 5(a)(1) due to the very reason that your hazard recognition, to that point, could not be shown. As a result of that inspection, the OSHA area director may have even sent you a nonbinding letter bringing the hazard to your attention.

The table has been set. If the hazard exists during a subsequent inspection, your arguments against employer recognition are nullified. If industry or employer recognition of the hazard cannot be established, recognition can still be established if it is concluded that any reasonable person would have recognized the hazard. OSHA's posture on this plan of attack is that this common sense theory shall be used only in flagrant cases.

Isolated Incidents

In an earlier chapter, I covered isolated incidents of unpreventable employee misconduct in much detail. I urge you to reread that section.

I repeat a part of it here and add comments because this defense is one of the most important. The basic premise is that it would be unfair and would not promote employee safety and health to penalize an employer for conditions that were unpreventable and are not likely to recur. This defense is introduced frequently and, if supported suitably, is a winner. As in my earlier explanations, personal protective equipment standards are very often involved. For a company to be sustained, it must show that it had established an adequate work rule designed to prevent the violation, had adequately communicated the rule to the employees, had taken steps to discover violations, and had effectively enforced the rules when violations had been discovered.

Where employees were exposed and the employer's instructions were insufficient to eliminate the hazard, even if the employees had complied with those instructions, OSHRC has ruled that the defense would not be allowed. The compliance officer will want to find out if the work rule was communicated to all of the relevant employees or just to some. Where there is evidence of widespread noncompliance by employees, both in numbers of employees and duration of exposure, there is a strong inference that the employer's efforts in detecting and correcting violations have been lacking.

The defense is much more apt to be sustained when there was a single incident of noncompliance by only one or a very few employees. The defense has been unsuccessful when there was inadequate supervision of employees. If the employer did not discipline employees after prior incidents of violative conduct, subsequent violations have not been considered unpreventable, and the defense has failed. Evidence of discipline following an inspection has been considered by OSHRC. However, there is a far greater chance that the defense will be upheld when there have been sanctions, such as reprimands and the docking of pay, prior to the inspection.

This is an affirmative defense, and the burden of establishing it has to rest with the employer. Keep in mind that an affirmative defense is any matter which, if established by the employer, will excuse the employer from a violation that has otherwise been proven by the compliance officer.

Impossibility of Compliance

Now comes the impossibility or infeasibility of compliance defense. The employer must demonstrate that compliance with the standard is not practical or reasonable under the particular circumstances and that alternative safety measures had been taken or are not available. This has been upheld when the employer was able to show that it was physically impossible to comply with the standard, when the compliance would have prevented the performance of work, and when compliance would not have reduced the hazard due to the nature of the work.

Reframing and new interpretations of this defense have evolved to the point that OSHA is required to show that the alternative means of protection is "feasible" and not merely "available." It must therefore show that the alternative means is a practical and realistic method, given the circumstances at the workplace, to protect the employer's employees and that the employer did not use it. Yet, OSHRC has also indicated that even when an employer cannot fully comply with literal terms of a standard, it must nevertheless comply to the extent that compliance is feasible.

The infeasibility defense has encompassed both technological and economic feasibility. Surprisingly, an employer can assert an economic infeasibility argument against complying with a standard if the employer demonstrates that it is extremely costly to comply with the order and that the employer cannot absorb this cost. Where these assertions would be made, the employer should apply for a variance before any OSHA inspection takes place. This argument does not apply simply because abatement is generally expensive. There is a major burden on the employer to make the case. In fact, it is only with mixed emotions and a duty to the reader that I report on this defense. Nevertheless, OSHRC has held that an employer that believes that compliance with a standard is inappropriate cannot challenge the wisdom of the standard through the judicial process. Instead, the employer should have sought a variance or petitioned to have the standard modified through the rulemaking process.

Noncompliance Is Safer Than Compliance

Another defense deals with the idea that compliance with a standard would result in a greater hazard than if noncompliance remained. The other two key elements of this argument are that alternative means of protecting employees are unavailable and that a variance application would be inappropriate. For the main element, the evidence of a greater hazard must be clear. The mere verbalized fears of employees along with an employer's unsupported opinion have been held to be inadequate. Similarly, it is not enough that compliance with the standard will cause momentary lapses of protection or inconvenience to employees. OSHRC has held that the term "*alternative means*" refers to means other than those specifically required or permitted under a cited standard and encompasses more than "equivalent means." I find that the functional distinction between "alternative" and "equivalent" is not immediately apparent. This may be an area for significant scrutiny by an attorney. A variance application would be inappropriate if the employer incorrectly assumes that its practice is safer than complying.

Multiemployer Worksites

The other well-known and not infrequently recognized affirmative defense relates to multiemployer worksites, which usually means construction. Although much of my material relates more closely to single employer fixed establishments, this defense must be elucidated as it is so commonly asserted and does result in victory more frequently than most other defenses. To have a legitimate defense, the employers, who in the relevant cases are nearly always subcontractors, must show that: they did not create the hazard; they did not have the responsibility or the authority to have the hazard corrected; they did not have the ability to correct or remove the hazard; the creating, the controlling, and/or the correcting employers, as appropriate, had been specifically notified of the hazards to which the employees were exposed; and they had instructed their employees to recognize the hazard and, where necessary, informed them how to avoid the dangers associated with it. Regarding the last element, exposing employers, where feasible, must have taken appropriate alternative means of protecting employees from the hazard, and when extreme circumstances justify it, exposing employers should remove their employees from the job to avoid citation and serious harm.

If an exposing employer meets all of these defenses, that employer should not have been cited. OSHA's guidelines state that, if all employers on a worksite with employees exposed to a hazard meet these conditions, then the citation shall be issued only to the employers who are responsible for creating the hazard and/or who are in the best position to correct the hazard or to ensure its correction. In such circumstances, the controlling employer and/or the hazard-creating employer shall be cited even though none of their employees are exposed to the violative condition. Continuing with OSHA's guidelines, penalties for such citations shall be appropriately calculated using the exposed employees of all employers.

Improper Inspection

One defense is that the compliance officer failed to follow legal procedures in conducting the inspection. I recommend that you obtain a copy of OSHA's internal guidelines for the conduct of inspections. The word "*internal*" does not preclude you from having a copy. The current relevant guidelines may be called the *Field Inspection Reference*

Manual, the *Field Operations Manual*, or something similar. The Agency has spent considerable time updating and renaming such materials. Additionally, you should have a copy of the Occupational Safety and Health Act and the germane sections of the code of federal regulations. Armed with all of these documents, you will be in a better position to assess whether or not the compliance officer conducted a legally proper inspection. To have grounds for a defense relating to conduct of the inspection, there must be clear evidence that the deviation from the guidelines had prejudicial effects.

Citation Classifications

You may challenge the citation classification (e.g., "serious," "willful," etc.) itself. In all cases, do not assume that you know what each classification means and what is presupposed by its designation. You should carefully read the legal definition of each. With OSHA's *Field Inspection Reference Manual* (or its equivalent) and the Occupational Safety and Health Act in hand, you can gain insight into the Agency's interpretations of the classifications and what documentation is required of/by the compliance officer.

"Serious" Violations

With "serious" violations, OSHA must prove that there is substantial probability that death or serious physical harm could result. The Agency is to consider the most serious injury or illness which could reasonably be expected to result from the type of accident or health hazard exposure that has been identified. Then, OSHA is to consider whether or not the results of that injury or illness could include death or serious physical harm. Serious physical harm, according to OSHA, can fall under either of two categories:

1. Impairment of the body in which part of the body is made functionally useless or is substantially reduced in efficiency on or off the job. Such impairment may be permanent or temporary, chronic or acute. Injuries involving such impairment would usually require treatment by a medical doctor.

2. Illnesses that could shorten life or significantly reduce physical or mental efficiency by inhibiting the normal function of a part of the body.

These definitions leave a lot of room for argument. In fact, going against the two-part test just listed, OSHRC has held that with respect to standards for toxic substances, if the standard was violated, then the violation is "serious." By that definition, OSHA does not have to prove that the level of exposure would lead to a serious disease.

It is of paramount importance to focus in on the "substantial probability that death or serious physical harm" could result. The criteria do not deal with possibility or mere probability. Then there is the "most serious injury or illness which could reasonably be expected to result." The criteria do not deal with the most severe possible injury or illness. As to "serious physical harm," you may be able to claim that the identified impairment injury would not render the body part "functionally useless" and that it would not be "substantially reduced in efficiency." Similarly, you may be able to fairly assert that the identified illness could not "shorten life" or "significantly reduce physical or mental efficiency by inhibiting the normal function of a part of the body." There are a lot of words with which to find issue.

"Repeated" Violations

OSHA must base "repeated" violations on previous citations where a substantially similar condition existed. In most cases, the similarity will be evident if the prior and present violations are for failure to comply with the same standard. There are times when that would be unreasonable and you can fairly easily defend against the classification. As an example, previously a citation was issued for a violation of 29 CFR 1910.132 for not requiring employees to wear safety-toe footwear. A recent inspection of the same establishment revealed a violation of 29 CFR 1910.132 for not requiring the wearing of head protection (i.e., hard hats). Although the same standard was involved, the hazardous conditions found were not substantially similar and therefore a "repeated" violation would not be appropriate.

Another example of the same standard being cited, but with a solid available defense against a "repeated," involves general machine guarding. A company could be cited for a violation of 29 CFR 1910.212(a)(1) for some type of in-running nip point on a machine that is not covered by a standard which specifies the machine by name. For instance, the machine is not a mechanical power press, table saw, calender, abrasive grinder, or one of the particular machines referenced by the standards dedicated to the textile industry. There are other specific machines mentioned in the standards as well. Now, a compliance officer observes another violation of 29 CFR 1910.212(a)(1), again regarding a machine that is not covered by a standard which specifies the machine by name; in other words, 212(a)(1) is the proper standard. However, in this situation, the hazard involves a pinch point on a completely different type of machine.

The machines do not look alike and their functions are very different—distinct, in fact. Your defense against the "repeated" should be clear and sustainable.

On the other hand, your defense cannot simply be that different standards were involved. You may have been cited for the lack of guarding on a horizontal power transmission belt under 29 CFR 1910.219(e)(1)(i). As a result of the next inspection, you are cited for the lack of guarding on a vertical power transmission belt on a different machine under 29 CFR 1910.219(e)(3)(i). The hazards were substantially similar, even if the standards were different and the machines bore little resemblance to each other. The defense should not hold up.

Although there are no statutory limitations upon the length of time that a citation may serve as a basis for a "repeated", there is an OSHA policy that is used to ensure uniformity. A citation will be issued as a "repeated" violation if the citation is issued within 3 years of the final order of the previous citation or if the citation is issued within 3 years of the final abatement date of that citation, whichever is later. OSHA has been known to err on this in at least two ways. They have cited when the 3-year limitation has lapsed, and they have cited before the original citation became a final order. In the second case, this is sometimes where the initial citation had been issued recently. At other times, even if the inspections were a few months apart, the OSHA area office was slow to process the first citation, and the compliance officer recommending the "repeated" had mistakenly believed or assumed that the first citation was issued and that a final order date had elapsed.

Another OSHA error has stemmed from the fact that the company was still in contest with the first citation. The inspection that resulted in that citation may have been conducted several months before the second inspection. It does not matter. If there was no final order in the first instance, there cannot be a "repeated." Look carefully into this time limitation.

A multifacility employer may be cited for a "repeated" if the violation recurred at any worksite at any of its locations nationwide. One glitch is that this is only done, barring exceptional circumstances, within the same two-digit standard industrial classification. If the facilities that are the subject of the violations are in totally different businesses, you may defend on that basis.

If a violation was cited and then abated, but OSHA found that it became "unabated" (e.g., the guard that was installed for abatement purposes had been removed), a "repeated" should be cited. A "repeated" may also be cited in a case where the subject of the violation did not even exist during the first inspection. Again, a machine serves as a good example. A company may have been cited for the unguarded machine and abated the situation. Two years later, at the same or at another location, a machine with a substantially similar hazard is put into operation.

There is a totally different kind of challenge that you may make to a classification of "repeated." If you can show that a compliance officer definitely saw the machine that was the subject of the "repeated" in operation when he or she first visited your establishment, but no citation was issued, you should have good grounds for recourse. It will be best if the compliance officer specifically discussed the machine with you and the rest of the inspection party. The more witnesses to the conversation, the better. You must be able to demonstrate that the machine condition and use, especially as related to safety, were identical or virtually so during the two inspections. The compliance officer may claim subtle differences in exposures, or related to different parts being worked, or to positions of the guard. He or she may claim that the machine was being serviced during the first inspection. The closer you can relate the overall scenario involving the machine on the two inspections, the better for your challenge. Although OSHA guidelines do not prohibit the citing of "repeated" in this type of case, if you can show that the compliance officer or different compliance officers really looked at the same situation (in plain English), you will have a viable defense.

Note that just because one compliance officer or even the same one missed a hazard/violation during one inspection but recognized it during a subsequent inspection does not necessarily form a basis for a claim that there should not be any type of citation. A hazard/violation cannot be allowed to exist forever, for instance, just because one compliance officer failed to recognize it one time. The point here has been limited to "repeated" violations, but it can also apply to "willful" violations.

"Failure-to-Abate" Violations

A "failure-to-abate" violation clearly differs from a "repeated." It exists when the employer has not corrected a violation for which a citation has been issued, the citation has become a final order of OSHRC, and the abatement date has passed. A "failure-to-abate" also exists when the employer has not complied with interim measures involved in a long-term abatement within the time given. OSHA must prove that the cited condition is the identical one for which the employer was originally cited and that the item of equipment or condition previously cited has never been brought into compliance. If, however, the violation was not continuous, the subsequent occurrence is a "repeated" violation. Proposed penalties attached to "repeated" violations and to "failure-to-abate" violations can both be extreme.

If the hazard/violation is noted for the second time, even several weeks after the initial observation, the "failure-to-abate" can be significantly more expensive because a penalty can be applied for each and every day the violation continued beyond the prescribed abatement date. I am referring to federal calculations; state calculations may differ, but probably not by much. Thus, you may well benefit from claiming that the violation was a "repeated" at the worst, but certainly not a "failure-to-abate." You may want to study OSHA's penalty formula because, when an employer has "repeated" the violation more than once, hefty coefficients are progressively attached to the gravity-based penalty. Thus, it is possible that, on rare occasions, a "repeated" violation can carry a larger penalty than a "failure-to-abate."

In most cases, you will want to show that, at some time between the inspections, the hazard was brought into compliance. A major caution is due here. Do not bring this up if you have no reason to suspect that a "failure-to-abate" or a "repeated" will be cited. If you do, you may talk the compliance officer into a "repeated," although he or she had previously only considered a "serious."

If you are in clear violation relating to a personal protective equipment situation, you can usually demonstrate why it is a "repeated," but only do this if you are trying to avoid a "failure-to-abate" that is definitely being considered. You can usually show that the personal protective equipment has been used since the first inspection or, perhaps more important legally, since receipt of the original citation. This type of argument can be made relating to impeded access to exits, electrical disconnects, and extinguishers. Are they blocked by different miscellaneous storage, or are they blocked by the exact same large piece of machinery? The implications are obvious.

One more example involves the violative storage of compressed gas cylinders. Oxygen and fuel gas may have been stored, as opposed to used or when attached to regulators on a cart, within 20 ft of each other and not separated by a 5 ft high noncombustible barrier. OSHA finds the same hazard/violation, but are those the same cylinders that were seen during the earlier inspection?

There is another defense against the "failure-to-abate" classification. The employer can challenge the original uncontested citation if the employer has not contested it previously. The employer can claim factual defenses to the earlier uncontested citations, or that the original citation lacked specificity, or that the incorrect standard was originally cited.

"Willful" Violations

"Willful" may be cited if the violation was committed voluntarily with either an intentional or purposeful disregard for the requirements of the Act or with plain indifference to employee safety. Thus, a violation that fits one of these definitions, noting that sometimes the word *"deliberate"* is also used, can result in the citing of a "willful." This distinguishes the classification from those where the violation was merely inadvertent, accidental, or arising from ordinary negligence.

The issue of willfulness focuses on the employer's state of mind and general attitude toward employee safety to a greater extent than for a nonwillful violation. On one hand, OSHA need not prove that the violation was committed with a bad purpose (e.g., malice) or an evil or criminal intent on the part of the employer. On the other hand, when formulating your defense, consider that it is not enough for OSHA simply to show carelessness or lack of diligence in discovering or eliminating a violation. For a "willful," it certainly is not sufficient for OSHA to show that an employer was aware of conduct or conditions constituting a violation; such evidence is necessary to establish any violation. Don't remove a "CARCINOGEN" label; that's asking for it.

A "willful" violation is often differentiated by a heightened awareness. A "willful" charge is not justified if an employer has made a good faith effort to comply with a standard or to eliminate a hazard, even though the employer's efforts are not entirely effective or complete. In addition, you will have a good argument against a "willful" if you can demonstrate that the employer had an evident, sincere, reasonable belief that the conduct conformed to the law. This can go a long way in negating a "willful" classification.

OSHA cannot assume that every act consciously done is done willfully. Keep in mind the extra ingredient that was referenced earlier: the element of intentional or purposeful disregard or plain indifference. The fact that the employer is aware of the requirements of the standard, standing alone, does not establish this element. OSHA must address, in effect, the motives of the employer. Without some outward manifestation of intent or overt act, the mere fact that the employer knew all elements of the standard is insufficient.

To assess the effectiveness of your particular defenses against a "willful," you have a need to understand how the Agency interprets its definitions. Then you will have a better idea of where the possible loopholes are, as well as where your defense may fall on deaf ears or, worse, get you into even hotter water. I will discuss the committing of intentional and knowing violations and the committing of violations with plain indifference. However, OSHRC has interpreted that OSHA must also show a certain state of mind on the part of the employer to sustain its burden of proof for a "willful."

The Agency must show evidence of a conscious disregard for the requirement and/or a cavalier attitude toward employee safety and health. OSHA considers that the employer committed an intentional and knowing violation if an employer representative was aware of the requirements of the Act, or the existence of an applicable standard or regulation, and was also aware of a condition or practice in violation of those requirements but did not abate the hazard. The Agency also interprets that definition to be met when an employer representative was not aware of the requirements of the Act or standards, but was aware of comparable state or local legal requirements, and was also aware of a condition or practice in violation of that requirement but did not abate the hazard.

OSHA considers that the employer committed a violation with plain indifference to the law if higher management officials were aware of an OSHA requirement applicable to the company's business but made little or no effort to communicate the requirement to lower level supervisors and employees. Also in this definition category is when company officials were aware of a continuing compliance problem but made little or no effort to avoid violations. This could be a passive situation. Now for the flip side. You had proactively summoned your loss control rep to inspect a machine. The rep did so and assured you that there was no hazard or violation. A defense is set.

Another example of plain indifference, according to the Agency, is when an employer representative was not aware of any legal requirement but was aware that a condition or practice was hazardous to the safety or health of employees and made little or no effort to determine the extent of the problem or to take the corrective action. Knowledge of a hazard may be gained from insurance company reports, safety committee or other internal reports, consultant reports, the occurrence of injuries or illnesses, media coverage, employee complaints, employee representative complaints, information gathered at seminars or from publications, and so on.

Finally in this category, in particularly flagrant situations, willfulness can be found despite lack of knowledge of either a legal requirement or the existence of a hazard if the circumstances show that the employer would have placed no importance on such knowledge even if he or she had possessed it or had no concern for the health or safety of employees. You can certainly mount a strong defense against that one in the great majority of cases.

Violations of Section 5(a)(1)

There is the defense concerning the improper citing of an alleged violation of section 5(a)(1) of the Act. In establishing a violation of section 5(a)(1), OSHA must show that a condition or activity in the employer's workplace presented a hazard to employees, that the cited employer or the employer's industry recognized the hazard, that the hazard was likely to cause death or serious physical harm, and that feasible means existed to eliminate or materially reduce the hazard. Considering the "death or serious physical harm" element, section 5(a)(1) is not to be cited for "other than serious" violations. A defense can be established when the general duty clause was invoked, even though a specific standard governing the employer's conduct exists. An OSHA area director may not issue a section 5(a)(1) citation simply because the compliance officer did not or could not locate the specific applicable standard. By the same token, he or she may not issue the citation because the existing applicable standard was not as stringent as the compliance officer or area director had preferred.

"Egregious" Penalty Policy

When scouring OSHA's "egregious" penalty policy (i.e., violation-by-violation or instance-by-instance penalties) for defenses, I strongly advise a hard look at recent case law. The "egregious," which is not a regular citation classification, has not fared too well. You should find support in OSHRC decisions. OSHA has used the word *"outrageous"* to encapsulate its real-world depiction of an "egregious" violation. The Agency opens its consideration of "egregious" when the employer is found in violation of an OSHA requirement of which he or she had actual knowledge at the time of the violation and intentionally, through conscious, voluntary action or inaction, made no reasonable effort to eliminate the known violation. The major additional factors include the following: the violations resulted in employee deaths, a worksite catastrophe, a large number of injuries or illnesses, or in persistently high rates of employee injuries or illnesses; the employer has an extensive history of prior violations of the Act; the employer has intentionally disregarded safety and health responsibilities; the employer's conduct, taken as a whole, amounts to clear bad faith in the performance of its duties under the Act; and the employer has committed a large number of violations so as to significantly undermine the effectiveness of any safety and health program that might be in place.

You may make general challenges to the fundamental fairness of the "egregious" policy, particularly because it is such a drastic departure from previous interpretations of what the Act allowed for penalty assessment. OSHA has definitely jumped the gun on some of their allegations of "egregious." One specific argument can address how instances are differentiated. If several employees are not wearing required personal protective equipment, the Agency can make a decent case that each of the employees constitutes one instance. Nevertheless, you can counter if you can back up a claim that your error was in not implementing an eye protection program. Therefore, you assert that the case really reflects one multiexposure instance. If the idea of a penalty for each instance has any validity that will be difficult to rebut, it is anchored in situations such as several unguarded machines scattered throughout the facility, with each operated by a different employee or mutually exclusive sets of employees. You can still contend that you lacked a machine guarding program, but this should be a weaker assertion than the one based on eye protection.

On the other hand, you may have a mezzanine that lacks perimeter protection, and since you have 10 employees that are exposed to falling off of the edge, OSHA views the situation as having 10 instances, with a hefty penalty to be proposed for each. Now you have a good case. The fact is that choosing one particular and common means of abatement (i.e., a railing system) would have protected all 10 of the employees. There is no logical way that this can be viewed as 10 instances, although OSHA has attempted to do so, and failed, in similar situations.

"Criminal Willful" Violations

Whenever an employer is accused of a "criminal willful," it would be foolish to lack professional legal representation. Hire an attorney. Period. There may be a plethora of defenses added to this classification because it is in the criminal arena as opposed to the civil arena. Those defenses should be researched by your attorney. For now, I stress that a plain reading of section 17(e) of the Act refers to "any employer who willfully violates any standard, rule, or order...[where] that violation caused death to any employee." The key here is that there cannot be a "criminal willful" just because a standard that was violated willfully, and is thus a serious enough matter in itself, was somehow related to the work being performed by the victim. In order to prove that the violation of the standard caused the employee's death, the Agency's interpretation is that there must be evidence in the file which clearly demonstrates that the violation of the standard was the cause of, or a contributing factor to, an employee's death. Additionally, violations of section 5(a)(1) of the Act are not to be used as the basis of a "criminal willful."

Penalty Calculation

You may choose to challenge the way penalties were calculated. You should obtain OSHA's up-to-the-minute penalty policy. It has evolved and will continue to do so. As an example, the Agency has always had in place a system whereby penalty reductions depending on company size have been given. This generally relates to the number of employees controlled by the employer. OSHA uses a table with specific cutoff points to determine what the break will be. The trend is to allow for even greater reductions, particularly with very small companies. A company with 10 or fewer employees might receive as much as 80 percent off of the gravity-based penalty. The gravity-based penalty is where the base amount is determined as related to severity and probability.

Reductions have also been available if the company has a good history, or lack of history, with OSHA. These reductions are not substantial and the "history" reduction concept may not remain.

The last of the standard reductions categories involves "good faith." OSHA has fine-tuned their criteria for these breaks, and the system may be modified several more times. Suffice to say that, under the likely scales, a company can receive a significant percentage off of the gravity-based penalty if it can demonstrate a high degree of completeness and effectiveness of the workplace's safety and health program. I have alluded to this concept when discussing the compliance officer's evaluation of the program. I do not print any tables or charts herein for any of the reduction factors because the system is in a state of flux. Just know that you can challenge the amount of proposed penalties if OSHA did not adhere to its system, but that reductions are not always applicable.

As an example, in the case of "willful," "repeated," and "failure-to-abate" violations, there is generally no "good faith" reduction. At times, that has also been the rule when "high gravity" or "serious" was alleged. In that context, "high gravity" refers to high severity and greater probability. Further, the coefficient that is applied to the gravity-based penalty for "repeated" violations is significantly increased when the company is large, as determined by the number of persons employed. Once again, I emphasize that I am not offering a chart for the calculations or special situations simply because the formula and system may well change.

It is not unusual for the Agency to err in its calculations. It has done so both in favor of the employer and to the detriment of the employer. This does not only relate to the reductions or the coefficient, which vary for "repeated" and historically as a fixed number for "willful," but for the gravity-based penalty as well. Consider the two elements that go into that base penalty. In the case of the severity assessment, you may challenge the compliance officer's evaluation. He or she may have determined that the severity could be high, whereas you feel that, even if it is possible, it is not reasonable to predict such a severe injury or illness. The arguments could be complex and tedious, and if pushed, medical professionals could be brought into the picture. Although the nature and severity of physical trauma can be questioned, matters concerning chemical burns and respirable hazards might be more ripe for challenge. Study the current OSHA definitions and decide if there is room for a defense.

The probability factor should also be considered. OSHA analyzes many factors in arriving at a probability assessment. The list includes such things as the following: number of employees exposed, frequency and duration of exposure (with duration a larger factor for employee overexposure to contaminants), employee proximity to the hazardous conditions, use of appropriate personal protective equipment, medical surveillance program, noise level (for other than noise-related standards), lighting, condition of the adjacent floor, existence of piecework or incentive work or numerous other stress factors, the weather (for outside work), and possibly the level of skill or experience of the exposed employees. If the penalty was based on a "greater" probability, see where you can argue that such an assessment was excessive. Point out mitigating circumstances to balance out or overcome what OSHA believes to be exacerbating circumstances.

Partial Abatement and Attempts at Abatement

Be sure to point out partial abatement and/or attempts at abatement. This can be particularly important when you are dealing with a "failure-to-abate" or "repeated," and it can represent an argument to totally reject the classification of a "willful" violation. When defending against the penalty calculation for a "failure-to-abate," realize that (in my experience with OSHA) a violation had to be at least 24 h out of compliance to qualify for this classification. Thus, if an item carried an abatement date of Monday, June 1, which means that abatement was required before midnight of Tuesday morning, and the compliance officer observed that the item, with continuing employee exposure verified, was not abated as of Tuesday, June 2, there should not be a finding of a "failure-to-abate." For that reason, many OSHA area offices try to schedule follow-up inspections on the Wednesday following the Monday upon which several of the abatement dates fell. By that Wednesday, a "failure-to-abate" is viable.

OSHA guidelines explain that the number of days unabated shall be counted from the day following the abatement date specified in the citation or in the final order. It will include all calendar days between that date and the date of the reinspection, excluding the date of reinspection. On the other hand, OSHRC has held that the calculation is to be based on working days. At the discretion of the area director, a smaller penalty may be proposed with the reasons for doing so (e.g., achievement of an appropriate deterrent effect) documented in the case file. When the compliance officer believes, and so documents in the case file, that the employer has made a good faith effort to correct the violation and had good reason to believe that it was fully abated, the area director may reduce or eliminate the daily proposed penalty that would otherwise be justified.

I do not recommend that you make halfhearted attempts at abatement, and figure that you can read my last two sentences to the compliance officer. Yet, I do recommend that you consider bringing them up if they are really relevant. On a similar line, when the citation has been partially abated, the area director may authorize a major reduction to the proposed penalty normally calculated. In kind, when a violation consists of multiple instances and the follow-up inspection reveals that only some instances of the violation have been corrected, the additional daily proposed penalty shall take into consideration the extent that the violation has been abated. The Agency may pro rate the penalty.

Two more comments are due regarding "failure-to-abate" dates. If the informal conference involves an alleged "failure-to-abate," the area director shall set a new abatement date in the informal settlement agreement (oops, I forgot to tell you that before), documenting for the case file the time that has passed since the original citation, the steps that the employer has taken to inform the exposed employees of their risk and to protect them from the hazards, and the measures that will have to be taken to correct the condition. In fact, OSHA guidelines even allow that, if the employer has exhibited good faith, a late petition for modification of the abatement date may be considered when there are extenuating circumstances. The idea for a defense of sorts comes into play if, for instance, OSHA did not make it clear during an informal conference that new dates could be worked into the settlement agreement. All of this sounds great, but for the most part a "failure-to-abate" is a "failure-to-abate"—the point being that you will simply be in deep trouble with the very real possibility of horrific financial consequences.

Prompt Issuance of Citations

OSHA must issue the citation(s) with reasonable promptness. In addition, there is a statute of limitations of 6 months following the occurrence of a violation in which a citation must be issued. I briefly addressed this issue when discussing citation receipt. OSHRC has upheld the issuance of citations when a compliance officer first discovered the violation during an inspection made more than 6 months after the unsafe condition's occurrence. Further, OSHRC does not view the occurrence to be when the unsafe condition first came into existence. With recordkeeping, specifically, OSHRC has determined that an uncorrected error or omission in an employer's OSHA-required injury/illness log may be cited 6 months from the time that the compliance officer discovered, or reasonably should have discovered, the facts necessary to issue a citation.

Tying together the test of reasonable promptness and the statute of limitations, it is interesting to note that OSHRC will generally stick with the 6-month deadline; that is, it seldom considers long gaps between inspection and citation issuance to be unreasonable. However, if the delay put the employer in a position whereby the ability to prepare a defense was severely hampered, OSHRC may assert that sufficient prejudice existed and vacate the citation.

As an example, an employee who was a witness to the alleged violation had left the employ of the company and could not be located by the time the citation arrived several weeks after the inspection. In another case, OSHA alleged that defective forklift trucks were in operation. By the time the citation arrived, which was about 1 month after the inspection, the vehicles had been returned to the lessor and could not be examined for the alleged defects. Further, if they had been examined, they would not necessarily have been in the same condition that they were in during the compliance officer's visit. In both situations, OSHRC found that citation issuance delay did unfairly prejudice the employer in the preparation of its case.

It appears that OSHRC will generally be more liberal with OSHA when it is clear that the Agency and the particular OSHA area office have completed a long and complex inspection. The cases that I have referenced are more anecdotal than anything approaching a trend. There is also a defense involving the improper service of a citation, which tolls the time for filing a notice of contest. This has to do with the method of delivery and to whom the citation(s) were delivered. The chances that there are grounds for such a defense are slim. If, for instance, a compliance officer simply dropped by the plant and handed an envelope to an employee who was working on the loading dock, there could be a foundation for a defense.

Lack of Jurisdiction and Improper Promulgation

It is possible that OSHA will inspect and issue citation(s) against a business that does not fall under the Agency's jurisdiction. In such cases, the employer can raise the lack of jurisdiction as a defense. If OSHA does not have jurisdiction, the citation will be vacated.

There are at least four well-defined situations that fall under this heading. To be covered by OSHA, at least as federally enforced, the employer must affect interstate commerce. This is a requirement that is very easily satisfied. Broad interpretations have interpreted this effect as including the employer competing with employers from another state, fuel in the employer's vehicle or even parts of the vehicle being from another state, and so on. Then there is preemption by another federal agency. This could mean, for instance, the Federal Railroad Administration, Coast Guard, Nuclear Regulatory Commission, or Mine Safety and Health Administration. There is a three-part test for preemption. Preemption requires that the employer be covered by another federal act concerning safety and health, the other federal agency exercises its authority, and the federal agency exercises its authority over the cited working conditions. When questions arise over which agency has jurisdiction, they are often resolved through memoranda of understanding between the agencies. Next is the case of employer bankruptcy in court jurisdiction. With bankruptcy matters, as reported earlier, the employer is obligated to comply if the business is still functioning and employees are still exposed to hazards and violations. Finally, there are some employers that are not covered by the Act; this was explained in the Introduction.

There is also the assertion that there was no employer-employee relationship. If an employer is a legal sole owner, as opposed to the owner of a theoretical one-person corporation, and has no employees, he or she is not covered by OSHA. If two or more individuals are equal partners, in a legal partnership as opposed to a corporation, and have no employees, they are not covered by OSHA.

This part of the defense can be folded into the lack of jurisdiction contention. The employer-employee question can also arise when an employer, even if generally covered by OSHA, appears to be in a violative situation, but the only persons exposed are employees that, supposedly, are not his or her direct employees. A full analysis of this question can be cumbersome. For the most part, OSHA, and at this stage OSHRC, would want to know if the employer hired, directed, supervised, controlled, scheduled, paid in any manner, or monitored the employees and whether or not those employees worked for that employer exclusively or nearly so.

One more factor revolves around the type of work performed by those individuals. The chief issue goes to whether or not the work that they do is technically oriented or otherwise sophisticated to a degree or of a type that the employer could not be fully expected to comprehend it. The point is that the employer may not be in a position to judge whether or not those employees are exposed to hazards. They may be performing a trade that is out of their expertise and whose hazards are not evident. The employer cannot just curtly and smugly declare that they are all "independent contractors."

A different type of defense is when you can claim that a standard had been improperly promulgated, but this one is a long shot. The issue is two-fold. First, does the standard have a rational relationship to the actual situation to be regulated? Second, was OSHA arbitrary and capricious in promulgating the standard, as it does not fall under the purpose of the Act?

An Overview

Knowledge of cracks in the otherwise tight enforcement system provides you with an education so that you at least have the tools with which to attempt to carve out a defense. Some of these cracks are legal, and some are out of benevolence. Nevertheless, when you bring up these special provisions—these opportunities for grace—mainly as related to "failure-to-abate," you may be enlightening the OSHA personnel. This is not to say that they lack diligence. It is to say that their mind-set generally does not incorporate opportunities to stray from the everyday procedures. You may, indeed, have to remind them of the discretion that they are allowed when justified, as outlined in their guidelines. Ideally, if there were arguments to make, you made them prior to the filing of a notice of contest. If not, or if those arguments fell upon seemingly deaf ears, those arguments now become true defenses.

In this chapter, I discussed several classic procedural, constitutional, affirmative, and substantive defenses. Some of them might be considered less sophisticated defenses, such as those relating to citation classification, penalty calculation, and similar topics. Throughout the chapters dealing with the inspection process and the citation remedies, I alluded to some of these defenses and highlighted numerous, specific safeguards to help you assure, as much as you can, that the compliance officer did not lose sight of the bottom line basics. In all cases, do not neglect to come full circle back to these basics when mulling over defenses. Forgetting the legal mumbo jumbo, while not neglecting its value, was there a real hazard, were employee(s) realistically exposed to the hazard, and if so, did the employer have good reason to have answers to those two questions?

The best defenses against OSHA citations should be preemptive. Strive to avoid the necessity of presenting legal arguments against citations. Make every effort to obviate your company's vulnerability to them. Protect your employees (this is their right, not a privilege) as you would your family. Vitalize your facility to achieve and maintain compliance. In so doing, you should reduce risk, loss, financial burden, legal jeopardy, and human suffering. The result will be a safer, more healthful, more productive, and more profitable business venture.

Whether in respect to providing a safe and healthful workplace or to taking meaningful steps to combat OSHA citations, resounding proclamations and verbal posturing will lack the substantial, telling impetus of well-formulated, positive actions. With that in mind, I respectfully suggest that you heed the wisdom of the following Apache proverb:

> *It is better to have less thunder in the mouth, and more lightning in the hand.*

Selected Safety and Health Organizations and Associations

Please note that not all of the listings below are involved *solely* with occupational safety and health.

Action on Smoking and Health
2013 H Street N.W.
Washington, D.C. 20006
phone 202/659-4310, fax 202/833-3921

American Academy of Industrial Hygiene
6015 West Saint Joseph, Suite 102
Lansing, Michigan 48917-3980
phone 517/321-5025, fax 517/321-4624

American Association of Occupational Health Nurses, Inc.
50 Lenox Pointe
Atlanta, Georgia 30324
phone 404/262-1162, fax 404/262-1165

American Board of Industrial Hygiene
6015 West Saint Joseph, Suite 102
Lansing, Michigan 48917-3980
phone 517/321-2638,
ABIH prefers no fax input

American Board for Occupational Health Nurses
201 East Ogden, Suite 114
Hinsdale, Illinois 60521-3652
phone 630/789-5799, fax 630/789-8901

American Chemical Society
1155 16th Street N.W.
Washington, D.C. 20036
phone 202/872-4600, fax 202/872-4615

American College of Occupational and Environmental Medicine
55 West Seegers Road
Arlington Heights, Illinois 60005
phone 847/228-6850, fax 847/228-1856

American Conference of Governmental Industrial Hygienists
Building D-7
1330 Kemper Meadow Drive
Cincinnati, Ohio 45240
phone 513/742-2020, fax 513/742-3355

American Hospital Association
One North Franklin
Chicago, Illinois 60606
phone 312/422-3000, fax 312/422-4796

American Industrial Health Council
2001 Pennsylvania Avenue N.W.
Suite 760
Washington, D.C. 20006
phone 202/833-2131, fax 202/833-2201

American Industrial Hygiene
Association
2700 Prosperity Avenue
Suite 250
Fairfax, Virginia 22031
phone 703/849-8888, fax 703/207-3561

American Institute of Chemical
Engineers
345 East 47th Street
New York, New York 10017
phone 800/242-4363, fax 212/752-3294

American Institute of Plant
Engineers
8180 Corporate Park Drive
Suite 305
Cincinnati, Ohio 45208
phone 513/489-2473, fax 513/247-7422

American Insurance Services Group,
Inc.
Engineering and Safety Service
85 John Street
New York, New York 10038
phone 212/669-0400, fax 212/669-0550

American National Standards
Institute
11 West 42nd Street
New York, New York 10038
642-4900, fax 212/764-3274

American Petroleum Institute
1220 L Street N.W.
Washington, D.C. 20005
phone 202/682-8000, fax 202/682-8232

American Society of Civil Engineers
1801 Alexander Bell Drive
Reston, Virginia 20191-4400
phone 703/295-6000, fax 703/295-6333

American Society of Heating,
Refrigerating and Air-Conditioning
Engineers
1791 Tullie Circle N.E.
Atlanta, Georgia 30329
phone 404/636-8400, fax 404/321-5478

American Society of Mechanical
Engineers
345 East 47th Street
New York, New York 10017
phone 212/705-7722, fax 212/705-7739

American Society of Safety Engineers
1800 East Oakton Street
Des Plaines, Illinois 60016
phone 847/699-2929, fax 847/296-3769

American Society for Testing and
Materials
100 Barr Harbor Drive
West Conshohocken, Pennsylvania
19428
phone 610/832-9500, fax 610/832-9555

American Subcontractors Association
1004 Duke Street
Alexandria, Virginia 22314
phone 703/684-3450, fax 703/836-3482

American Welding Society
Post Office Box 351040
550 LeJeune Road N.W.
Miami, Florida 33135
phone 305/443-9353, fax 305/443-7559

Asbestos Information
Association/North America
1745 Jefferson Douglas Highway
Crystal Square 4, Suite 406
Arlington, Virginia 22202
phone 7-3/412-1150, fax 703/412-1152

Asbestos Institute
1002 Sherbrooke Street W.
Suite 1750
Montreal, Quebec, Canada H3A3L6
phone 514/844-3956, fax 514/844-1381

Asphalt Institute
Research Park Drive
Post Office Box 14052
Lexington, Kentucky 40512-4052
phone 606/288-4960, fax 606/288-4999

Associated Builders and Contractors
1300 N. 17th Street
Rosslyn, Virginia 22209
phone 703/812-2000, fax 703/812-8200

Associated General Contractors of America
1957 F Street, N.W.
Washington, D.C. 20006-5194
phone 202/393-2040, fax 202/347-4004

Association for Repetitive Motion Syndromes
Post Office Box 514
Santa Rosa, California 95402
phone 707/571-0397, no fax

Association of University Programs in Occupational Health and Safety
c/o University of Washington
Department of Environmental Health
Post Office Box 357234
Seattle, Washington 98195-7234
phone 206/543-6991, fax 206/543-9616

A. M. Best Company
Best's Safety Directory
Ambest Road
Oldwick, New Jersey 08858
phone 908/439-2200, fax 908/439-3296

Board of Certified Hazard Control Management
8009 Carita Court
Bethesda, Maryland 20817
phone 301/984-8969, fax 301/984-1516

Board of Certified Healthcare Safety Professionals
8009 Carita Court
Bethesda, Maryland 20817
phone 301/984-8969, fax 301/984-1516

Board of Certified Safety Professionals
208 Burwash Avenue
Savoy, Illinois 61874
phone 217/359-9263, fax 217/359-0055

Building Officials Code Administrations, International
4051 W. Flossmoor Road
Country Club Hills, Illinois 60478
phone 708/799-2300, fax 708/799-4981

Bureau of National Affairs, Inc.
Occupational Safety & Health Reporter
1231 25th Street N.W.
Washington, D.C. 20037
phone 202/452-4200, fax 202/452-5331

Canadian Centre for Occupational Safety & Health
250 Main Street East
Hamilton, Ontario, Canada L8N 1H6
phone 800/668-4284, fax 905/572-2206

Canadian Society of Safety Engineering
P.O. Box 294
Kleinburg, Ontario, Canada L0J 1C0
phone 905/893-1689, fax 905/893-2392

CCH Incorporated
Employment Safety & Health Guide
4025 West Peterson Avenue
Chicago, Illinois 60646
phone 800/835-5224; fax 800/224-8299

Center for Chemical Process Safety
c/o American Institute of Chemical
Engineers
345 East 47th Street, 12th Floor
New York, New York 10017
phone 212/705-7319, fax 212/838-8274

Center for Ergonomics Research
Miami University
Department of Psychology
104 Benton Hall
Miami, Ohio 45056
phone 513/529-2414, fax 513/529-2420

Chemical Manufacturers Association,
Inc.
1300 Wilson Boulevard
Arlington, Virginia 22209
phone 703/741-5000, fax 703/741-6000

Chemtrec Center Non-Emergency
Services
c/o Chemical Manufacturers
Association, Inc.
1300 Wilson Boulevard
Arlington, Virginia 22209
phone 800/262-8200, fax 703/741-6089

Chlorine Institute
2001 L Street N.W., No. 506
Washington, D.C. 20036
phone 202/775-2790, fax 202/223-7225

Compressed Gas Association
1725 Jefferson Davis Hwy.,
Suite 1004
Arlington, Virginia 22202
phone 703/412-0900, fax 703/412-0128

Council of American Building
Officials
5203 Leesburg Pike, Suite 708
Falls Church, Virginia 22041
phone 703/931-4533, fax 703/379-1546

Factory Mutual Engineering Corp.
1151 Boston-Providence Turnpike
Norwood, Massachusetts 02062
phone 781/762-4300, fax 781/762-9375

Fire Retardant Chemicals Association
851 New Holland Avenue
Box 3535
Holland, Pennsylvania 17604
phone 717/291-5616, fax 717/295-4538

Hazardous Materials Advisory
Council
1110 Vermont Ave., N.W., Suite 250
Washington, D.C. 20005-3406
phone 202/728-1460, fax 202/728-1459

Human Factors and Ergonomics
Society
Post Office Box 1369
Santa Monica, California 90406-1369
phone 310/394-1811, fax 310/394-2410

Industrial Chemical Research
Association
2547 Monroe Street
Dearborn, Michigan 48124
phone 313/563-0360, fax 313/563-1448

Industrial Safety Equipment
Association, Inc.
1901 North Moore Street
Suite 808
Arlington, Virginia 22209
phone 703/525-1695, fax 703/528-2148

Institute of Industrial Engineers
25 Technology Park
Norcross, Georgia 30092
phone 770/449-0461, fax 770/263-8532

Inter-American Safety Council
(Consejo Interamericano de
Seguridad)
33 Park Place
Englewood, New Jersey 07631
phone 201/871-0004, fax 201/871-
2074

International Association of
Industrial Accident Boards and
Commissions
1575 Aviation Court Parkway
Suite 512
Daytona Beach, Florida 33114
phone 904/252-2915, fax 904/258-
9965

International Ergonomics Association
Central Advisory Group on
Ergonomics
Personnel Affairs Department
Netherlands Railways, NL-3511
Utrecht, Netherlands
phone 30-354-455, fax 30-357-639

International Loss Control Institute
4546 Atlanta Highway
Post Office Box 1898
Loganville, Georgia 30249
phone 404/466-2208, fax 404/466-
4318

International Occupational Safety
and Health Information Centre
International Labour Organization
4 Rue des Morillons, CH-1211
Geneva 22, Switzerland
phone 22-799-6740, fax 22-798-6253

International Society for Respiratory
Protection
Post Office Box 158
Jonesborough, Tennessee 37659
phone 615/753-1388, fax 615/753-
8645

Lead Industries Association Inc.,
295 Madison Avenue
New York, New York 10017
phone 212/578-4750, fax 212/684-
7714

Mason Contractors Association of
America
1550 Spring Road, Suite 320
Oak Brook, Illinois 60251-1363
phone 630/782-6767, fax 630/782-
6786

Metal Treating Institute
1550 Roberts Drive
Jacksonville Beach, Florida 32250-
3222
phone 904/249-0448, fax 904/249-
0459

National Association of Home Builders
1201 15th Street, N.W.
Washington, D.C. 20005
phone 202/822-0200, fax 202/822-
0559

National Association of Manufacturers
1331 Pennsylvania Avenue, N.W.
Suite 1500N
Washington, D.C. 20004-1790
phone 202/637-3000, fax 202/637-
3182

National Fire Protection Association
1 Batterymarch Park
Quincy, Massachusetts 02269
phone 800/344-3555, fax 617/770-
0700

National Restaurant Association Risk
and Safety Managers
c/o National Restaurant Association
150 North Michigan Avenue, Suite
2000
Chicago, Illinois 60601
phone 312/853-2525, fax 312/853-
2548

National Roofing Contractors
Association
1025 W. Higgins Road, Suite 600
Rosemont, Illinois 60018
phone 800/323-9545, fax 847/299-
1183

National Safe Workplace Institute
3008 Bishops Ridge
Monroe, North Carolina 28110
phone 704/289-6061, fax 704/289-
6766

National Safety Council
1121 Spring Lake Drive
Itasca, Illinois 60143-3201
phone 630/285-1121, fax 630/285-1315

National Safety Management Society
12 Pickens Lane
Weaverville, North Carolina 28787
phone 704/645-5229, fax 704/645-5229

National Society of Professional Engineers
1420 King Street
Alexandria, Virginia 22314
phone 703-684-2800, fax 703/836-4875

Rubber Manufacturers Association
1400 K Street, N.W., Suite 900
Washington, D.C. 20005
phone 202/682-4800, fax 202/682-4854

Safety Equipment Institute
1901 North Moore Street
Suite 808
Arlington, Virginia 22209
phone 703/525-3354, fax 703/528-2148

Safety Equipment Manufacturers
Agents Association
Post Office Box 30310
Cincinnati, Ohio 45230
phone 513/624-3535, fax 513/231-1456

Safety, Health, and Environmental
Resource Center International
Central Missouri State University
Homphreys Building
Suite 202
Warrensburg, Missouri 64093
phone 816/543-4744, fax 816/543-4959

Scaffold Industry Association
14039 Sherman Way
Van Nuys, California 91405-2599
phone 818/782-2012, fax 818/786-3027

Sheet Metal Occupational Health
Institute
601 North Fairfax St., Suite 250
Alexandria, Virginia 22314
phone 703/739-7132, fax 703/739-7134

Society of Fire Protection Engineers
One Liberty Square
Boston, Massachusetts 02109
phone 617/482-0686, fax 617/482-8184

Society of Manufacturing Engineers
One SME Drive
Dearborn, Michigan 48121
phone 313/271-1500, fax 313/271-2861

Society for Occupational and
Environmental Health
6728 Old McLean Village Drive
McLean, Virgina 22101
phone 703/556-9222, fax 703/556-8729

The Society of the Plastics Industry
1275 K Street, N.W., Suite 400
Washington, D.C. 20005
phone 202/371-5200, fax 202/371-1022

System Safety Society
5 Export Drive
Suite A
Sterling, Virginia 20164
phone 703/450-0310, fax 703/450-1745

Underwriters Laboratories, Inc.
333 Pfingsten Road
Northbrook, Illinois 60062
phone 847/272-8800, fax 847/272-8129

Voluntary Protection Programs
Participants' Association (VPPPA)
7600-B Leesburg Pike, Suite 440
Falls Church, Virginia 22043
phone 703/761-1146, fax 703/761-1148

Selected Federal Safety and Health Agencies

Agency for Toxic Substances and Disease Registry
U.S. Department of Health and Human Services
1600 Clifton Road
Atlanta, Georgia 30333
phone 404/639-0500

Bureau of Labor Statistics
U.S. Department of Labor
Occupational Safety and Health Statistics
441 G Street N.W.
Washington, D.C. 20212
phone 202/523–1382

Centers for Disease Control and Prevention
U.S. Department of Health and Human Services
1600 Clifton Avenue N.E.
Atlanta, Georgia 30333
phone 404/329–3311

Crew System Ergonomics Information Analysis Center, AL/CFH/CSERIAC
2255 H Street, Building 248
Wright-Patterson AFB, Ohio 45433-7022
phone 937/255-4842

Environmental Protection Agency
401 M Street S.W.
Washington, D.C. 20460
phone 202/260–2090

Federal Highway Administration
Office of Motor Carriers
U.S. Department of Transportation
400 7th Street S.W.
Washington, D.C.
phone 202/366–4000

Federal Railroad Administration
U.S. Department of Transportation
400 7th Street S.W.
Washington, D.C. 20590
phone 202/366–4000

Mine Safety and Health Administration
U.S. Department of Labor
4015 Wilson Boulevard
Arlington, Virginia 22203
phone 703/235–1452

National Archives and Records Administration
Multimedia and Publications Distribution Center
Customer Services Section
8700 Edgeworth Drive
Capitol Heights, Maryland 20743–3701
phone 301/763–1896

National Institute for Occupational
Safety and Health
U.S. Department of Health and
Human Services
Publications Dissemination
4676 Columbia Parkway
Cincinnati, Ohio 45226
phone 513/533–8287

National Institute of Standards and
Technology
U.S. Department of Commerce
National Engineering Laboratory
Route I-270 and Quince Orchard
Road
Gaithersburg, Maryland 20899
phone 301/975–2000

National Institutes of Health
U.S. Department of Health and
Human Services
900 Rockville Pike
Bethesda, Maryland 20205
phone 301/496–5787

National Library of Medicine
8600 Rockville Pike
Bethesda, Maryland 20014
phone 301/496–4000

National Technical Information
Service
U.S. Department of Commerce
5285 Port Royal Road
Springfield, Virginia 22161
phone 703/487–4650

Nuclear Regulatory Commission
Washington, D.C. 20555 (no street
address necessary)
phone 301/415–7000

Occupational Safety and Health
Administration (OSHA)
National Office
U.S. Department of Labor
200 Constitution Avenue N.W.
Washington, D.C. 20210
phone 202/219–9308

Occupational Safety and Health
Administration (OSHA)
Publications Office
Room N–3101
Third Street and Constitution
Avenue N.W.
Washington, D.C. 20210
phone 202/219–4667

Occupational Safety and Health
Administration (OSHA)
Training Institute
1555 Times Drive
Des Plaines, Illinois 60018
phone 847/297–4810

Occupational Safety and Health
Review Commission
1825 K Street N.W.
Washington, D.C. 20006
phone 202/643–7943

Office of Drug and Alcohol Policy and
Compliance
U.S. Department of Transportation
400 7th Street S.W., Room 10317
Washington, D.C. 20590
phone 202/366–3784

Research and Special Programs
Administration
Office of Hazardous Materials
Safety
U.S. Department of Transportation
400 7th Street S.W.
Washington, D.C. 20590
phone 202/366–4000

Superintendent of Documents
U.S. Government Printing Office
732 North Capitol Street N.W.
Washington, D.C. 20402
phone 202/512–1800

U.S. Fire Administration
Federal Emergency Management
Agency
16825 South Seton Avenue
Emmitsburg, Maryland 21727
phone 301/447–1018

Selected Safety and Health Internet Mailing Lists

To subscribe to any of the lists in the following group, send this message to the listserv: subscribe [name of list] [your real name].

Name of List	Listserv Address	General Subject Area
BIOSAFTY	listserv@mitvma.mit.edu	safe handling of biohazards
CHEMED-L	listserv@uwf.cc.uwf.edu	chemical safety
CMTS-L	listserv@cornell.edu	chemical management and tracking
DISPATCH	majordomo@tcomeng.com	police, fire, EMS telecommunications
EMERG-L	listserv@vm.marist.edu	emergency services
FIRENET	listserv@life.anu.edu.au	firefighting and emergency response
FIRST-AID	first-aid request @rabble.uow.edu.au	casual and professional users of first-aid
HEALTHRE	listserv@ukcc.uky.edu	health-care reform
HELPNET	listserv@vm1.nodak.edu	network emergency response planning
LEPC	listproc@moose.uvm.edu	hazardous materials emergency response
RISKNET	listproc@mcfeeley.cc.utexas.edu	safety and health
SYSTEM-SAFETY	listserv@listserv.gsfc.nasa.gov	system safety

To subscribe to any of the lists in the following group, send this message to the listserv: subscribe [name of list][your e-mail address].

Name of List	Listserv Address	General Subject Area
CCM-L	ccm-request@eja.anes.hscsyr.edu	critical care
DMATNEWS	listserv@mediccom.norden1.com	disaster medical assistance teams
EMC-PSTC	majordomo@ieee.org	product safety
EMED-L	emed-1@itsa.ucsf.edu	for health-care professionals only
HAZMATMED	listserv@mediccom.norden1.com	hazardous materials emergency response
SSAVETT_NCEMSF	majordomo@indiana.edu	National Collegiate EMS Foundation

Selected Safety and Health Web Sites

AFL-CIO
  http://www.aflcio.org

Agency for Toxic Substances and Disease Registry
  http://atsdr1.atsdr.cdc.gov:8080/atsdrhome.html

American Conference of Governmental Industrial Hygienists
  http://www.acgih.org

American Health Information Management Association
  http://www.ahima.org/

American Industrial Hygiene Association
  http://www.aiha.org

American National Standards Institute
  http://web.ansi.org

American Society for Testing and Materials
  http://www.astm.org

American Society of Safety Engineers
  http://www.ASSE.org

Asbestos Institute
  http://www.odyssee.net/~ai/

BNA Communications, http://zeus.bna.com/bnac/

Bureau of Labor Statistics
  http://stats.bls.gov

Brookhaven National Laboratory, Safety and Environmental
Protection Division
  http://sun10.sep.bnl.gov/seproot.html

Canadian Centre for Occupational Health and Safety (CCOHS)
http://www.ccohs.ca

Chemical Manufacturers Association
http://www.cmahq.com/index.html

Chemistry Resources on the Internet
http://www.rpi.edu/dept/chem/cheminfo/chemres.html

Duke University Occupational & Environmental Medicine
http://occ-env-med.mc.duke.edu/oem

Enviro-Net MSDS Index
http://www.environ-net.com/msds/msds.html

Environmental Health Center/National Safety Council
http://envirolink.org

Environmental Protection Agency
http://www.epa.gov

Ergo Web
http://tucker.mech.utah.edu

FedWorld
http://www.fedworld.gov

Government Information Xchange
http://www.info.gov/info.html

Government Printing Office
http://www.thorplus.lib.purdue.edu/gpo/

Human Factors and Ergonomic Society
http://www.hfes.vt.edu/HFES/

IAPPS (International Assoc. of Personal Protection Specialists)
http://www.mps.ohio-state.edu/cgi-bin/hpp?Iapps_home/html

Index of Occupational Safety and Health Resources on the Internet
http://turva.me.tut.fi/oshweb/

Institute for Research in Construction
http://www.cisti.nrc.ca/irc/irccontents.html

Martindale's Health Science Guide '96
http://www-sci.lib.uci.edu/~martindale/HSGuide.html

Mine Safety and Health Administration
http://www.msha.gov

MSDS On-line from University of Utah
gopher://atlas.chem.utah.edu/70/11/MSDS

MSU Radiation, Chemical & Biological Safety
http://www.orcbs.msu.edu

National Fire Protection Association
http://www.nfpa.com/

National Institute for Occupational Safety and Health
http://www.cdc.govniosh/homepage.html

National Safety Council
http://www.nsc.org

National Universty of Singapore BioMed Web Server (natural toxins database
http://biomed.nus.sg

Occupational Injury and Illness Rates
http://www.osha.gov/oshstats/bls.text

OSHA
http://www.osha.gov

OSHA Consultation Service
http://www.osha.gov/oshaprogs/consult.html

OSHA Federal Register
http://www.osha-s/c.gov/OCIS/TOC_fed_reg.html

OSHA Salt Lake Technical Center (Federal Register and OSHA regulations)
http://www.osha-slc.gov

OSHA Standards & Related Documents
http://www.osha-slc.gov/OCIS/standards

OSHA Statistics & Data
http://www.osha.gov/oshstats/index.html

Periodic Table
http://www.cchem.berkeley.edu/Table/index.html

RISKWeb
http://riskweb.bus.utexas.edu/riskweb.html

Rocky Mountain Center for Occupational and Environmental Health
http://rocky.utah.edu

RSI/UK focuses on Repetitive Strain Injury and related topics
http://www.demon.co.uk/rsi

Safety Information Resources on the Internet
http://ds.internic.net/cgi-bin/enthtml/environment/siri.b

Safety Net Yellow Pages
http://www.tiac.net/users/dploss/html/olddir.html

Safety Online
http://www.safetyonline.net

SIC Search
http://www.osha.gov/oshstats/sicser.html

Stanford University glossary of MSDS terms
http://www-nanonet.stanford.edu/NanoFab/safety/
S8glossary.html

University of Illinois, Urbana, Division of Environmental Health and Safety
http://phanton.ehs.uiuc.edu/Dehs.html

University of London Ergonomics and Human Computer Interaction
http://www.ergohci.ucl.ac.uk

University of Virginia EPA Chemical Substances Factsheets
gopher://ecosys.drdr.virginia.edu/11/library/gen/toxics

U.S. Centers for Disease Control
http://www.cdc.gov

U.S. Department of Energy—Office of Environment, Safety and Health
http://tis.eh.doe.gov/

U.S. Department of Health and Human Services
http://www.os.dhhs.gov

U.S. EPA Center for Exposure Assessment Modeling
http://ftp.epa.gov/epa_ceam/www.html/ceam_home.html

U.S. FEMA
http://www.fema.gov

U.S. Food and Drug Administration
http://vm.cfsan.fda.gov/index.html

U.S. National Institutes of Health
http://www.nih.gov/

U.S. National Library of Medicine
http://www.nlm.nih.gov/

UVA's Video Display Ergonomics page
http://www.virginia.edu/~enhealth/ERGONOMICS/toc.html

Where to Find MSDS on the Internet (from the University of Kentucky)
http://www.uky.edu/ArtsSciences/Chemistry/Resources/MSDS.HTML

World Health Organization
http://www.who.ch

Self-Inspection Checklist

Disclaimer

This checklist is not intended to be all inclusive. Further, I recommend that you add items that are applicable to your operations and delete items that are not applicable. This list does not include certain major topics that are almost exclusive to the construction industry. These include such subjects as excavations, scaffolds, construction vehicles, tunneling, demolition, concrete and masonry construction, traffic in construction areas, and various fall protection scenarios. I recommend that you give special attention to these subjects, as needed. This checklist is not to be considered as a substitute for OSHA standards. You should refer to those standards for complete and detailed requirements. If your workplace is in a geographical area where the OSHA program is under state jurisdiction, consider that the state's standards may be more stringent.

Note that some topics appear under more than one subject heading. In some of those cases, I have noted the related heading.

Abrasive Wheel Machinery

	Yes	No
1. Does each machine have proper guarding?	☐	☐
2. Are unused runs of sanding belts guarded against accidental contact?	☐	☐
3. Are nip points, where sanding belts run onto a pulley, guarded?	☐	☐
4. Are guards in proper position?	☐	☐
5. Are tongue guards in proper position?	☐	☐
6. Is eye protection provided and used?	☐	☐

7. Are signs posted reminding operators to wear
   protective eye wear? ☐ ☐
8. Are machines properly grounded? ☐ ☐
9. Are machines properly anchored? ☐ ☐
10. Is the work rest used and kept adjusted to
    within $\frac{1}{8}$ in of the wheel? ☐ ☐
11. Is the adjustable tongue on the top side of the grinder
    used and kept adjusted to within $\frac{1}{4}$ in of the wheel? ☐ ☐
12. Do side guards cover the spindle, nut, flange,
    and 75 percent of the wheel diameter? ☐ ☐
13. Are bench and pedestal grinders permanently
    mounted? ☐ ☐
14. Is the maximum RPM rating of each abrasive wheel
    compatible with the RPM rating of the grinder motor? ☐ ☐
15. Are fixed or permanently mounted grinders
    connected to their electrical supply system with
    metallic conduit or other permanent wiring method? ☐ ☐
16. Does each grinder have an individual on and off
    control switch? ☐ ☐
17. Is each electrically operated grinder effectively
    grounded? ☐ ☐
18. Before new abrasive wheels are mounted,
    are they visually inspected and ring tested? ☐ ☐
19. Are dust collectors and powered exhausts provided
    on grinders used in operations that produce
    large amounts of dust? ☐ ☐
20. Are splash guards mounted on grinders that
    use coolant to prevent the coolant from reaching
    employees? ☐ ☐
21. Is cleanliness maintained around grinders? ☐ ☐

Anhydrous Ammonia Emergency Plan

	Yes	No

1. Is there an anhydrous ammonia emergency
   plan in place? ☐ ☐
2. Does the plant provide training for employees
   in ammonia emergency procedures? ☐ ☐
3. Has each employee gone through the
   training course? ☐ ☐
4. Is the training documented? ☐ ☐
5. Is a deluge shower (or equivalent means of full
   body flushing) easily accessible? ☐ ☐
6. Are there at least two suitable gas masks in
   readily accessible locations? ☐ ☐

Back Safety (see also Ergonomics)

	Yes	No
1. Does the plant provide proper training (including for conditioning) in back safety?	☐	☐
2. Has each employee gone through the training course?	☐	☐
3. Is the training documented?	☐	☐
4. Are proper lifting techniques being used?	☐	☐

Bloodborne Pathogens

	Yes	No
1. Do you have any employees who could be reasonably anticipated, as the result of performing their job duties, to face contact with blood and other potentially infectious materials? *If so, proceed.*	☐	☐
2. Does your plant have a written exposure control plan?	☐	☐
3. Has employee training been provided?	☐	☐
4. Is safety equipment provided?	☐	☐
5. Are employee medical and training records maintained?	☐	☐
6. Does the employee training program on the bloodborne pathogens standard contain the following elements:		
a. an accessible copy of the standard and an explanation of its contents?	☐	☐
b. a general explanation of the epidemiology and symptoms of bloodborne diseases?	☐	☐
c. an explanation of the modes of transmission of bloodborne pathogens?	☐	☐
d. an explanation of the employer's exposure control plan and the means by which employees can obtain a copy of the written plan?	☐	☐
e. an explanation of the appropriate methods for recognizing tasks and the other activities that may involve exposure to blood and other potentially infectious materials?	☐	☐
f. an explanation of the use and limitations of methods that will prevent or reduce exposure (including appropriate engineering controls, work practices, personal protective equipment, and the concept of universal precautions)?	☐	☐
g. information on the types, proper use, location, removal, handling, decontamination, and disposal of personal protective equipment?	☐	☐
h. an explanation of the basis for selection of personal protective equipment?	☐	☐

    i. information on the hepatitis B vaccine and the
company policy to provide HEP vaccinations if an
exposure event is reasonably anticipated or if an
exposure event occurs (unless declined in writing
by an employee)? ☐   ☐

    j. information on the appropriate actions to take
and persons to contact in an emergency involving
blood or other potentially infectious materials? ☐   ☐

    k. an explanation of the procedure to follow if an
exposure incident occurs, including the methods
of reporting the incident and the medical follow-up
that will be made available? ☐   ☐

    l. information on postexposure evaluations
and follow-up? ☐   ☐

    m. an explanation of signs, label, and color coding? ☐   ☐

Compressed Gas Cylinders (see also Welding, Cutting, and Brazing)

	Yes	No
1. Are cylinders with a water weight capacity over 30 lb equipped with means for connecting a valve protector device or with a collar or recess to protect the valve?	☐	☐
2. Are cylinders legibly marked to clearly identify the gas contained?	☐	☐
3. Are compressed gas cylinders stored in areas which are protected from external heat sources such as flame impingement, intense radiant heat, electric arcs, or high temperature lines?	☐	☐
4. Are cylinders located or stored in areas where they will not be damaged by passing or falling objects or subject to tampering by unauthorized persons?	☐	☐
5. Are cylinders stored or transported in a manner to prevent them from creating a hazard by tipping, falling, or rolling?	☐	☐
6. Are cylinders containing liquefied fuel gas stored or transported in a position so that the safety relief device is always in direct contact with the vapor space in the cylinder?	☐	☐
7. Are valve protectors always placed on cylinders when the cylinders are not in use or connected for use?	☐	☐
8. Are all valves closed off before a cylinder is moved when the cylinders are not in use or connected for use?	☐	☐

9. Are all valves closed off before a cylinder is moved, when the cylinder is empty, and at the completion of each job? ☐ ☐
10. Are low-pressure fuel-gas cylinders checked periodically for corrosion, general distortion, cracks, or any other defect that might indicate a weakness or render it unfit for service? ☐ ☐
11. Does the periodic check of low-pressure fuel-gas cylinders include a close inspection of the cylinders, bottom? ☐ ☐

Compressors and Air Receivers

	Yes	No

1. Is every receiver equipped with a pressure gauge and with one or more automatic spring-loaded safety valves? ☐ ☐
2. Is the total relieving capacity of the safety valve capable of preventing pressure in the receiver from exceeding the maximum allowable working pressure of the receiver by more than 10 percent? ☐ ☐
3. Is every air receiver provided with a drainpipe and valve at the lowest point for the removal of accumulated oil and water? ☐ ☐
4. Are compressed air receivers periodically drained of moisture and oil? ☐ ☐
5. Are all safety valves tested frequently and at regular intervals to determine whether they are in good operating condition? ☐ ☐
6. Is there a current operating permit required by state or local authorities? ☐ ☐
7. Is the inlet of air receivers and piping systems kept free of accumulated oil and carbonaceous materials? ☐ ☐

Compressors and Compressed Air

	Yes	No

1. Are compressors equipped with pressure relief valves and pressure gauges? ☐ ☐
2. Are compressor air intakes installed and equipped so as to ensure that only clean uncontaminated air enters the compressor? ☐ ☐
3. Are air filters installed on the compressor intake? ☐ ☐
4. Are compressors operated and lubricated in accordance with the manufacturer's recommendations? ☐ ☐

5. Are safety devices on compressed air systems checked frequently? ☐ ☐

6. Before any repair work is done on the pressure system of a compressor, is the pressure bled off and the system locked out? ☐ ☐

7. Are signs posted to warn of the automatic starting feature of the compressors? ☐ ☐

8. Is the belt drive (and blade, if applicable) system totally enclosed to provide protection for the front, back, top, and sides? ☐ ☐

9. Is it strictly prohibited to direct compressed air toward a person? ☐ ☐

10. Are employees prohibited from using compressed air for cleaning purposes, except where reduced to less than 30 psi dead-ended? ☐ ☐

11. Are employees prohibited from using compressed air to clean off their bodies or clothing? ☐ ☐

12. When using compressed air for cleaning, do employees wear protective chip guarding and personal protective equipment? ☐ ☐

13. Are safety chains or other suitable locking devices used at couplings of high pressure hose lines where a connection failure would create a hazard? ☐ ☐

14. Before compressed air is used to empty containers of liquid, is the safe working pressure of the container checked? ☐ ☐

15. When compressed air is used with abrasive blast cleaning equipment, is the operating valve a type that must be held open manually? ☐ ☐

16. When compressed air is used to inflate auto tires, is a clip-on chuck and an inline regulator preset to 40 psi required? ☐ ☐

17. Is it prohibited to use compressed air to clean-up or move combustible dust if such action could cause the dust to be suspended in the air and cause a fire or explosion hazard? ☐ ☐

18. Are hose trip hazards abated (e.g., by reels)? ☐ ☐

Confined Space Entry (Permit Required)

	Yes	No
1. Is there a confined space entry program in place?	☐	☐
2. Does the plant provide training for employees in confined space entry?	☐	☐
3. Has each employee gone through the training course?	☐	☐

4. Is the training documented? ☐ ☐
5. Do the employees use a permit before entering a permit confined space? ☐ ☐
6. Are these permits kept on file? ☐ ☐
7. Are all permits signed? ☐ ☐
8. Are all confined spaces locked and tagged out before they are entered? ☐ ☐
9. Are all doors blocked before the confined space is entered? ☐ ☐
10. Are all lines bled before the confined space is entered? ☐ ☐
11. Are all lines blocked before the confined space is entered? ☐ ☐
12. Is air quality checked before the confined space is entered? ☐ ☐
13. Is air quality monitored continuously during the time any employee is in the confined space? ☐ ☐
14. Are there at least two employees assigned to each confined space job? ☐ ☐
15. Is rescue equipment available at the plant including:
    a. self-contained breathing apparatus? ☐ ☐
    b. safety harness? ☐ ☐
    c. safety line? ☐ ☐
    d. safety hoist, tripod, or winch? ☐ ☐

16. Is air monitoring equipment calibrated as required? ☐ ☐
17. Are the calibration certificates kept on file? ☐ ☐
18. Is there a program set up for outside contractors? ☐ ☐
19. Are confined spaces thoroughly emptied of any corrosive or hazardous substances, such as acids or caustics, before entry? ☐ ☐
20. Are all lines to a confined space containing inert, toxic, flammable, or corrosive materials valved off and blanked or disconnected and separated before entry? ☐ ☐
21. Is it required that all impellers, agitators, or other moving equipment inside confined spaces be locked out if they present a hazard? ☐ ☐
22. Is either natural or mechanical ventilation provided prior to confined space entry? ☐ ☐
23. Are appropriate atmospheric tests performed to check for oxygen deficiency, toxic substances, and explosive concentrations in the confined space before entry? ☐ ☐
24. Is adequate illumination provided for the work to be performed in the confined space? ☐ ☐

25. Is the atmosphere inside the confined space frequently tested or continuously monitored during the conduct of work?   □   □

26. Is there an assigned safety standby employee outside of the confined space, when required, whose sole responsibility is to watch the work in progress, sound an alarm if necessary, and render assistance?   □   □

27. Is the standby employee appropriately trained and equipped to handle an emergency?   □   □

28. Is the standby employee or other employees prohibited from entering the confined space without lifelines and respiratory equipment if there is any question as to the cause of an emergency?   □   □

29. Is approved respiratory equipment required if the atmosphere inside the confined space cannot be made acceptable?   □   □

30. Is all portable electrical equipment used inside confined spaces either grounded and insulated, and where applicable, equipped with ground fault protection?   □   □

31. Before gas welding or burning is started in a confined space, are hoses checked for leaks, are compressed gas bottles forbidden inside the confined space, are torches lighted only outside of the confined area, and is the confined area tested for an explosive atmosphere each time prior to taking a lighted torch into the confined space?   □   □

32. If employees will be using oxygen-consuming equipment such as salamanders, torches, furnaces, etc. in a confined space, is sufficient air provided to assure combustion without reducing the oxygen concentration of the atmosphere below 19.5 percent by volume?   □   □

33. Whenever combustion-type equipment is used in a confined space, are provisions made to ensure the exhaust gases are vented outside of the enclosure?   □   □

34. Is each confined space checked for decaying vegetation or animal matter which may produce methane?   □   □

35. Is the confined space checked for possible industrial waste which could contain toxic properties?   □   □

36. If the confined space is below the ground and near areas where motor vehicles will be operating, is it possible for vehicle exhaust or carbon monoxide to enter the space?   □   □

Control of Harmful Substances by Ventilation

	Yes	No
1. Has in-plant air quality been checked and compared to OSHA PELs?	☐	☐
2. Have engineering controls (hoods, ducts, fans, etc.) been provided as required?	☐	☐
3. Is the volume and velocity of air in each exhaust system sufficient to gather the dusts, fumes, mists, vapors, or gases to be controlled and to convey them to a suitable point of disposal?	☐	☐
4. Are exhaust inlets, ducts, and plenums designed, constructed, and supported to prevent collapse or failure of any part of the system?	☐	☐
5. Are clean-out ports or doors provided at intervals not to exceed 12 ft in all horizontal runs of exhaust ducts?	☐	☐
6. Where two or more different types of operations are being controlled through the same exhaust system, will the combination of substances being controlled constitute a fire, explosion, or chemical reaction hazard in the duct?	☐	☐
7. Is adequate make-up air provided to areas where exhaust systems are operating?	☐	☐
8. Is the source point for make-up air located so that only clean, fresh air, which is free of contaminates, will enter the work environment?	☐	☐
9. Where two or more ventilation systems are serving a work area, is their operation such that one will not offset the functions of the other?	☐	☐
10. Is the work area's ventilation system appropriate for the work being performed?	☐	☐
11. Are spray painting operations done in spray rooms or booths equipped with an appropriate exhaust system?	☐	☐
12. Is employee exposure to welding fumes controlled by ventilation, use of respirators, exposure time, or other means?	☐	☐
13. Are grinders, saws, and other machines that produce respirable dusts vented to an industrial collector or central exhaust system?	☐	☐
14. Are all local exhaust ventilation systems designed and operating properly (e.g., airflow and volume appropriate for the application, ducts not plugged, and belts not slipping)?	☐	☐

Dip Tanks Containing Flammable or Combustible Liquids

	Yes	No
1. Are all dip tanks with more than 150 gal capacity or 10 ft^2 in liquid surface area equipped with a properly trapped overflow pipe leading to a safe location outside of the building?	☐	☐
2. Are all dip tanks with more than 150 gal capacity or 4 ft^2 in liquid surface area protected with at least one of the following automatic extinguishing facilities: water spray system, foam system, carbon dioxide system, dry chemical system, or automatic dip tank cover?	☐	☐
3. Do you prohibit open flames, spark producing devices, or heated surfaces having a temperature sufficient to ignite vapors from being in any vapor area?	☐	☐

Electrical Equipment

	Yes	No
1. Are ground fault circuit interrupters installed for each receptacle in wet areas or areas subject to splashing, dripping, or leaking liquids?	☐	☐
2. Is only properly listed and labeled equipment used and maintained as designed?	☐	☐
3. Are all employees required to report as soon as practicable any obvious hazard to life or property observed in connection with electrical equipment or lines?	☐	☐
4. Are employees instructed to make preliminary inspections and/or appropriate tests to determine what conditions exist before starting work on electrical equipment or lines?	☐	☐
5. When electrical equipment or lines are to be serviced, maintained, or adjusted, are necessary switches opened, locked out, and tagged whenever possible?	☐	☐
6. Are portable electrical tools and equipment grounded or of the double insulated type?	☐	☐
7. Are electrical appliances such as vacuum cleaners, polishers, vending machines, etc. grounded?	☐	☐
8. Do extension cords have a grounding conductor?	☐	☐
9. Are multiple plug adapters prohibited?	☐	☐
10. Are ground fault circuit interrupters installed on each temporary 15 or 20 A, 120 V AC circuit at locations where construction, demolition, modifications, alterations, or excavations are being performed?	☐	☐

11. Are all temporary circuits protected by suitable disconnecting switches or plug connectors at the junction with permanent wiring? ☐ ☐
12. Do you have electrical installations in hazardous dust or vapor areas? If so, do they meet the National Electrical Code (NEC) for hazardous locations? ☐ ☐
13. Is exposed wiring and cords with frayed or deteriorated insulation repaired or replaced promptly? ☐ ☐
14. Are splices, joints, and free ends of conductors covered with insulation equivalent to that of conductors or with insulating devices suitable for the purpose? ☐ ☐
15. Have you assured that flexible electrical cords are not concealed, run through holes in walls, ceilings, or floors, or run through doorways, windows, or similar openings? ☐ ☐
16. Are all flexible electrical cords connected to devices and fittings so that strain relief is provided to prevent pull from being directly transmitted to joints or terminal screws? ☐ ☐
17. Are clamps or other securing means provided on flexible cords or cables at plugs, receptacles, tools, equipment, etc., and is the cord jacket securely held in place? ☐ ☐
18. Are all cord, cable, and raceway connections intact and secure? ☐ ☐
19. Do you prohibit the use of knockout type electrical boxes (designed for mounting in a permanent location) at the end of suspended cords and at the end of extension cords? ☐ ☐
20. In wet or damp locations, are electrical tools and equipment appropriate for the use or location or otherwise protected? ☐ ☐
21. Is the location of electrical power lines and cables (overhead, underground, underfloor, other side of walls, etc.) determined before digging, drilling, or similar work is begun? ☐ ☐
22. Are metal measuring tapes, ropes, handlines, or similar devices with metallic thread woven into the fabric prohibited where they could come in contact with energized parts of equipment or circuit conductors? ☐ ☐
23. Is the use of metal ladders prohibited in areas where the ladder or the person using the ladder could come in contact with energized parts of equipment, fixtures, or circuit conductors? ☐ ☐

24. Are all disconnecting switches and circuit breakers labeled to indicate their use or equipment served? ☐ ☐

25. Are disconnecting means always opened before fuses are replaced? ☐ ☐

26. Do all interior wiring systems include provisions for grounding metal parts of electrical raceways, equipment, and enclosures? ☐ ☐

27. Are electrical raceways and enclosures securely fastened in place? ☐ ☐

28. Are all energized parts of electrical circuits and equipment guarded against accidental contact by approved cabinets or enclosures? ☐ ☐

29. Is sufficient access and working space provided and maintained about all electrical equipment to permit ready and safe operations and maintenance? ☐ ☐

30. Are all unused openings, including conduit knockouts, in electrical enclosures and fittings closed with appropriate covers, plugs, or plates? ☐ ☐

31. Are electrical enclosures such as switches, receptacles, junction boxes, etc. provided with tight-fitting covers or plates? ☐ ☐

32. Are disconnecting switches for electrical motors in excess of 2 hp capable of opening the circuit when the motor is in a stalled condition without exploding? (Switches must be horsepower rated equal to or in excess of the motor hp rating.) ☐ ☐

33. Is low-voltage protection provided in the control device of motors driving machines or equipment which could cause probable injury from inadvertent starting? ☐ ☐

34. Is each motor disconnecting switch or circuit breaker located within sight of the motor control device? ☐ ☐

35. Is each motor located within sight of its controller, is the controller disconnecting means capable of being locked in the open position, or is a separate disconnecting means installed in the circuit within sight of the motor? ☐ ☐

36. Is the controller for each motor in excess of 2 hp rated in horsepower equal to or in excess of the rating of the motor it serves? ☐ ☐

37. Are employees who regularly work on or around energized electrical equipment or lines instructed in the cardiopulmonary resuscitation (CPR) methods? ☐ ☐

38. Are employees prohibited from working alone on energized lines or equipment over 600 V? ☐ ☐

39. Are documented safety-related work practices in place to prevent electrical shock or other injuries resulting from either direct or indirect electrical contact when work is performed near or on equipment or circuits that are or may be energized? ☐ ☐

Elevated Surfaces

	Yes	No

1. Are signs posted, when appropriate, showing the elevated surface load capacity? ☐ ☐
2. Are surfaces elevated 4 ft or more above the floor or ground provided with standard guardrails? ☐ ☐
3. Are all elevated surfaces (beneath which people or machinery could be exposed to falling objects) provided with standard 4-in toeboards? ☐ ☐
4. Is a permanent means of access and egress provided to elevated storage and work surfaces? ☐ ☐
5. Is required headroom provided where necessary? ☐ ☐
6. Is material on elevated surfaces piled, stacked, or racked in a manner to prevent it from tipping, falling, collapsing, rolling, or spreading? ☐ ☐

Elevators

	Yes	No

1. Are all elevators regularly inspected, as per local and state law? ☐ ☐
2. Are all hoistway openings protected by doors or gates interlocked with controls so that the car cannot be started until all gates or doors are closed and so that gates or doors cannot be opened when the car is not at a landing? ☐ ☐
3. Are all elevators equipped with an emergency alarm? ☐ ☐

Emergency Action Plan and Emergency Response Plan (see also Hazardous Waste Operations)

	Yes	No

1. Have an emergency action plan and an emergency response plan been prepared for the plant? ☐ ☐
2. Has employee training been conducted? ☐ ☐
3. Have a plant emergency coordinator and a chain of command been established and assigned? ☐ ☐

4. Has special personal protective equipment (as needed) been provided? ☐ ☐
5. Have evacuation routes, including preferred assignments, and procedures been designated? ☐ ☐
6. Have regrouping areas been designated and procedures for accounting for employees and others been established? ☐ ☐
7. Have drills and practice evacuations been performed with postevent evaluation? ☐ ☐
8. Are procedures in place for shutdown of critical plant operations and, in special cases, for the performance of critical plant operations? ☐ ☐
9. Has the preferred means of reporting emergencies, which may differ depending on the type of emergency, been established? ☐ ☐
10. Are all alarm systems regularly checked to assure that they are functioning properly? ☐ ☐
11. Is emergency communications equipment readily available? ☐ ☐
12. Have you addressed every one of the following possibilities: fire, explosion, earthquake, hurricane, tornado, flood, chemical spills, toxic vapors, nuclear radiation, bomb threats, and workplace violence? ☐ ☐
13. Is there a clear plan for carrying out rescue and medical duties? ☐ ☐

Environmental Controls (general perspective—relates to several other headings)

	Yes	No
1. Can a less harmful method or product be used?	☐	☐
2. If forklifts and other vehicles are used in buildings or other enclosed areas, are the carbon monoxide levels kept below the maximum acceptable concentration?	☐	☐
3. Are proper precautions taken when handling asbestos and other fibrous materials?	☐	☐
4. Are wet methods used, when practicable, to prevent the emission of airborne asbestos fibers, silica dust, and similar hazardous materials?	☐	☐
5. Are engineering controls examined and maintained or replaced on a scheduled basis?	☐	☐
6. Is vacuuming (in some cases requiring a HEPA filter) with appropriate equipment used whenever possible rather than blowing or sweeping dust?	☐	☐

7. Is all water provided for drinking, washing, and cooking potable? ☐ ☐

8. Are all outlets for water that is not suitable for drinking clearly identified? ☐ ☐

9. Are employees' physical capacities assessed before being assigned to jobs requiring heavy work? ☐ ☐

10. Are employees working on streets and roadways where they are exposed to the hazards of traffic required to wear brightly colored (traffic orange) warning vests? ☐ ☐

11. Are exhaust stacks and air intakes so located that contaminated air will not be recirculated within a building or other enclosed area? ☐ ☐

12. Is equipment producing ultraviolet radiation properly shielded? ☐ ☐

Ergonomics (see also Back Safety)

Yes **No**

By engineering (first choice), administrative controls, and/or training and monitoring, have you made every reasonable effort to assure that employees avoid the following actions:

1. Working in a position where one is supporting the weight of his or her upper body by leaning on the hard, sharp edges of a work surface such as a workbench? ☐ ☐

2. Working with one's elbows above his or her shoulders? ☐ ☐

3. Performing a repetitive motion that requires rotating one's forearm with the wrist bent, such as turning the handle of a screwdriver? ☐ ☐

4. Performing a repetitive motion while holding one's wrist in a flexed or extended position or tilted toward one's thumb or little finger? (*Note:* Try to maintain a neutral wrist position, that is, the position the wrist assumes when it is hanging relaxed at one's side.) ☐ ☐

5. Holding on to an object such as a toolbox handle with the tips of the fingers? (*Note:* Use the entire hand, including the thumb and fingers wrapped around the handle, to grasp it.) ☐ ☐

6. Using tools that have hard, sharp edges that would come into contact with one's hands? ☐ ☐

7. Using tools with handles that are too short? (*Note:* The handle should fit in the palm of one's hand and be cradled by all the fingers.) ☐ ☐

8. Working with gloves that are either too large or too small? (*Note:* Those that are too large could reduce the ability to manipulate things and may cause one to use too much force while wearing them; those that are too small could put pressure on the median nerve, contributing to carpal tunnel syndrome.) ☐ ☐
9. Applying too much force to accomplish a job? ☐ ☐
10. Gripping a tool too tightly? ☐ ☐
11. Using the palm of one's hand like a hammer to pound on something or against a tool to increase the force? ☐ ☐
12. Working for prolonged periods of time doing the same thing? ☐ ☐
13. Working with cold hands? (*Note:* This reduces one's sense of feel and may cause a person to use too much force without knowing it.) ☐ ☐
14. Using tools with bare metal handles and grips? (*Note:* Better choices, if possible, are those with padded, textured grips.) ☐ ☐
15. Using a pinch grip to hold on to something? (*Note:* A pinch grip uses the thumb and fingers opposite each other to grasp something such as one would use to pick a penny up off the floor.) ☐ ☐
16. Holding parts of one's body in a stationary awkward position for long periods of time? (*Note:* To experience how fatiguing this really is, hold your arms out in front of you without moving them. You will soon experience discomfort.) ☐ ☐
17. Using tools that subject the hands and fingers to prolonged periods of significant vibration? ☐ ☐
18. Lifting and holding excessive weights, particularly with great frequency, with legs straight (no bend at the knees) when turning the upper body away from the stationary feet and/or when the load is not close to the body? ☐ ☐
19. Lifting or holding excessive weights, particularly when the load is slippery, sharp, easily capable of shifting, has an unusual center of gravity, and/or lacks well-designed and positioned hand holds or grip areas? ☐ ☐

Exit Doors

	Yes	No
1. Are doors which are required to serve as exits designed and constructed so that the way of exit travel is obvious and direct?	☐	☐

2. Are windows which could be mistaken for exit doors made inaccessible by means of barriers or railings? □ □
3. Are exit doors operable from the direction of exit travel without the use of a key or any special knowledge or effort when the building is occupied? □ □
4. Is a revolving, sliding, or overhead door prohibited from servicing as a required exit door? □ □
5. Where panic hardware is installed on a required exit door, will it allow the door to open by applying a force of 15 lb or less in the direction of the exit traffic? □ □
6. Are doors on cold storage rooms provided with an inside release mechanism which will release the latch and open the door even if it's padlocked or otherwise locked on the outside? □ □
7. Where exit doors open directly onto any street, alley, or other area where vehicles may be operated, are adequate barriers and warnings provided to prevent employees from stepping into the path of traffic? □ □
8. Are doors that swing in both directions and are located between rooms where there is frequent traffic provided with viewing panels in each door? □ □

Exiting or Egress

	Yes	No
1. Are all exits marked with an exit sign and illuminated by a reliable light source?	□	□
2. Are the directions to exits, when not immediately apparent, marked with visible signs?	□	□
3. Are doors, passageways, or stairways that are neither exits nor access to exits and which could be mistaken for exits appropriately marked *Not an Exit, To Basement, Storeroom,* etc.?	□	□
4. Are exit signs provided with the word *Exit* in lettering at least 6 in high and the stroke of the lettering at least ¾-in wide?	□	□
5. Are exit doors side-hinged?	□	□
6. Are all exits kept free of obstructions?	□	□
7. Are at least two means of egress provided from elevated platforms, pits, or rooms where the absence of a second exit would increase the risk of injury from hot, poisonous, corrosive, suffocating, flammable, or explosive substances?	□	□
8. Are there sufficient exits to permit prompt escape in case of emergency?	□	□

9. Are special precautions taken to protect employees during construction and repair operations? □ □
10. Is the number of exits from each floor of a building and the number of exits from the building itself appropriate for the building occupancy load? □ □
11. Are exit stairways which are required to be separated from other parts of a building enclosed by at least 2-h fire-resistive construction in buildings more than four stories in height and not less than 1-h fire-resistive construction elsewhere? □ □
12. Where ramps are used as part of required exiting from a building, is the ramp slope limited to 1 ft vertical and 12 ft horizontal? □ □
13. Where exiting will be through frameless glass doors, glass exit doors, storm doors, etc., are the doors fully tempered and do they meet the safety requirements for human impact? □ □

Fire Protection (see also Flammable and Combustible Materials)

	Yes	No
1. Is there a portable fire extinguisher program in place?	□	□
2. Does the plant provide employee training in portable fire extinguishers?	□	□
3. Has each employee who is expected to use the extinguishers gone through the training course?	□	□
Is the training documented?	□	□
Is the training repeated annually?	□	□
4. Are all fire extinguishers mounted properly?	□	□
5. Are all fire extinguishers checked on a monthly basis?	□	□
6. Have all fire extinguishers had annual inspection within the last 12 months?	□	□
7. Are all fire extinguishers currently properly charged?	□	□
8. Are all extinguishers readily accessible, adequately conspicuous, and of the correct type/class for anticipated fires?	□	□
9. Is there a master inventory list describing the location and type of each fire extinguisher at the plant?	□	□
10. Is there a sprinkler system at the plant?	□	□

11. Does the sprinkler system meet OSHA requirements or, if installed prior to the effective date of the standard (September 12, 1980), those of the National Fire Protection Association (NFPA) or the National

Board of Fire Underwriters (FBFU) standard in effect at the time of the system's installation? ☐ ☐
12. Is the sprinkler system given an acceptance check annually? ☐ ☐
Is documentation of checks on file at the plant? ☐ ☐
13. Is the sprinkler system currently in acceptable working order? ☐ ☐
14. Is your local fire department well acquainted with your facilities, its location, and specific hazards? ☐ ☐
15. If you have a fire alarm system, is it certified as required? ☐ ☐
16. If you have a fire alarm system, is it tested at least annually? ☐ ☐
17. If you have interior standpipes and valves, are they inspected regularly? ☐ ☐
18. If you have outside private fire hydrants, are they flushed at least once a year and on a routine preventive maintenance schedule? ☐ ☐
19. Are fire doors and shutters in good operating condition? ☐ ☐
20. Are fire doors and shutters unobstructed and protected against obstructions, including their counterweights? ☐ ☐
21. Are fire door and shutter fusible links in place? ☐ ☐
22. Are all fusible links for fire doors, sprinklers, and washing/dip tanks free of paint and similar coating? ☐ ☐
23. Is there a written emergency action plan, including escape procedures and routes, critical plant operations, employee accounting following an emergency evacuation, rescue and medical duties, means of reporting emergencies, and persons to be contacted for information and clarification? ☐ ☐

First Aid, Medical Services, and Medical Records (see also Bloodborne Pathogens)

	Yes	No
1. Is there a hospital, clinic, or infirmary for medical care in near proximity of your workplace?	☐	☐
2. If medical and first-aid facilities are not in proximity of your workplace, is at least one employee (preferably two) on each shift currently qualified to render first aid and CPR?	☐	☐
3. Is their certification being updated annually?	☐	☐

4. Have all employees who are expected to respond to medical emergencies as part of their work:

    a. received first-aid training? □ □

    b. had hepatitis B vaccination made available to them? □ □

    c. had appropriate training or procedures to protect themselves from bloodborne pathogens, including universal precautions? □ □

    d. have available and understand how to use appropriate personal protective equipment? □ □

5. Are there first-aid supplies available to all employees 24 h a day? □ □

6. Is the first-aid cabinet properly maintained? □ □

7. Is it inspected, at minimum, on a monthly basis? □ □

8. Where employees have had an exposure incident involving bloodborne pathogens, did you provide an immediate postexposure medical evaluation and follow-up? □ □

9. Are medical personnel readily available for advice and consultation on matters of employees' health? □ □

10. Are emergency phone numbers posted? □ □

11. Are first-aid kits easily accessible to each work area, are necessary supplies available, and are they periodically inspected and replenished as needed? □ □

12. Have first-aid kit supplies been approved by a physician, indicating that they are adequate for a particular area or operation? □ □

13. Are means provided for quick drenching or flushing of the eyes and body in areas where corrosive liquids or materials are handled? □ □

14. Have employees been provided with information on, and the availability of, employee medical records, the person responsible for the records, and employees' rights of access? □ □

Fixed Ladders

	Yes	No
1. Are all ladders more than 20 ft in length equipped with a cage or, as an alternative, are ladder-climbing devices always used?	□	□
2. Are ladder platforms provided for all ladders more than 20 ft in length?	□	□

3. Do side rails extend at least 42 in above the landing? ☐ ☐
4. Are all side rails easy to grip and hold with a hand? ☐ ☐

Flammable and Combustible Materials

                                              Yes   **No**

1. Does the quantity of flammable or combustible liquids located outside of an inside storage room or storage cabinet *not* exceed 120 gal, or 60 gal if all the liquid is in a single portable tank? ☐ ☐
2. Are combustible scrap, debris, and waste materials (e.g., oily rags, etc.) stored in covered metal receptacles and removed from the worksite promptly? ☐ ☐
3. Is proper storage practiced to minimize the risk of fire, including spontaneous combustion? ☐ ☐
4. Are approved containers and tanks used for the storage and handling of flammable and combustible liquids? ☐ ☐
5. Are the liquids drawn from or transferred into containers within a building drawn only through a closed piping system, from safety cans, by means of a device drawing through the top, or by gravity through an approved self-closing valve? ☐ ☐
6. Are all connections on drums and combustible liquid piping, vapor, and liquid tight? ☐ ☐
7. Are all flammable liquids kept in closed containers when not in use (e.g., parts cleaning tanks, pans, etc.)? ☐ ☐
8. Are bulk drums of flammable liquids grounded and bonded to containers during dispensing? ☐ ☐
9. Do storage rooms for flammable and combustible liquids have explosion-proof lights? ☐ ☐
10. Do storage rooms for flammable and combustible liquids have mechanical or gravity ventilation? ☐ ☐
11. Are inside storage rooms of fire-resistive construction, with self-closing fire doors at all openings? ☐ ☐
12. Are openings from inside storage rooms to other rooms or buildings provided with noncombustible, liquid-tight raised sills or ramps at least 4 in in height or a floor at least 4 in below the surrounding floor or equivalent construction to assure that spills will not flow out of the room? ☐ ☐
13. If a mechanical exhaust system is used in the inside storage room, is it controlled by a switch outside of the door that controls the ventilation and the lighting and that has a pilot light? ☐ ☐

14. Is all spark producing equipment, including hand tools, prohibited from inside storage rooms and from being used otherwise in near proximity to possible vapor ignition areas? ☐ ☐

15. Is liquefied petroleum gas stored, handled, and used in accordance with safe practices and standards? ☐ ☐

16. Are *No Smoking* signs posted on liquefied petroleum gas tanks? ☐ ☐

17. Are liquefied petroleum storage tanks guarded to prevent damage from vehicles? ☐ ☐

18. Are all solvent wastes and flammable liquids kept in fire-resistant, covered containers until they are removed from the worksite? ☐ ☐

19. Is vacuuming used (as approved for a particular application or operation) whenever possible rather than blowing or sweeping combustible dust? ☐ ☐

20. Are firm separators placed between containers of combustibles or flammables when stacked one upon another to assure their support and stability? ☐ ☐

21. Are fuel gas cylinders and oxygen cylinders separated from each other by 20 ft or by 5-ft-high noncombustible barriers while in storage? ☐ ☐

22. Are fire extinguishers selected and provided for the types of materials in areas where they are to be used?

    Class A   Ordinary combustible material fires.
    Class B   Flammable liquid, gas, or grease fires.
    Class C   Energized electrical equipment fires.

23. Are appropriate fire extinguishers mounted within 75 ft of outside areas containing flammable liquids and within 10 ft of any inside storage area for such materials? ☐ ☐

24. Are extinguishers free from obstructions or blockage? ☐ ☐

25. Are all extinguishers serviced, maintained, and tagged at intervals not to exceed 1 year? ☐ ☐

26. Where sprinkler systems are permanently installed, are the nozzle heads so directed or arranged that water will not be sprayed into operating electrical switchboards and equipment? ☐ ☐

27. Are *No Smoking* signs posted where appropriate in areas where flammable or combustible materials are used or stored? ☐ ☐

28. Are safety cans, including spring-loads, emergency vents, and flash arrestors, used for dispensing flammable or combustible liquids at the point of final use? ☐ ☐

29. Are all spills of flammable or combustible liquids cleaned up promptly? □ □
30. Are storage tanks adequately vented to prevent the development of excessive vacuum or pressure as a result of filling, emptying, or atmosphere-temperature changes? □ □
31. Are storage tanks equipped with emergency venting that will relieve excessive internal pressure caused by fire exposure? □ □
32. Are no smoking rules enforced in areas involving the storage and use of hazardous materials? □ □

Floor and Wall Openings

           Yes   **No**

1. Are floor openings guarded by a cover, a guardrail, or equivalent on all sides, except at the entrance to stairways or ladders? □ □
2. Do all open platforms and floors 4 ft or more above the adjacent floor, or at any height if above dangerous equipment, have adequate perimeter protection? □ □
3. Are toeboards installed around the edges of permanent floor openings, where persons may pass below the opening? □ □
4. Are skylight screens of such construction and mounting that they will withstand a load of at least 200 lb? □ □
5. Is the glass in the windows, doors, glass walls, etc., which are subject to human impact, of sufficient thickness and type for the condition of use? □ □
6. Are grates or similar type covers over floor openings such as floor drains of such design that foot traffic or rolling equipment will not be affected by the grate spacing? □ □
7. Are unused portions of service pits and similar openings either covered, protected by guardrails (or equivalent), or tended? □ □
8. Are manhole covers, trench covers, and similar covers, plus their supports, designed to carry a truck rear axle load of at least 20,000 lb when located in roadways and subject to vehicle traffic? □ □
9. Are floor or wall openings in fire-resistive construction provided with doors or covers compatible with the fire rating of the structure and provided with self-closing features when appropriate? □ □

Fueling

	Yes	No
1. Is it prohibited to fuel an internal combustion engine with a flammable liquid while the engine is running?	☐	☐
2. Are fueling operations done in such a manner that the likelihood of spillage will be minimal?	☐	☐
3. When spillage occurs during fueling operations, is the spilled fuel washed away completely, evaporated, or are other measures taken to control vapors before restarting the engine?	☐	☐
4. Are fuel tank caps replaced and secured before starting the engine?	☐	☐
5. In fueling operations, is there always metal contact between the container and the fuel tank?	☐	☐
6. Are fueling hoses of a type designed to handle the specific type of fuel?	☐	☐
7. Is it prohibited to handle or transfer gasoline in open containers?	☐	☐
8. Are open lights, open flames, sparking, or arcing equipment prohibited near fueling or transfer of fuel operations?	☐	☐
9. Is smoking prohibited in the vicinity of fueling operations?	☐	☐
10. Are fueling operators prohibited in buildings or other enclosed areas that are not specifically ventilated for this purpose?	☐	☐
11. Where fueling or transfer of fuel is done through a gravity flow system, are the nozzles of the self-closing type?	☐	☐

General Work Environment

	Yes	No
1. Are floor surfaces kept dry or are appropriate means taken to assure the surfaces are slip resistant?	☐	☐
2. Are all spilled hazardous materials or liquids, including blood and other potentially infectious materials, cleaned up immediately and according to proper procedures?	☐	☐
3. Is all regulated waste, including (but not limited to) oils, acids, solvents, and substances covered under the bloodborne pathogens standard (29 CFR 1910.1030), discarded according to federal, state, and local regulations?	☐	☐

4. Are accumulations of combustible dust routinely removed from elevated surfaces, including the overhead structure of buildings, etc.? □ □

5. Is combustible dust cleaned up with a vacuum system to prevent the dust from going into suspension? □ □

6. Is metallic or conductive dust prevented from entering or accumulating on or around electrical enclosures or equipment? □ □

7. Are covered metal waste cans used for oily and paint-soaked waste? □ □

8. Are all oil and gas fired devices equipped with flame failure controls that will prevent flow of fuel if pilots or main burners are not working? □ □

9. Are paint spray booths, dip tanks, etc. cleaned regularly? □ □

10. Are at least the minimum number of toilets and washing facilities provided? □ □

11. Are an adequate number of water fountains reasonably available? □ □

12. Are all work areas adequately illuminated? □ □

13. Are all aisles clear of clutter, oil, water, tripping hazards, loose boards, holes, projections (including open furniture drawers/doors), and uneven surfaces? □ □

14. Are locker rooms, restrooms, and lunchrooms kept in sanitary, safe, and healthful condition, and are all work and common areas maintained in a relatively clean and orderly condition? □ □

15. Are employees prohibited from consuming food or beverage in toilet rooms or in any area exposed to a toxic material? □ □

16. Is the parking lot kept in good repair? □ □

17. Are permanent aisles marked? □ □

18. Is pest harborage eliminated or effectively controlled? □ □

19. Are all surfaces free from protruding nails, splinters, sharp corners, and similar hazards? □ □

Hand Tools and Equipment

                                                                    Yes   No

1. Are all tools and equipment (company or employee owned), that are used by employees at their workplace, maintained in good condition? □ □

2. Are hand tools such as chisels, punches, etc., which develop mushroomed heads during use, reconditioned or replaced as necessary? □ □

3. Are broken or splintered handles on hammers, axes, and similar equipment replaced promptly? ☐ ☐
4. Are worn or bent wrenches replaced regularly? ☐ ☐
5. Are appropriate handles used on files and similar tools? ☐ ☐
6. Are employees made aware of the hazards caused by faulty or improperly used hand tools? ☐ ☐
7. Are appropriate safety glasses, face shields, etc. worn while using hand tools or equipment which might produce flying materials or be subject to breakage? ☐ ☐
8. Are jacks checked periodically to assure they are in good operating condition? ☐ ☐
9. Are tool handles wedged tightly in the head of all tools? ☐ ☐
10. Are tool cutting edges kept sharp so the tool will move smoothly without binding or skipping? ☐ ☐
11. Are tools stored in dry, secure locations where they won't be tampered with? ☐ ☐
12. Are sharp edges on abandoned tools (e.g., razor knives) retracted, holstered, or otherwise covered? ☐ ☐

Hazard Communication (Chemical)

	Yes	No

1. Is there a hazard communication program in place? ☐ ☐
2. Does the plant have an employee training program that includes the following elements:
   a. An explanation of what an MSDS is and how to use and obtain one? ☐ ☐
   b. MSDS contents for each hazardous substance or class of substances? ☐ ☐
   c. Explanation of the standard and written program? ☐ ☐
   d. Identification of where an employee can see the employer's written hazard communication program and where hazardous substances are present in their work areas? ☐ ☐
   e. The physical and health hazards of substances in the work area and specific protective measures, including personal protective equipment, work practices, and emergency procedures, to be used? ☐ ☐
   f. Details of the hazard communication program, including how to use the labeling system and MSDSs? ☐ ☐
   g. Hazardous chemical properties, including visual appearance, odor, and methods, that can be used to detect the presence or release of hazardous chemicals? ☐ ☐

3. Has each employee gone through hazard communication training? ☐ ☐

4. Is the training documented? ☐ ☐
5. When questioned, can the employees explain:
   a. What the program involves? ☐ ☐
   b. What a MSDS is? ☐ ☐
   c. Where the MSDSs are? ☐ ☐
   d. What information is on a MSDS? ☐ ☐
   e. To what hazards they may be exposed? ☐ ☐
   f. How to read a label? ☐ ☐
   g. How to protect themselves against the hazards? ☐ ☐
   h. How to evaluate when they are in significant
      danger? ☐ ☐

6. Is refresher training given when a new hazardous
   substance is introduced into the plant, when new
   information is received about an existing substance,
   or when an existing substance is used in a new way? ☐ ☐
7. Is there a material-safety data sheet (MSDS) for each
   hazardous substance at the plant? ☐ ☐
8. Are all MSDSs placed in a MSDS book? ☐ ☐
9. Is the MSDS book complete, accessible, and readily
   available for all employees? ☐ ☐
10. Is there a table of contents by both chemical and
    common name at the beginning of the MSDS book? ☐ ☐
11. Are all containers marked with labels? ☐ ☐
12. Is there a program in place for outside contractors? ☐ ☐
13. Is it documented? ☐ ☐
14. Is there a list of hazardous substances used in your
    workplace? ☐ ☐
15. Is there a written hazard communication program
    dealing with material-safety data sheets (MSDS),
    labeling, employee training, nonroutine tasks, and
    contractors? ☐ ☐
16. Is each container for a hazardous substance (e.g.,
    vats, bottles, storage tanks, etc.) labeled with product
    identity and an appropriate hazard warning
    (communication of the specific health hazards and
    physical hazards)? ☐ ☐
17. Do the substance identifications on labels match (i.e.,
    correspond to) those of the material-safety data sheets? ☐ ☐

**Hazardous Chemical Exposure (see also
Respiratory Protection)**

                                                        Yes   No

1. Are employees trained in the safe handling practices
   of hazardous chemicals (e.g., acids, caustics, etc.)? ☐ ☐

2. Are employees aware of the potential hazards involving various chemicals stored or used in the workplace, including (but not limited to) acids, bases, caustics, solvents, epoxies, phenols, etc.? ☐ ☐

3. Is employee exposure to chemicals kept within acceptable levels? ☐ ☐

4. Are eye wash fountains (and safety showers, depending on the degree and amount of potential exposure) provided in areas where corrosive chemicals are handled? ☐ ☐

5. Are all containers (e.g., vats, storage tanks, etc.) labeled as to their contents and with appropriate hazard warnings? ☐ ☐

6. Are all employees required to use appropriate personal protective clothing and equipment (e.g., gloves, eye protection, respirators, etc.) when handling chemicals? ☐ ☐

7. Are flammable or toxic chemicals kept in closed containers when not in use? ☐ ☐

8. Are chemical piping systems clearly marked as to their content? ☐ ☐

9. Where corrosive liquids are frequently handled in open containers or drawn from storage vessels or pipelines, is adequate means readily available for neutralizing or disposing of spills or overflows properly and safely? ☐ ☐

10. Have standard operating procedures been established and are they followed when cleaning up chemical spills? ☐ ☐

11. Where needed for emergency use, are respirators stored in a convenient, clean, and sanitary location? ☐ ☐

12. Are respirators intended for emergency use adequate for the various uses for which they may be needed? ☐ ☐

13. Are employees prohibited from eating in areas where hazardous chemicals are present? ☐ ☐

14. Is personal protective equipment provided, used, and maintained whenever necessary? ☐ ☐

15. Are there written standard operating procedures for the selection and use of respirators where needed? ☐ ☐

16. If you have a respirator protection program, are your employees instructed on the correct usage and limitations of the respirators? ☐ ☐

17. Are the respirators NIOSH-approved for this particular application? ☐ ☐

18. Are they regularly inspected, cleaned, sanitized, and maintained? ☐ ☐

19. If hazardous substances are used in your processes, do you have a medical or biological monitoring system in operation? ☐ ☐

20. Are you familiar with the permissible exposure limits of airborne contaminants and physical agents used in your workplace? ☐ ☐
21. Have control procedures been instituted for hazardous materials where appropriate (e.g., respirators, ventilation systems, handling practices, etc.)? ☐ ☐
22. Whenever possible, are hazardous substances handled in properly designed and exhausted booths or similar locations? ☐ ☐
23. Do you use dilution or local exhaust ventilation systems to control dusts, vapors, gases, fumes, smoke solvents, or mists which may be generated in your workplace? ☐ ☐
24. Is ventilation equipment provided for removal of contaminants from such operations as production grinding, buffing, spray painting, and/or vapor degreasing, and is it operating properly? ☐ ☐
25. Do employees complain about dizziness, headaches, nausea, irritation, or other problems of discomfort when they use solvents or other chemicals? ☐ ☐
26. Is there a dermatitis problem? ☐ ☐
27. Do employees complain about dryness, irritation, or sensitization of the skin? ☐ ☐
28. Have you considered the use of an industrial hygienist or environmental health specialist to evaluate your operation? ☐ ☐
29. If internal combustion engines are used, is carbon monoxide kept within acceptable levels? ☐ ☐
30. Is vacuuming used, rather than blowing or sweeping dusts whenever possible for cleanup? ☐ ☐
31. Are materials which give off toxic asphyxiant, suffocating, or anesthetic fumes stored in remote or isolated locations when not in use? ☐ ☐

Hazardous Waste Operations (Cleanup and Other Emergency Responses)

	Yes	No
1. Is there a documented emergency response plan for all potential emergencies involving hazardous wastes?	☐	☐
2. Are all employees who are expected to respond to emergencies involving hazardous wastes (most often, leaks and spills) trained for the required number of hours in how to perform their jobs safely?	☐	☐
3. Is safety equipment provided?	☐	☐
4. Are containers of hazardous wastes labeled?	☐	☐

Health and Safety Program (see Part 1 of this book for in-depth discussion)

	Yes	No

1. Is your safety and health program concisely documented, readily available, well-organized, precisely worded, and clearly communicated to each employee? ☐ ☐

2. Is that program characterized by employee involvement/participation, management leadership, hazard prevention and control, workplace analysis, and comprehensive training? ☐ ☐

3. Do you insist on accountability and evaluation of management specifically in regard to safety and health matters? ☐ ☐

4. Is one person clearly responsible for the overall activities of the safety and health program? ☐ ☐

5. Do you have a safety and health program for all contractors and other visitors entering your grounds? ☐ ☐

6. Do you have easy access to all necessary documentation and resources, such as OSHA standards, other standards, governmental agencies, consultants, testing devices (or those who can perform the tests), audiovisuals (including computer software and CD ROM), technical manuals, company rules, material-safety data sheets, injury/illness records, training records, etc. ☐ ☐

7. Do you have a labeling and logging system (nonredundant number and/or letter) for each tool, machine, ladder, fire alarm, fire extinguisher, fire hose, electrical disconnect, chemical tank, furnace, sprinkler control, forklift truck, other vehicle, hoist, crane, sling deluge shower, eye fountain, first-aid kit, MSDS station, hazardous waste cleanup station, and so on? ☐ ☐

8. Do you have a labeling and logging system (nonredundant number and/or letter) for each room, area, exit, stairway, department, building, shack, operator's station, spray booth, loading dock, office, closet, mezzanine, column, and so on? ☐ ☐

9. Do you have a working procedure for handling in-house employee complaints regarding safety and health? ☐ ☐

10. Do you have a highly focused medical management system? ☐ ☐

11. Do you have a full array of documented preventive maintenance programs? ☐ ☐

12. Is there an active safety and health committee at the plant? ☐ ☐
13. Does it meet, at the minimum, monthly? ☐ ☐
14. Are the meetings documented and minutes disseminated to committee members? ☐ ☐
15. Is a copy sent to a vice president (or person of equivalent or higher rank)? ☐ ☐
16. Are minutes of meetings posted? ☐ ☐
17. Is a safety tour taken during each meeting? ☐ ☐
18. Does the safety and health committee consist of both labor and management representatives (with more employees than management)? ☐ ☐
19. Does management act on all recommendations? ☐ ☐
20. Does safety and health committee membership include a person in high authority and a person with high knowledge of the technical and maintenance issues that can arise? ☐ ☐
21. Are all accidents (even those not resulting in injury or illness) that occurred since the last safety and health committee meeting discussed in detail with the goal to uncover the cause(s) and to preclude or significantly reduce the chance of a similar event? ☐ ☐
22. When plans are made at safety and health committee meetings to abate, control, or eliminate hazards, are priorities, deadlines, methods, and persons responsible clearly identified? ☐ ☐
23. Are accident investigations conducted by line supervision for all injuries, illnesses, and near misses? ☐ ☐
24. Have you adequately addressed the needs of disabled employees? (*Note:* Steps to take include visual *and* audible alarms, buddies to assist individuals in evacuating the building, ergonomic redesign of work stations or tools, and so on.) ☐ ☐
25. Is there a progressive disciplinary procedure for safety violators? (*Note:* Do not overlook horseplay.) ☐ ☐
26. Are employees in fact disciplined for safety violations? ☐ ☐
27. If so, is the disciplinary action documented? ☐ ☐
28. Is the disciplinary action enforced and applied consistently? ☐ ☐

Heat Stress

	Yes	No
1. Is there a heat stress safety program in place?	☐	☐
2. Does the plant provide proper training for employees in heat stress acclimatization, prevention and safety?	☐	☐

3. Has each employee gone through the training course? ☐ ☐
4. Is the training documented? ☐ ☐
5. In high heat areas, are engineering controls such as spot cooling, air conditioning, shielding for radiant heat sources, elimination of steam leaks, or equipment modification in place? ☐ ☐
6. Are employees screened before assignment to areas of high heat to determine if their health condition might make them more susceptible to heat stress? ☐ ☐
7. Is cool drinking water readily available? ☐ ☐

Hoists, Cranes, and Auxiliary Equipment

	Yes	No
1. Is there a crane and hoist safety program in place?	☐	☐
2. Does the plant have proper training for employees in crane and hoist safety?	☐	☐
3. Has each employee gone through the training course?	☐	☐
4. Is the training documented?	☐	☐
5. Is there a documented preventive maintenance program including monthly inspection and repair of hoists, hooks, wire ropes, and other components?	☐	☐
6. Are inspections documented?	☐	☐
7. Do cranes and hoists have annual inspections by an outside contractor?	☐	☐
8. Are lifting slings, chains, and hooks inspected monthly?	☐	☐
9. Are lifting slings, chains, and hooks tagged?	☐	☐
10. Are all hooks equipped with safety latches?	☐	☐
11. Are all safety latches in proper working order?	☐	☐
12. Do you assure that no modification or rerating is done without approval of a qualified engineer or equipment manufacturer?	☐	☐
13. Are all pendant controls plainly marked to indicate functions and direction of travel?	☐	☐
14. Are all U-bolt wire clips in use on wire ropes installed so that the U-bolt is in contact with the short (i.e., nonload-carrying) end of the rope?	☐	☐
15. Do you prohibit the use of chains, rope slings, and wire rope that is kinked, twisted, bird-caged, or similarly damaged?	☐	☐
16. Are bridge trucks equipped with rail sweeps?	☐	☐
17. Are bridges and trolleys equipped with bumpers or the equivalent protection?	☐	☐
18. Are slings checked daily, and are defective slings immediately removed from service?	☐	☐

19. Are chain slings permanently affixed with identification stating size, grade, capacity, and reach? ☐ ☐

20. Is each overhead electric hoist equipped with a limit device to stop the hook at its highest and lowest points of safe travel? ☐ ☐

21. Will each hoist automatically stop and hold any load up to 125 percent of its rated load if its actuating force is removed? ☐ ☐

22. Is the rated load of each hoist legibly marked and visible to the operator? ☐ ☐

23. Are stops provided at the safe limits of travel for trolley hoists? ☐ ☐

24. Are the controls of hoists plainly marked to indicate the direction of travel or motion? ☐ ☐

25. Is each cage-controlled hoist equipped with an effective warning device? ☐ ☐

26. Are close-fitting guards or other suitable devices installed on hoists to assure that hoist ropes will be maintained in the sheave grooves? ☐ ☐

27. Are all hoist chains or ropes of sufficient length to handle the full range of movement of the application while still maintaining two full wraps on the drum at all times? ☐ ☐

28. Are nip points or contact points between hoist ropes and sheaves which are permanently located within 7 ft of the floor, ground, or working platform guarded? ☐ ☐

29. Is it prohibited to use chains or rope slings that are kinked or twisted? ☐ ☐

30. Is the operator instructed to avoid carrying loads over people? ☐ ☐

Laboratory Chemical Hygiene Plan

	Yes	No
1. Does your plant have a laboratory that is used for functions other than just quality control?	☐	☐
2. If so, has a chemical hygiene plan been prepared?	☐	☐
3. Has employee training been conducted?	☐	☐

Lift Trucks

	Yes	No
1. Is there a lift truck program in place?	☐	☐
2. Does your plant have training for employees in forklift safety?	☐	☐

3. Has each employee gone through the training course? ☐ ☐
   Is the training documented? ☐ ☐
4. Do all operators carry their operators' license? ☐ ☐
5. Are forklift trucks on a P.M. schedule? ☐ ☐
6. Are inspection reports kept on file? ☐ ☐
7. Do all lift trucks have properly maintained and
   functioning:

   a. Lights? ☐ ☐
   b. Horn? ☐ ☐
   c. Backup light? ☐ ☐
   d. Backup bell or horn? ☐ ☐
   e. Fire extinguisher? ☐ ☐
   f. Overhead guard? ☐ ☐
   g. Emergency brakes? ☐ ☐
   h. Regular brakes? ☐ ☐
   i. Tires? ☐ ☐
   j. Steering? ☐ ☐

8. Are all forklift rules obeyed? ☐ ☐
9. Are wheel chocks available for trucks? ☐ ☐
10. Are wheel chocks used? ☐ ☐
11. Are forklifts examined before use and after each
    shift, with defects reported immediately? ☐ ☐
12. Do you prohibit lifting personnel on the forks, unless
    a firmly secured safety platform with adequate
    perimeter/fall protection is in use? ☐ ☐
13. When traveling, are forks only raised enough to
    safely clear the road or floor surface? ☐ ☐
14. Are dock boards or bridge plates used when transferring
    materials between docks and trucks or rail cars? ☐ ☐
15. Are portable dock boards anchored or equipped with
    devices that will prevent their slipping? ☐ ☐

Lockout/Tagout (Control of Hazardous Energy)

	Yes	No

1. Is there a lockout/tagout (LO/TO) program in place? ☐ ☐
2. Are proper lockout devices available for:

   a. All types of valves? ☐ ☐
   b. All electrical shutoffs? ☐ ☐

3. Are blocks provided for doors, gates, conveyors, etc.? ☐ ☐
4. Is there a control on the locks and keys to assure
   that each lock used can only be removed by the
   employee placing it? ☐ ☐

5. Is each valve, switch, etc. that must be locked out
   marked in some manner? ☐ ☐
6. Is current locked-out equipment done so properly? ☐ ☐
7. Is there a program for contractors? ☐ ☐
   Is there documentation on contractor compliance? ☐ ☐
8. Is there a machine-specific, detailed, written LO/TO
   program in place? ☐ ☐
9. Are the specific instructions for locking out each
   individual piece of equipment posted on or near that
   piece of equipment? ☐ ☐
10. Is there documented LO/TO training for each
    authorized, affected, and other person? ☐ ☐
11. Is all machinery or equipment capable of movement
    required to be deenergized or disengaged and blocked
    or locked out during cleaning, servicing, adjusting, or
    setting up operations, whenever required? ☐ ☐
12. When the power disconnecting means for equipment
    does not also disconnect the electrical control circuit:

    a. Are the appropriate electrical enclosures identified? ☐ ☐
    b. Is means provided to assure the control circuit
       can also be disconnected and locked out? ☐ ☐

13. Is the locking out of control circuits in lieu of locking
    out main power disconnects prohibited? ☐ ☐
14. Are all equipment control valve handles provided
    with a means for locking out? ☐ ☐
15. Does the lockout procedure require that stored
    energy (e.g., mechanical, hydraulic, air, etc.) be
    released or blocked before equipment is locked out
    for repairs (the ZES principle)? ☐ ☐
16. Are appropriate employees provided with individually
    keyed personal safety locks? ☐ ☐
17. Are employees required to keep personal control of
    their key(s) while they have safety locks in use? ☐ ☐
18. Is it required that only the employee exposed to the
    hazard place or remove the safety lock? ☐ ☐
19. Is it required that employees check the safety of the
    lockout by attempting a start up after making sure
    no one is exposed? ☐ ☐
20. Are employees instructed to always push the control
    circuit stop button prior to reenergizing the main
    power switch? ☐ ☐
21. Is there a means provided to identify any or all
    employees who are working on locked-out equipment
    by their locks or accompanying tags? ☐ ☐

22. Are a sufficient number of accident prevention signs or tags and safety padlocks provided for any reasonably foreseeable repair emergency? ☐ ☐
23. When machine operations, configuration, or size requires the operator to leave his or her control station to install tools or perform other operations, and that part of the machine could move if accidentally activated, is such an element required to be separately locked or blocked out? ☐ ☐
24. In the event that equipment or lines cannot be shut down, locked out, and tagged, is a safe job procedure established and rigidly followed? ☐ ☐

Machine Safeguarding

	Yes	No
1. Is there a training program to instruct employees on safe methods of machine operation?	☐	☐
2. Is there adequate supervision to ensure that employees are following safe machine operating procedures?	☐	☐
3. Is there a regular program of safety inspection of machinery and equipment?	☐	☐
4. Is all machinery and equipment kept clean and properly maintained?	☐	☐
5. Is sufficient clearance provided around and between machines to allow for safe operations, set up and servicing, material handling, and waste removal?	☐	☐
6. Is equipment and machinery securely placed and anchored when necessary to prevent tipping or other movement that could result in personal injury?	☐	☐
7. Is there a power shut-off switch within reach of the operator's position at each machine?	☐	☐
8. Can electric power to each machine be locked out for maintenance, repair, or security?	☐	☐
9. Are the noncurrent-carrying metal parts of electrically operated machines bonded and grounded?	☐	☐
10. Are foot-operated switches guarded or arranged to prevent accidental actuation by personnel or falling objects?	☐	☐
11. Are hand controls guarded or arranged to prevent accidental activation by personnel?	☐	☐
12. Are manually operated valves and switches controlling the operation of equipment and machines clearly identified and readily accessible?	☐	☐

13. Are all emergency stop buttons colored red and of the mushroom configuration or otherwise easily accessed and operated? ☐ ☐
14. Are all pulleys and belts properly guarded? ☐ ☐
15. Are all moving chains, gears, and couplings properly guarded? ☐ ☐
16. Are splash guards mounted on machines that use coolant to prevent the coolant from reaching employees? ☐ ☐
17. Are methods provided to protect the operator and other employees in the machine area from hazards created at the point of operation, ingoing nip points, rotating parts, pinch points, reciprocating tiny parts, flying chips, and sparks? ☐ ☐
18. Are machinery guards secure and so arranged that they do not offer a hazard in their use? ☐ ☐
19. If special hand tools are used for placing and removing material, do they protect the operator's hands? ☐ ☐
20. Are revolving drums, barrels, and containers required to be guarded by an enclosure that is interlocked with the drive mechanism so that revolution cannot occur unless the guard enclosure is in place and so guarded? ☐ ☐
21. Do arbors and mandrels have firm and secure bearings, and are they free from play? ☐ ☐
22. Are provisions made to prevent machines from automatically starting when power is restored after a power failure or shutdown? ☐ ☐
23. Are machines constructed so as to be free from excessive vibration when the largest size tool is mounted and run at full speed? ☐ ☐
24. If machinery is cleaned with compressed air, is air pressure controlled and is personal protective equipment or other safeguards utilized to protect operators and other workers from eye and body injury? ☐ ☐
25. Are fan blades protected with a guard having openings no larger than $\frac{1}{2}$ in, when operating within 7 ft of the floor? ☐ ☐
26. Are saws used for ripping equipped with antikickback devices and spreaders? ☐ ☐
27. Are radial arm saws so arranged that the cutting head will gently return to the back of the table when released? ☐ ☐

28. Are radial saws so arranged that no portion of the blade can extend beyond the front or rear ends of the table? ☐ ☐

29. Are radial saws equipped with guards that completely enclose the upper portion of the blade and with floating guards for the lower exposed portions of the blade? ☐ ☐

30. Are table/circular saws equipped with a hood guard over the portion of the blade above the table, so mounted that the hood will automatically adjust itself to the thickness of the material being cut and remain in contact with that material? ☐ ☐

31. Are band saw wheels fully enclosed? ☐ ☐

32. Are all portions of the band saw blade, except for the working surface of the blade, enclosed or guarded? ☐ ☐

33. Are guards in place to protect against the hazards of rotating lathe chuck jaws? ☐ ☐

34. Are lathe chuck keys and flywheel turnover bars spring-loaded, or as an alternative, are other means used to assure that the rotation of the chuck and flywheel cannot begin when the key or bar can become a projectile? ☐ ☐

Mandated (and Other) Posters and Signs

	Yes	No

1. Is the Job Safety & Health Protection (OSHA) notice posted? ☐ ☐

2. Are signs posted to indicate the locations of exits, extinguishers, fire hoses, manual pull alarms, emergency eye fountains, deluge showers, hazardous waste clean-up stations, first-aid supplies, emergency shut-offs, and other critical equipment or areas?

3. Is the emergency evacuation plan posted? ☐ ☐

4. Are emergency telephone numbers posted where they can be readily found in case of emergency? ☐ ☐

5. Where employees may be exposed to any toxic substances or harmful physical agents, has appropriate information concerning employee access to medical and exposure records and material safety data sheets been posted or otherwise made readily available to affected employees? ☐ ☐

6. Are signs posted concerning crain/hoist capacities, floor loading, and various clearances? ☐ ☐

7. Are signs posted to warn of the hazards of biological dangers, and exposure to X-ray, microwave, or other harmful radiation or substances? (*Note:* Be sure to include substance identity and appropriate hazard warnings, where required by the hazard communication standard.) ☐ ☐
8. Are signs posted to warn of the hazards of vapor ignition (by smoking, open flames, etc.), permit-required confined spaces, falling parts, hot surfaces, head bumps (e.g., low overheads), step-downs, vehicular traffic, slippery floors, and similar conditions? ☐ ☐
9. Is all required information, with relevant sections of OSHA standards, posted regarding exposure, including the hazards of lead and noise? ☐ ☐
10. Is the summary of occupational illnesses and injuries posted? ☐ ☐

Noise and Hearing Conservation

	Yes	No

1. Are there areas in the workplace where continuous noise levels exceed 85 dBA? ☐ ☐
2. Is there an ongoing preventive health and hearing conservation program to educate employees in safe levels of noise, exposures, effects of noise on their health, and the use of personal protection? ☐ ☐
3. Is there a hearing conservation program in place? ☐ ☐
4. Has each employee gone through the training course? ☐ ☐
5. Is the training documented? ☐ ☐
6. Have work areas where noise levels make voice communication between employees difficult and where the use of personal protective equipment is required been identified and posted? ☐ ☐
7. Are noise levels being measured using a sound level meter or an octave band analyzer, and are records being kept? ☐ ☐
8. Have engineering controls been used to reduce excessive noise levels? ☐ ☐
9. Where engineering controls are determined not to be feasible, are administrative controls (e.g., worker rotation) used to minimize individual employee exposure to noise? ☐ ☐
10. Is a variety of approved hearing protective equipment (noise attenuating devices) available to every employee working in noisy areas? ☐ ☐

11. Have you tried isolating noisy machinery from the rest of your operation? ☐ ☐
12. If you use ear protectors, are employees properly fitted and instructed in their use? ☐ ☐
13. Are employees in high-noise areas given periodic audiometric testing to ensure that you have an effective hearing protection system? ☐ ☐
14. Are all employees wearing their hearing protection in designated areas? ☐ ☐
15. Have all employees received base line audiograms? ☐ ☐
16. Are employees getting annual audiograms? ☐ ☐
17. Are employees getting the results of their audiograms? ☐ ☐
18. Are current noise levels in the plant equivalent to those on record? ☐ ☐

Open Surface Tanks

	Yes	No

1. Where ventilation is used to control potential exposure to employees, is it adequate to reduce the concentration of the air contaminant to a degree that hazards to employees do not exist? ☐ ☐
2. Where there is a danger of splashing, are employees required to wear tight-fitting chemical goggles or an effective face shield, impervious gloves, impervious boots, and other general body protection? ☐ ☐
3. Have all employees working at or near open surface tank operations been instructed as to the hazards of their respective jobs and in the personal protection and first-aid procedures applicable to these hazards? ☐ ☐
4. Are means provided for quick drenching or flushing of the eyes and body? ☐ ☐

Over-the-Road Trucks

	Yes	No

1. Are trucks and trailers secured from movement (by chocks or equivalent means) during loading and unloading operations? ☐ ☐
2. Do all trucks have fire extinguishers? ☐ ☐
   Are they properly charged? ☐ ☐
   Are they properly inspected? ☐ ☐
3. Do all the exterior lights work? ☐ ☐

4. Are road flares and triangles in the trucks? ☐ ☐
5. Does the horn work? ☐ ☐
6. Are the seat belts operable and worn at all times? ☐ ☐
7. Are drivers properly licensed and qualified? ☐ ☐
8. Are hours of service monitored and logged? ☐ ☐
9. Is the two-way radio, telephone, or other means of communication in proper working order? ☐ ☐
10. Are tie-down straps or their equivalent, and trailer or other storage flooring in good condition? ☐ ☐
11. Are required placards and markings (for hazardous materials and capacity) in place? ☐ ☐
12. Are regular safety inspections conducted to check the condition of critical items including horn, lights, mirrors, windshields, turn signals, regular braking system, emergency braking system, tires, steering, fluids, placards and markings, and any other items as mandated by state and federal law? ☐ ☐

Personal Protective Equipment and Clothing
(see also Noise and Hearing Conservation;
Respiratory Protection)

	Yes	No

1. Are appropriate ANSI-approved tight-fitting chemical goggles, or face shields over safety glasses with side shields, provided and worn where there is a danger of flying particles, corrosive materials, or other relevant hazards? ☐ ☐
2. Are ANSI-approved safety glasses with side shields required to be worn at all times in areas where there is a risk of eye injuries such as punctures, abrasions, contusions, or burns? ☐ ☐
3. Are employees who need corrective lenses (i.e., glasses or contacts) in working environments having harmful exposures required to wear only ANSI-approved safety glasses, protective goggles, or use other medically approved precautionary procedures? ☐ ☐
4. Are protective gloves, aprons, shields, or other means required and worn, where employees could be cut or where they are exposed to toxic chemicals (e.g., corrosives), blood or other infectious materials, or very hot/cold surfaces/substances? ☐ ☐
5. Are hard hats provided and worn where the danger of falling objects exists? ☐ ☐

6. Are hard hats inspected periodically for damage to the shell and suspension system? ☐ ☐

7. Is appropriate foot protection (e.g., safety toes, metatarsal guards, impervious boots, etc.) required where there is the related risk of foot injuries from hot, corrosive, poisonous substances; falling objects; crushing, rolling, or penetrating actions? ☐ ☐

8. Are approved respirators provided for regular or emergency use where needed? ☐ ☐

9. Is all protective equipment maintained in sanitary condition and ready for use? ☐ ☐

10. Where special equipment is needed for electrical workers, is it available? ☐ ☐

11. Where food or beverages are consumed on the premises, are they consumed in areas where there is no exposure to toxic material, blood, or other potentially infectious materials? ☐ ☐

12. Is protection against the effects of occupational noise exposure provided when sound levels exceed those of the OSHA noise standard? ☐ ☐

13. Are adequate work procedures, protective clothing, and equipment provided and used when cleaning up spilled toxic or otherwise hazardous materials or liquids? ☐ ☐

14. Are there appropriate procedures in place for disposing of or decontaminating personal protective equipment contaminated with, or reasonably anticipated to be contaminated with, pertinent toxic materials, blood, or other potentially infectious materials? ☐ ☐

15. Is there written documentation and/or certification that an assessment of personal protective equipment needs has been performed? ☐ ☐

16. Is there written documentation and/or certification of employee training regarding the use and care of personal protective equipment? ☐ ☐

17. Are employees exposed to rotating parts, in-running rolls, and material running onto rolls precluded from having loose, droopy, billowy clothing, jewelry, and unrestrained long hair that can be pulled into danger areas? (*Note:* Do not place long hair in conventional ponytails; it should be restrained up and close to the head.) ☐ ☐

Pipe Markers

	Yes	No
1. Are all pipes marked as to their contents?	☐	☐
2. Are pipes marked at point of shut-off valves and lockout points?	☐	☐
3. When nonpotable water is piped through a facility, are outlets or taps posted to alert employees that it is unsafe and not to be used for drinking, washing, or other personal use?	☐	☐
4. When hazardous substances are transported through above ground piping, is each pipeline identified at points where confusion could introduce hazards to employees?	☐	☐
5. When pipelines are identified by color painting, are all visible parts of the line so identified?	☐	☐
6. When pipelines are identified by color painted bands or tapes, are the bands or tapes located at reasonable intervals and at each outlet, valve, or connection?	☐	☐
7. When pipelines are identified by color, is the color code posted at all locations where confusion could introduce hazards to employees?	☐	☐
8. When the contents of the pipelines are identified by name or name abbreviation, is the information readily visible on the pipe near each valve or outlet?	☐	☐
9. When pipelines carrying hazardous substances are identified by tags, are the tags constructed of durable materials, is the message carried clearly and permanently distinguishable, and are tags installed at each valve or outlet?	☐	☐
10. When pipelines are heated by electricity, steam, or other external source, are suitable warning signs or tags placed at unions, valves, or other serviceable parts of the system?	☐	☐

Portable Ladders

	Yes	No
1. Are all ladders maintained in good condition, are joints between steps and side rails tight, are all hardware and fittings securely attached, and are movable parts operating freely without binding or undue play?	☐	☐
2. Are nonslip safety feet provided on each ladder?	☐	☐

3. Are ladder rungs and steps free of grease and oil? ☐ ☐
4. Is it prohibited to place a ladder in front of doors opening toward the ladder except when the door is blocked open, locked, or guarded? ☐ ☐
5. Is it prohibited to place ladders on boxes, barrels, or other unstable bases to obtain additional height? ☐ ☐
6. Are employees instructed to face the ladder when ascending or descending? ☐ ☐
7. Are employees prohibited from using ladders that are broken, missing steps, rungs, or cleats, have broken side rails, or other faulty equipment? ☐ ☐
8. Are employees instructed not to use the top step of ordinary stepladders as a step? ☐ ☐
9. When portable rung ladders are used to gain access to elevated platforms, roofs, etc., does the ladder always extend at least 3 ft above the elevated surface? ☐ ☐
10. Is it required when portable rung or cleat type ladders are used that the base is so placed that slipping will not occur or that it is lashed or otherwise held in place? ☐ ☐
11. Are portable metal ladders legibly marked with signs reading *Caution—Do Not Use Around Electrical Equipment* or equivalent wording? ☐ ☐
12. Are employees prohibited from using ladders as guys, braces, skids, gin poles, or for other than their intended purposes? ☐ ☐
13. Are employees instructed only to adjust extension ladders while standing at a base, not while standing on the ladder or from a position above the ladder? ☐ ☐
14. Are ladders inspected for damage? ☐ ☐
15. Are the rungs of ladders uniformly spaced at 12 in, center to center? ☐ ☐

Portable (Power-Operated) Tools and Equipment

	Yes	No

1. Are grinders, saws, and similar equipment provided with appropriate safety guards? ☐ ☐
2. Are power tools used with the correct shield, guard, or attachment recommended by the manufacturer? ☐ ☐
3. Are portable circular saws equipped with guards above and below the base shoe? ☐ ☐
4. Are circular saw guards checked to assure they are not wedged up, thus leaving the lower portion of the blade unguarded? ☐ ☐

5. Are rotating or moving parts of equipment guarded
   to prevent physical contact? ☐ ☐
6. Are all cord-connected, electrically operated tools
   and equipment effectively grounded or of the
   approved double insulated type? ☐ ☐
7. Are effective guards in place over belts, pulleys,
   chains, and sprockets on equipment such as
   concrete mixers, air compressors, etc.? ☐ ☐
8. Are portable fans provided with full guards or
   screens having openings $\frac{1}{2}$ in or less? ☐ ☐
9. Is hoisting equipment available and used for lifting
   heavy objects, and are hoist ratings and characteristics
   appropriate for the task? ☐ ☐
10. Are ground fault circuit interrupters provided on
    all temporary electrical 15 and 20 A circuits used
    during periods of construction? ☐ ☐
11. Are pneumatic and hydraulic hoses on power-operated
    tools checked regularly for deterioration or damage? ☐ ☐

Powder-Actuated Tools

	Yes	No

1. Are employees who operate powder-actuated tools trained
   in their use, and do they carry a valid operator's card? ☐ ☐
2. Is each powder-actuated tool stored in its own
   locked container when not being used? ☐ ☐
3. Is a sign at least 7 in by 10 in with boldface type
   reading **Powder-Actuated Tool in Use**
   conspicuously posted when the tool is being used? ☐ ☐
4. Are powder-actuated tools left unloaded until they
   are actually ready to be used? ☐ ☐
5. Are powder-actuated tools inspected for obstructions
   or defects each day before use? ☐ ☐
6. Do powder-actuated tool operators have and use
   appropriate personal protective equipment such as hard
   hats, safety goggles, safety shoes, and ear protectors? ☐ ☐

**Process Safety Management (PSM)
of Highly Hazardous Chemicals**

	Yes	No

1. Do you use chemicals covered by the PSM standard? ☐ ☐
2. Are they stored on site in quantities that exceed their
   designated threshold quantity (TQ)? ☐ ☐
3. Has a hazard analysis been conducted? ☐ ☐

4. Have risk reduction measures been identified and implemented? ☐ ☐
5. Have emergency response plans been prepared? ☐ ☐
6. Have procedures for conducting PSM incident investigations been prepared? ☐ ☐
7. Is there a written plan of action regarding employee participation, and have the employees and their representatives been consulted on the conduct and development of process hazards analyzes and on the development of the other elements of process safety management where required by law? ☐ ☐
8. Was a compilation of written process safety information completed prior to conducting a process hazard analysis? ☐ ☐
9. Does the process hazard analysis appropriately address the complexity of the company's processes, identifying, evaluating, and controlling the hazards involved in the process? ☐ ☐
10. Have written operating procedures been developed and implemented to provide clear instructions for safely conducting activities involved in each covered (by law) process consistent with process safety information? ☐ ☐
11. Has each employee presently involved in operating a process and each employee before being involved in operating a newly assigned process been trained in an overview of the process and in the operating procedures? ☐ ☐
12. When selecting a contractor, have you obtained and evaluated information regarding the contract employer's safety performance and programs? ☐ ☐
13. Has the contract employer assured that each contract employee has been trained in the work practices necessary to safely perform his or her job? ☐ ☐
14. Have you performed a prestart-up safety review for new facilities and for modified facilities when the modification is significant enough to require a change in the process safety information? ☐ ☐
15. Have written procedures been established and implemented to maintain the ongoing integrity of process equipment? ☐ ☐
16. Have written procedures been established and implemented to manage changes to process chemicals, technology, equipment, procedures, and changes to facilities that affect a covered process? ☐ ☐

Recordkeeping

	Yes	No
1. Are all occupational injuries and illnesses, except minor ones requiring only first aid, entered properly on the OSHA log and summary?	☐	☐
2. Is a supplementary OSHA incident record completed for each of those cases?	☐	☐
3. Is a first-aid log kept?	☐	☐
4. Are copies of all safety-related paperwork, including for required inspection/maintenance records for such items as cranes and mechanical power presses, kept on file at the plant?	☐	☐
5. Are employee medical records and records of employee exposure to hazardous substances or harmful physical agents up-to-date and in compliance with current OSHA standards?	☐	☐
6. Are employee training records kept and accessible for review by employees when required by OSHA standards?	☐	☐
7. Have arrangements been made to maintain required records for the legal period of time for each specific type record (some records must be maintained for at least 40 years)?	☐	☐
8. Are operating permits and records up-to-date for such items as elevators, air pressure tanks, liquefied petroleum gas tanks, etc.?	☐	☐

Respiratory Protection

	Yes	No
1. Does the plant provide respirators?	☐	☐
2. Is there an emergency respirator program in place?	☐	☐
3. Has a written respirator program been prepared?	☐	☐
4. Has a workplace hazard analysis been conducted?	☐	☐
5. Does the plant provide training for employees in respirator selection, use, and limitations?	☐	☐
6. Has each employee gone through the training course?	☐	☐
7. Have qualitative or quantitative fit tests been provided, as required?	☐	☐
8. Is the training documented?	☐	☐
9. Are respirators stored, maintained, and cleaned properly?	☐	☐
10. Are respirators readily accessible and NIOSH-approved for the particular application?	☐	☐
11. Are respirators inspected regularly?	☐	☐

12. Have respirator users been medically evaluated, and
the results documented?   ☐   ☐

Smoking

	Yes	No

1. Do you prohibit smoking in your facility and on
your grounds?   ☐   ☐
2. If not, is smoking allowed only in closed break rooms
with dedicated ventilation.   ☐   ☐

Spray Finishing Operations

	Yes	No

1. Is adequate ventilation assured before spray
operations are started?   ☐   ☐
2. Is mechanical ventilation provided when spraying
operations are done in enclosed areas?   ☐   ☐
3. When mechanical ventilation is provided during
spraying operations, is it so arranged that it will not
circulate the contaminated air?
4. Is the spray area free of hot surfaces?   ☐   ☐
5. Is the spray area at least 20 ft from flames, sparks,
operating electrical motors, and other ignition sources? ☐   ☐
6. Are portable lamps used to illuminate spray areas
suitable for use in a hazardous location?   ☐   ☐
7. Is approved respiratory equipment provided and
used when appropriate during spraying operations?   ☐   ☐
8. Do solvents used for cleaning have a flash point to
100°F or more?   ☐   ☐
9. Are fire control sprinkler heads kept clean?   ☐   ☐
10. Are *No Smoking* signs posted in spray areas, paint
rooms, paint booths, and paint storage areas?   ☐   ☐
11. Is the spray area kept clean of combustible residue? ☐   ☐
12. Is infrared drying apparatus kept out of the spray
area during spraying operations?   ☐   ☐
13. Is the spray booth completely ventilated before using
the drying apparatus?   ☐   ☐
14. Is the electric drying apparatus properly grounded?   ☐   ☐
15. Are lighting fixtures for spray booths located outside
of the booth and is the interior lighted through sealed,
clear panels?   ☐   ☐
16. Are the electric motors for exhaust fans placed
outside booths or ducts?   ☐   ☐
17. Are belts and pulleys inside the booth fully enclosed? ☐   ☐

18. Do ducts have access doors to allow cleaning? ☐ ☐
19. Do all drying spaces have adequate ventilation? ☐ ☐

Stairs and Stairways

	Yes	No

1. Are there standard stair rails or handrails on all
   stairways having four or more risers? ☐ ☐
2. Are all stairways at least 22 in wide? ☐ ☐
3. Do stairs have landing platforms not less that 30 in
   in the direction of travel that extend 22 in in width
   at every 12 ft or less of vertical rise? ☐ ☐
4. Do stairs angle no more than 50 and no less than 30°? ☐ ☐
5. Are stairs of hollow-pan type treads and landings
   filled to the top edge of the pan with solid material? ☐ ☐
6. Are step risers on stairs uniform from top to bottom? ☐ ☐
7. Are steps on stairs and stairways designed or provided
   with a surface that renders them slip resistant? ☐ ☐
8. Are stairway handrails located between 30 and 34 in
   above the leading edge of stair treads? ☐ ☐
9. Is there at least 3 in of clearance between the
   stairway handrails and the wall or surface they are
   mounted on? ☐ ☐
10. Where doors or gates open directly on a stairway, is
    there a platform provided so the swing of the door does
    not reduce the width of the platform to less than 21 in? ☐ ☐
11. Are stairway handrails capable of withstanding a
    load of 200 lb, applied within 2 in of the top edge in
    any downward or outward direction? ☐ ☐
12. Where stairs or stairways exit directly into any area
    where vehicles may be operated, are adequate barriers
    and warnings provided to prevent employees from
    stepping into the path of traffic? ☐ ☐
13. Do stairway landings have a dimension measured in
    the direction of travel that is at least equal to the
    width of the stairway? ☐ ☐
14. Is the vertical distance between stairway landings
    limited to 12 ft or less? ☐ ☐

Storage

1. Are all stored items blocked, interlocked, limited
   in height, or otherwise arranged so that they are secure
   against dangerous sliding, collapse, displacement, or
   unsafe retrieval? ☐ ☐

2. Are floor load ratings posted where required? ☐ ☐
3. Are damaged pallets and skids immediately repaired
   or put out of service? ☐ ☐
4. Are pallets, brooms, rakes, and similar items stored
   so as not to be inadvertently flipped toward the body,
   by being stepped on? ☐ ☐
5. Are papers and other combustibles stored away from
   light bulbs, space heaters, and other hot or sparking
   surfaces? ☐ ☐

Substance Abuse

	Yes	No

1. Do you train supervisors to identify employees who may
   be abusing drugs or alcohol? (*Note:* Of course, supervisors
   are not exempt from substance abuse themselves.) ☐ ☐
2. Do you have a written substance abuse policy? ☐ ☐
3. Is there access to an Employee Assistance Program? ☐ ☐

Walkways

	Yes	No

1. Are aisles and passageways kept clear? ☐ ☐
2. Are walking and working surfaces maintained free
   of tripping hazards from protruding nails, splinters,
   holes, projections, loose or uneven surfaces, and
   similar hazards? ☐ ☐
3. Are aisles and walkways marked as appropriate? ☐ ☐
4. Are wet surfaces covered with nonslip materials? ☐ ☐
5. Are holes in the floor, sidewalk, or other walking
   surface repaired properly, covered, or otherwise
   made safe? ☐ ☐
6. Is there safe clearance for walking in aisles where
   motorized or mechanical handling equipment
   is operating? ☐ ☐
7. Are materials or equipment stored so as not to
   protrude into walkways? (*Note:* This includes items
   stored above floor level in racks and shelves.) ☐ ☐
8. Are spilled materials cleaned up immediately? ☐ ☐
9. Are changes of direction or elevations readily
   identifiable? ☐ ☐
10. Are aisles or walkways that pass near moving or
    operating machinery, welding operations, or similar
    operations arranged so employees will not be
    subjected to potential hazards? ☐ ☐

11. Is adequate headroom provided for the entire length of any aisle or walkway? ☐ ☐
12. Are standard guardrails provided wherever aisle or walkway surfaces are elevated 4 ft or more above any adjacent floor or the ground? ☐ ☐
13. Are bridges provided over conveyors and similar hazards? ☐ ☐

Welding, Cutting, and Brazing

	Yes	No

1. Are only authorized and trained personnel permitted to use welding, cutting, or brazing equipment? ☐ ☐
2. Do all operators have a copy of the appropriate operating instructions, and are they directed to follow them? ☐ ☐
3. Are compressed gas cylinders regularly examined for obvious signs of defects, deep rusting, or leakage? ☐ ☐
4. Is care used in the handling and storage of cylinders, safety valves, relief valves, etc. to prevent damage? ☐ ☐
5. Are precautions taken to prevent the mixture of air or oxygen with flammable gases, except at a burner or in a standard torch? ☐ ☐
6. Are only approved apparatus (e.g., torches, regulators, pressure-reducing valves, acetylene generators, manifolds) used? ☐ ☐
7. Are cylinders kept away from sources of heat? ☐ ☐
8. Are cylinders kept away from elevators, stairs, exits, loading docks, and gangways? ☐ ☐
9. Is it prohibited to use cylinders as rollers or supports? ☐ ☐
10. Are empty cylinders appropriately marked and their valves closed (while understanding that no cylinder is to be considered truly empty unless it has been thoroughly purged)? ☐ ☐
11. Are signs reading *Danger—No Smoking, Matches, or Open Lights,* or the equivalent, posted? ☐ ☐
12. Are cylinders, cylinder valves, couplings, regulators, hoses, and apparatus kept free of oily or greasy substances? ☐ ☐
13. Is care taken not to drop or strike cylinders? ☐ ☐
14. Unless secured on special trucks, are regulators removed and valve-protection caps put in place before moving cylinders? ☐ ☐
15. Do cylinders without fixed hand wheels have keys, handles, or nonadjustable wrenches on stem valves when in service? ☐ ☐

16. Are liquefied gases stored and shipped valve-end up with valve covers in place? ☐ ☐

17. Are provisions made never to crack a fuel-gas cylinder valve near sources of ignition? ☐ ☐

18. Before a regulator is removed, is the valve closed and gas released from the regulator? ☐ ☐

19. Is red used to identify acetylene and other fuel-gas hoses, green for oxygen hoses, and black for inert gas and air hoses? ☐ ☐

20. Are pressure-reducing regulators used only for the gas and pressures for which they are intended? ☐ ☐

21. Is open circuit (no load) voltage of arc welding and cutting machines as low as possible and not in excess of the recommended limits? ☐ ☐

22. Under wet conditions, are automatic controls for reducing no load voltage used? ☐ ☐

23. Is grounding of the machine frame and safety ground connections of portable machines checked periodically? ☐ ☐

24. Are electrodes removed from the holders when not in use? ☐ ☐

25. Is it required that electric power to the welder be shut off when no one is in attendance? ☐ ☐

26. Is suitable fire extinguishing equipment available for immediate use? ☐ ☐

27. Is a welder forbidden to coil or loop welding electrode cable around his or her body? ☐ ☐

28. Are wet machines thoroughly dried and tested before being used? ☐ ☐

29. Are work and electrode lead cables frequently inspected for wear and damage, and are they replaced when needed? ☐ ☐

30. Do means for connecting cable lengths have adequate insulation? ☐ ☐

31. When the object to be welded cannot be moved and fire hazards cannot be removed, are shields used to confine heat, sparks, and slag? ☐ ☐

32. Are fire watchers assigned when welding or cutting is performed in locations where a serious fire might develop? ☐ ☐

33. Are combustible floors kept wet, covered with damp sand, or protected by fire-resistant shields? ☐ ☐

34. When floors are wet down, are personnel protected from possible electrical shock? ☐ ☐

35. When welding is done on metal walls, are precautions taken to protect combustibles on the other side? ☐ ☐
36. Before hot work is begun, are used drums, barrels, tanks, and other containers so thoroughly cleaned that no substances remain that could explode, ignite, or produce toxic vapors? ☐ ☐
37. Is it required that eye protection helmets, hand shields, and goggles meet appropriate standards? ☐ ☐
38. Are employees exposed to the hazards created by welding, cutting, or brazing operations protected with personal protective equipment and clothing (and/or curtains to protect against flash burns for employees in the area)? ☐ ☐
39. Is a check made for adequate ventilation in and where welding or cutting is performed? ☐ ☐
40. When working in confined places, are environmental monitoring tests taken and means provided for quick removal of welders in case of an emergency? ☐ ☐

Workplace Violence Policy and Prevention Program

	Yes	No

1. Do you have a site-specific, "zero-tolerance" workplace violence policy and prevention program (periodically reviewed) in place? ☐ ☐
2. Do you have a site-specific plan (periodically reviewed) in place to react to workplace violence? ☐ ☐
3. Do you consider the potential for workplace violence, when hiring, by carefully checking employment history and otherwise screening applicants? ☐ ☐
4. Do you provide access to an Employee Assistance Program? (*Note:* Consider, too, the value of wellness programs.) ☐ ☐
5. Do you have a specific plan in place, designed to avoid workplace violence, when employees are terminated or disciplined, or other negative personnel actions are to be taken against them? (*Note:* Consider timing, including season, day of the week, and hour of the day; minimizing humiliation and maintaining dignity; the presence of two members of management; added security measures if a violent response is reasonably anticipated; and counseling and/or outplacement.) ☐ ☐

6. Do you train supervisors to identify employees who appear to be under excessive stress (or who display warning signs of potential violent behavior), to respond to those employees with appropriate sensitivity, and to confidentially report concerns to appropriate personnel as designated by company policy? □ □

7. Do you train employees in courteous behavior toward customers, clients, and visitors? □ □

8. Do you have a plan in place regarding how employees should react to hostile customers, clients, and visitors? (*Note:* This may include a coded signal, of a verbal and/or other type, to alert staff.) □ □

9. Do you have a program in place to clearly identify and denounce harassment and intimidation (in particular, taking all threats of violence seriously), to promptly report and thoroughly investigate all instances, and to impose suitable disciplinary measures? □ □

10. Do you provide training in thoughtful (nonvisceral) conflict intervention and resolution, tension reduction, threat assessment, crisis management, and nonviolent response? (*Note:* At times it is highly advisable to institute "cooling-off" periods and keep individuals from concurrently occupying the same area, and at times it is a wise strategy to bring them together in a safe, controlled setting with skilled mediators.) □ □

11. Do you provide training in interpersonal communication skills (verbal and nonverbal), active listening, and acceptance of (and response to) criticism? □ □

12. Do you have a well-communicated, equitably applied system for the airing of grievances? □ □

13. Do you foster a supportive, harmonious work environment, and a climate of mutual trust and respect? □ □

14. Do you apprise employees of pending or potential significant modifications in the workplace (particularly when directly affecting their jobs), well in advance of the changes occurring, and/or the last opportunity for consideration of employee-propounded options to those changes? □ □

15. Do you actively solicit employee input, including on how to improve working conditions and which alternatives to potential negative changes may exist? (*Note:* This includes encouraging employee participation in problem-solving, personal and team responsibility, and a reasonable degree of job control, to aid in reducing frustration.) □ □

16. Do you maintain well-lit and uncluttered business premises, and assure high visibility inside and outside the facility? (*Note:* This can include removing bushes or structures behind which a person could conceal himself or herself near entrances or exits, limiting the height of shelving, and installing mirrors.) □ □

17. Do you have security guards, monitoring devices, company identification tags/badges (that include the name of the authorized individual), pass cards, "buzz-in" doors, strict sign-in policies, and/or a logical layout to help control access to the facility or to particular areas of concern? (*Note:* In some situations, bullet-resistant enclosures are recommended.) □ □

18. Do you provide drop safes and minimize the amount of cash that is on hand, and post signs to that effect? □ □

19. Do you have silent alarms? (*Note:* In some situations, personal alarms are recommended.) □ □

20. Do you train employees to avoid resistance during robberies? □ □

21. Do you minimize situations in which employees work alone? □ □

22. Do you have survelliance cameras and manual emergency alarms in elevators, stairwells, parking garages, and similar remote or isolated locations where employees (and others) may be alone, and do you provide trained escorts to lessen the risk? □ □

23. Do police patrol the perimeter of your facility when employees work late at night or early in the morning? □ □

24. Do you strictly prohibit personal firearms and other weapons from the premises? □ □

25. Do you train relevant employees in how to respond to (and document information regarding) telephone callers making bomb threats (or similar declarations), and what actions the receiver should take immediately following the termination of the call? (*Note:* Where there are reasons to exercise extraordinary vigilance regarding the possibility of mail/parcel bombs, consider training relevant employees in how to recognize mail/parcels that should raise suspicion, and how to handle such items.) □ □

OSHA Offices*

OSHA Regional Offices

Region I
(CT,* MA, ME, NH, RI, VT*)
John F. Kennedy Federal Building
Room E-340
Boston, MA 02203
(617) 565-9860

Region II
(NJ, NY,* PR,* VI*)
201 Varick Street
Room 670
New York, NY 10014
(212) 337-2330

Region III
(DC, DE, MD,*, PA, VA,* WV)
Gateway Building
3535 Market Street, Room 2100
Philadelphia, PA 19104
(215) 596-1201

Region IV
(AL, FL, GA, KY,* MS, NC, SC,* TN*)
Atlanta Federal Center
61 Forsyth St. SW, Room 6T50
Atlanta, GA 30303
(404) 562-2300

Region V
(IL, IN,* MI,* MN,* OH, WI)
230 South Dearborn Street
Room 3244
Chicago, IL 60604
(312) 353-2220

Region VI
(AR, LA, NM,* OK, TX)
525 Griffin Square Building
Room 602
Dallas, TX 75202
(214) 767-4731

Region VII
(IA,* KS, MO, NE)
City Center Square
1100 Main Street, Suite 800
Kansas City, MO 64105
(816) 426-5861

Region VIII
(CO, MT, ND, SD, UT,* WY*)
Suite 1690
1999 Broadway
Denver, CO 80202-5716
(303) 391-5858

---

*These states and territories operate their own OSHA-approved job safety and health programs (Connecticut and New York plans cover public employees only). States with approved programs must have a standard that is identical to, or at least as effective as, the federal standard.

Region IX
(American Samoa, AZ,* CA,*
Guam, HI,* NV,* Trust
Territories of the Pacific)
71 Stevenson Street
Room 420
San Francisco, CA 94105
(415) 975-4310

Region X
(AK,* ID, OR,* WA*)
Federal Office Building
1111 Third Avenue, Suite 715
Seattle, WA 98101-3212
(206) 553-5930

States with Approved Plans

Commissioner
Alaska Department of Labor
1111 West 8th Street
Room 306
Juneau, AK 99801
(907) 465-2700

Director
Industrial Commission of Arizona
800 W. Washington
Phoenix, AZ 85007
(602) 542-5795

Director
California Department of Industrial
Relations
45 Fremont Street
San Francisco, CA 94105
(415) 972-8835

Director
Connecticut Department of Labor
Division of Occupational Safety and
Health
38 Wolcott Hill Road
Wethersfield, CT 06109-1114
(860) 566-4550

Director
Hawaii Department of Labor and
Industrial Relations
830 Punchbowl Street
Honolulu, HI 96813
(808) 586-8844

Commissioner
Indiana Department of Labor
State Office Building
402 West Washington Street
Room W195
Indianapolis, IN 46204
(317) 232-2378

Commissioner
Iowa Division of Labor Services
1000 E. Grand Avenue
Des Moines, IA 50319
(515) 281-3447

Secretary
Kentucky Labor Cabinet
1049 U.S. Highway, 127 South
Frankfort, KY 40601
(502) 564-3070

Commissioner
Maryland Division of Labor and
Industry
Department of Licensing and
Regulation
1100 North Eutaw St., Room 613
Baltimore, MD 21201-2206
(410) 767-2215

Director
Michigan Department of Consumer
and Industry Services
7150 Harris Drive
Lansing, MI 48909
(517) 373-7230

Commissioner
Minnesota Department of Labor and
Industry
443 Lafayette Road
St. Paul, MN 55155
(612) 296-2342

Director
Nevada Division of Industrial
Relations
400 West King Street
Carson City, NV 89710
(702) 687-3032

Secretary
New Mexico Environment
Department
Occupational Health and Safety
Bureau
1190 St. Francis Drive
P.O. Box 26110
Santa Fe, NM 87502
(505) 827-2850

Commissioner
New York Department of Labor
State Campus
Building 12, Room 130
Albany, NY 12240
(518) 457-2741

Commissioner
North Carolina Department of
Labor
319 Chapanoke Road
Raleigh, NC 27603
(919) 662-4585

Administrator
Oregon Occupational Safety and
Health Division
Department of Consumer and
Business Services
Room 430
350 Winter Street, NE
Salem, OR 97310
(503) 378-3272

Secretary
Puerto Rico Department of Labor
and Human Resources
Prudencio Rivera Martinez Building
505 Munoz Rivera Avenue
Hato Rey, PR 00918
(787) 754-2119

Director
South Carolina Department of
Labor
Koger Office Park
Kingstree Building
110 Centerview Drive
P.O. Box 11329
Columbia, SC 29210
(803) 896-4300

Commissioner
Tennessee Department of Labor
710 James Robertson Parkway
Gateway Plaza, 2nd Floor
Nashville, TN 37243-0659
(615) 741-2582

Commissioner
Industrial Commission of Utah
160 East 300 South, 3rd Floor
P.O. Box 146600
Salt Lake City, UT 84114-6650
(801) 530-6898

Commissioner
Vermont Department of Labor and
Industry
National Life Building—Drawer 20
120 State Street
Montpelier, VT 05620
(802) 828-2288

Commissioner
Virgin Islands Department of Labor
2131 Hospital Street, Box 890
Christiansted
St. Croix, VI 00820-4666
(809) 773-1994

Commissioner
Virginia Department of Labor and
Industry
Powers-Taylor Building
13 South 13th Street
Richmond, VA 23219
(804) 786-2377

Director
Washington Department of Labor
and Industries
General Administration Building
P.O. Box 44001
Olympia, WA 98504-4001
(360) 902-4200

Safety Administrator
Wyoming Dept. of Employment
Workers' Safety and Compensation
Division
Herschler Building, 2nd Floor East
122 West 25th Street
Cheyenne, WY 82002
(307) 777-7786

OSHA Consultation Project Directory

Alabama
Safe State Program
University of Alabama
425 Martha Parham West
P.O. Box 870388
Tuscaloosa, AL 35487
(205) 348-3033

Alaska
Department of Labor
Occupational Safety and Health
301 Eagle Street
P.O. Box 107022
Anchorage, AK 99510
(907) 269-4957

Arizona
Consultation and Training
Division of Occupational Safety and
Health
Industrial Commission of Arizona
800 West Washington
Phoenix, AZ 85007-9070
(602) 542-5795

Arkansas
OSHA Consultation
Arkansas Department of Labor
10421 West Markham
Little Rock, AK 72205
(501) 682-4522

California
CAL/OSHA Consultation Service
Department of Industrial Relations
45 Fremont Street, Room 1260
San Francisco, CA 94105
(415) 972-8515

Colorado
Occupational Safety and Health
Section
Colorado State University
115 Environmental Health Building
Fort Collins, CO 80523
(970) 491-6151

Connecticut
Division of Occupational Safety and
Health

Connecticut Department of Labor
38 Wolcott Hill Road
Wethersfield, CT 06109-1114
(860) 566-4550

Delaware
Occupational Safety and Health
Division of Industrial Affairs
Delaware Department of Labor
4425 Market Street
Wilmington, DE 19802
(302) 761-8219

District of Columbia
Office of Occupational Safety and
Health
District of Columbia Department of
Employment Services
950 Upshur Street, N.W.
Washington, DC 20011
(202) 576-6339

Florida
7(c)(1) Onsite Consultation Program
Division of Industrial Safety
Florida Department of Labor and
Employment Security
2002 St. Augustine Road
Building E, Suite 45
Tallahassee, FL 32399-0663
(850) 922-8955

Georgia
7(c)(1) Onsite Consultation Program
Georgia Institute of Technology
O'Keefe Building, Room 22
Atlanta, GA 30332
(404) 894-2643

Guam
OSHA Onsite Consultation
Guam Department of Labor
P.O. Box 9970
Tamuning, GU 96931
(671) 475-0136

Hawaii
Consultation and Training Branch
Department of Labor and Industrial
Relations
830 Punchbowl Street
Honolulu, HI 96813
(808) 586-9100

Idaho
Safety and Health Consultation
Program
Boise State University
Department of Health Studies
1910 University Drive, ET-338A
Boise, ID 83725
(208) 385-3283

Illinois
Industrial Services Division
Department of Commerce and
Community Affairs
State of Illinois Center
100 W. Randolph St., Suite 3-400
Chicago, IL 60601
(312) 814-2337

Indiana
Division of Labor
Bureau of Safety, Education and
Training
402 West Washington, Room W195
Indianapolis, IN 46204-2287
(317) 232-2688

Iowa
7(c)(1) Consultation Program
Iowa Bureau of Labor
1000 East Grand Avenue
Des Moines, IA 50319
(515) 281-5352

Kansas
Kansas 7(c)(1) Consultation
Program
Kansas Department of Human
Resources
512 South West 6th Street
Topeka, KS 66603-3150
(913) 296-7476

Kentucky
Division of Education and Training
Kentucky Labor Cabinet
1049 U.S. Highway 127, South
Frankfort, KY 40601
(502) 564-6896

Louisiana
7(c)(1) Consultation Program
Louisiana Dept. of Labor
OWC-OSHA Consultation

P.O. Box 94094
Baton Rouge, LA 70804-9094
(504) 342-9601

Maine
Division of Industrial Safety
Maine Bureau of Labor
State House Station 82
Augusta, ME 04333
(207) 624-6460

Maryland
7(c)(1) Consultation Services
Division of Labor and Industry
312 Marshall Avenue, Room 600
Laurel, MD 20707
(410) 880-4970

Massachusetts
7(c)(1) Consultation Program
Division of Industrial Safety
Massachusetts Department of Labor
and Industries
1001 Watertown Street
West Newton, MA 02165
(617) 727-3982

Michigan
Department of Consumer and
Industry Services
7150 Harris Drive
Lansing, MI 48909
(517) 322-1809

Minnesota
Department of Labor and Industry
Consultation Division
443 Lafayette Road
St. Paul, MN 55155
(612) 297-2393

Mississippi
Mississippi State University Center
for Safety and Health
2906 N. State Street, Suite 201
Jackson, MS 39216
(601) 987-3981

Missouri
Onsite Consultation Program
Division of Labor Standards
Department of Labor and Industrial
Relations

3315 West Truman Boulevard
Jefferson City, MO 65109
(573) 751-3403

Montana
Department of Labor and Industry
Bureau of Safety
P.O. Box 1728
Helena, MT 59624-1728
(406) 444-6418

Nebraska
Division of Safety, Labor and Safety
Standards
Nebraska Department of Labor
State Office Building, Lower Level
301 Centennial Mall
South Lincoln, NE 68509-5024
(402) 471-4717

Nevada
Division of Preventive Safety
Department of Industrial Relations
2500 W. Washington, Suite 106
Las Vegas, NV 89106
(702) 486-5016

New Hampshire
Onsite Consultation Program
New Hampshire Division of Public
Health Services
6 Hazen Drive
Concord, NH 03301-6527
(603) 271-2024

New Jersey
New Jersey Department of Labor
Division of Public Safety and
Occupational Safety and Health
225 E. State St., 8th Fl. W.
P.O. Box 953
Trenton, NJ 08625-0953
(609) 292-3923

New Mexico
New Mexico Environment Dept.
Occupational Health and Safety
Bureau
525 Camino de Los Marquez
Suite 3
P.O. Box 26110
Santa Fe, NM 87502
(505) 827-4230

New York
Division of Safety and Health
State Campus
Building 12, Room 130
Albany, NY 12240
(518) 457-2238

North Carolina
North Carolina Consultative
Services
North Carolina Department of
Labor
319 Chapanoke Road, Suite 105
Raleigh, NC 27603-3432
(919) 662-4644

North Dakota
Division of Environmental
Engineering
North Dakota State Department of
Health
1200 Missouri Avenue, Room 304
P.O. Box 5520
Bismark, ND 58504-5520
(701) 328-5188

Ohio
Division of Onsite Consultation
Department of Industrial Relations
145 S. Front Street
Columbus, OH 43216
(614) 644-2246

Oklahoma
OSHA Division
Oklahoma Department of Labor
4001 North Lincoln Blvd.
Oklahoma City, OK 73105-5212
(405) 528-1500

Oregon
7(c)(1) Consultation Program
Department of Consumer and
Business Services
Labor and Industries Building
350 Winter Street, N.E., Room 430
Salem, OR 97310
(503) 378-3272

Pennsylvania
Indiana University of Pennsylvania
Safety Sciences Department
205 Uhler Hall
Indiana, PA 15705-1087
(724) 357-2561

Puerto Rico
Occupational Safety and Health
Office
Puerto Rico Department of Labor
and Human Resources
Prudencio Rivera Martinez Building
505 Munoz Rivera Avenue, 21st Fl.
Hato Rey, PR 00918
(787) 754-2171

Rhode Island
Division of Occupational Health
Rhode Island Department of Health
3 Capital Hill
Providence, RI 02908
(401) 277-2438

South Carolina
7(c)(1) Onsite Consultation Program
Licensing and Regulation, SCDOL
3500 Forest Drive
P.O. Box 11329
Columbia, SC 29204
(803) 734-9614

South Dakota
Engineering Extension
Onsite Technical Division
South Dakota State University
West Hall
907 Harvey Dunn Street
P.O. Box 510
Brookings, SD 57007-0510
(605) 688-4101

Tennessee
OSHA Consultative Services
Tennessee Department of Labor
710 James Robertson Pkwy.,
3rd Floor
Nashville, TN 37243-0659
(615) 741-7036

Texas
Texas Workers' Compensation
Commission
Health and Safety Division
Southfield Building
4000 South I H 35
Austin, TX 78704
(512) 440-3834

Utah
Utah Labor Commission
Consultation Services

160 East 300 South
Salt Lake City, UT 84114-6650
(801) 530-6868

Vermont
Division of Occupational Safety and
Health
Vermont Department of Labor and
Industry
National Life Building—Drawer 20
120 State Street
Montpelier, VT 05602
(802) 828-2765

Virgin Islands
Division of Occupational Safety and
Health
Virgin Islands Department of Labor
3012 Golden Rock
Christiansted
St. Croix, VI 00840
(809) 772-1315

Virginia
Virginia Department of Labor and
Industry
Occupational Safety and Health
Training and Consultation
13 S. 13th Street
Richmond, VA 23219
(804) 786-6359

Washington
Washington Department of Labor
and Industries
Division of Industrial Safety and
Health
P.O. Box 44643
Olympia, WA 98504
(360) 902-5638

West Virginia
West Virginia Department of Labor
Capitol Complex Building #3
1800 E. Washington St., Room 319
Charleston, WV 25305
(304) 558-7890

Wisconsin (Health)
Wisconsin Department of Health
and Human Services
Section of Occupational Health
1414 E. Washington Ave., Room 112
Madison, WI 53703
(608) 266-8579

Wisconsin (Safety)
Wisconsin Department of Industry
Labor and Human Relations
Bureau of Safety Inspection
401 Pilot Court, Suite C
Waukesha, WI 53186
(414) 521-5063

Wyoming
Occupational Health and Safety
State of Wyoming
Herschler Building, 2nd Floor East
122 West 25th Street
Cheyenne, WY 82002
(307) 777-7786

OSHA Area Offices

US Department of Labor—OSHA
2047 Canyon Road—Todd Mall
Birmingham, AL 35216-1981
(205) 731-1534

US Department of Labor—OSHA
3737 Government Blvd.
Suite 100
Mobile, AL 36693-4309
(334) 441-6131

US Department of Labor—OSHA
National Bank of Alaska Building
Suite 407
301 Northern Lights Blvd.
Anchorage, AK 99503-7571
(907) 271-5152

US Department of Labor—OSHA
3221 North 16th Street, Suite 100
Phoenix, AZ 85016
(602) 640-2007

US Department of Labor—OSHA
TCBY Building, Suite 450
425 West Capitol Avenue
Little Rock, AR 72201
(501) 324-6291

US Department of Labor—OSHA
101 El Camino Plaza, Suite 105
Sacramento, CA 95815
(916) 566-7470

US Department of Labor—OSHA
5675 Ruffin Road, Suite 330
San Diego, CA 92123
(619) 557-2909

US Department of Labor—OSHA
1391 North Speer Blvd., Suite 210
Denver, CO 80204-2552
(303) 844-5285

US Department of Labor—OSHA
7935 E. Prentice Avenue, Suite 209
Englewood, CO 80111-2714
(303) 843-4500

US Department of Labor—OSHA
1057 Broad Street, 4th Floor
Bridgeport, CT 06604
(203) 579-5579

US Department of Labor—OSHA
Federal Office Building
450 Main Street, Room 613
Hartford, CT 06103
(860) 240-3152

US Department of Labor—OSHA
1 Rodney Square
920 King Street, Suite 402
Wilmington, DE 19801
(302) 573-6115

US Department of Labor—OSHA
820 First Street NE, Suite 440
Washington, DC 20002
(202) 523-1452

US Department of Labor—OSHA
Executive Court
8040 Peters Road
Building H-100
Fort Lauderdale, FL 33324
(954) 424-0242

US Department of Labor—OSHA
Ribault Building
1851 Executive Center Dr., Suite 227
Jacksonville, FL 32207
(904) 232-2895

US Department of Labor—OSHA
5807 Breckenridge Pkwy., Suite A
Tampa, FL 33610-4249
(813) 626-1177

US Department of Labor—OSHA
450 Mall Blvd., Suite J
Savannah, GA 31406
(912) 652-4393

US Department of Labor—OSHA
2400 Herodian Way, Suite 250
Smyrna, GA 30080-2968
(770) 984-8700

US Department of Labor—OSHA
Building 7, Suite 110
La Vista Perimeter Office Park
Tucker, GA 30084-4154
(770) 493-6644

US Department of Labor—OSHA
300 Ala Moana Blvd., Room 5-146
Honolulu, HI 96850
(808) 541-2685

US Department of Labor—OSHA
1150 N. Curtis Road, Suite 201
Boise, ID 83706
(208) 321-2960

US Department of Labor—OSHA
1600 167th Street, Suite 9
Calumet City, IL 60409
(708) 891-3800

US Department of Labor—OSHA
O'Hare Lake Place
2360 E. Devon Avenue, Suite 1010
Des Plaines, IL 60018
(847) 803-4800

US Department of Labor—OSHA
11 Executive Drive, Suite 11
Fairview Heights, IL 62208
(618) 632-8612

US Department of Labor—OSHA
344 Smoke Tree Business Park
North Aurora, IL 60542
(630) 896-8700

US Department of Labor—OSHA
2918 West Willow Knolls Road
Peoria, IL 61614
(309) 671-7033

US Department of Labor—OSHA
46 East Ohio Street, Room 423
Indianapolis, IN 46204
(317) 226-7290

US Department of Labor—OSHA
210 Walnut Street, Room 815
Des Moines, IA 50309
(515) 284-4794

US Department of Labor—OSHA
8600 Farley, Suite 105
Overland Park, KS 66212
(913) 385-7380

US Department of Labor—OSHA
300 Epic Center
301 N. Main
Wichita, KS 67202
(316) 269-6644

US Department of Labor—OSHA
John C. Watts Fed. Bldg., Room 108
330 W. Broadway
Frankfort, KY 40601-1992
(502) 227-7024

US Department of Labor—OSHA
9100 Bluebonnet Center Blvd.
Suite 201
Baton Rouge, LA 70809
(504) 389-0474

US Department of Labor—OSHA
40 Western Avenue, Room121
Augusta, ME 04330
(207) 622-8417

US Department of Labor—OSHA
U.S. Federal Building
202 Harlow Street, Room 211
Bangor, ME 04401
(207) 941-8177

US Department of Labor—OSHA
300 West Pratt Street, Room 1110
Baltimore, MD 21201
(410) 962-2840

US Department of Labor—OSHA
639 Granite Street, 4th Floor
Braintree, MA 02184
(781) 565-6924

US Department of Labor—OSHA
Valley Office Park
13 Branch Street
Methuen, MA 01844
(978) 565-8110

US Department of Labor—OSHA
1145 Main Street, Room 108
Springfield, MA 01103-1493
(413) 785-0123

US Department of Labor—OSHA
801 South Waverly Rd., Suite 306
Lansing, MI 48917-4200
(517) 377-1892

US Department of Labor—OSHA
Federal Courts Bldg.
300 South 4th Street, Suite 1205
Minneapolis, MN 55415
(612) 664-5460

US Department of Labor—OSHA
3780 I-55 North, Suite 210
Jackson, MS 39211-6323
(601) 965-4606

US Department of Labor—OSHA
6200 Connecticut Avenue, Suite 100
Kansas City, MO 64120
(816) 483-9531

US Department of Labor—OSHA
911 Washington Avenue, Room 420
St. Louis, MO 63101
(314) 425-4249

US Department of Labor—OSHA
2900 4th Avenue N., Suite 303
Billings, MT 59101
(406) 247-7494

US Department of Labor—OSHA
Overland Wolf Building, Room 100
6910 Pacific Street
Omaha, NE 68106
(402) 221-3182

US Department of Labor—OSHA
705 North Plaza, Room 204
Carson City, NV 89701
(702) 885-6963

US Department of Labor—OSHA
279 Pleasant Street, Suite 201
Concord, NH 03301
(603) 225-1629

US Department of Labor—OSHA
1030 Saint Georges Avenue
Plaza 35, Suite 205
Avenel, NJ 07001
(908) 750-3270

US Department of Labor—OSHA
500 Route 17 South, 2nd Floor
Hasbrouck Heights, NJ 07604
(201) 288-1700

US Department of Labor—OSHA
Marlton Executive Park
701 Route 73 South, Building 2
Suite 120
Marlton, NJ 08053
(609) 757-5181

US Department of Labor—OSHA
299 Cherry Hill Road, Suite 304
Parsippany, NJ 07054
(973) 263-1003

US Department of Labor—OSHA
Western Bank Building, Suite 820
505 Marquette Avenue, N.W.
Albuquerque, NM 87102-2160
(505) 248-5302

US Department of Labor—OSHA
401 New Karner Road, Suite 300
Albany, NY 12205-3809
(518) 464-4338

US Department of Labor—OSHA
42-40 Bell Blvd., 5th Floor
Bayside, NY 11361
(718) 279-9060

US Department of Labor—OSHA
5360 Genesee Street
Bowmansville, NY 14026
(716) 684-3891

US Department of Labor—OSHA
6 World Trade Center, Room 881
New York, NY 10048
(212) 466-2482

US Department of Labor—OSHA
3300 Vickery Road,
North Syracuse, NY 13212
(315) 451-0808

US Department of Labor—OSHA
660 White Plains Road, 4th Floor
Tarrytown, NY 10591-5107
(914) 524-7510

US Department of Labor—OSHA
990 Westbury Road
Westbury, NY 11590
(516) 334-3344

US Department of Labor—OSHA
Century Station Federal Office
Building
300 Fayetteville Street Mall
Room 438
Raleigh, NC 27601-9998
(919) 856-4770

US Department of Labor—OSHA
Federal Office Building
3rd & Rosser, Room 348
P.O. Box 2439
Bismarck, ND 58502
(701) 250-4521

US Department of Labor—OSHA
36 Triangle Park Drive
Cincinnati, OH 45246
(513) 841-4132

US Department of Labor—OSHA
Federal Office Building, Room 899
1240 East Ninth Street
Cleveland, OH 44199
(216) 522-3818

US Department of Labor—OSHA
Federal Office Building, Room 620
200 N. High Street
Columbus, OH 43215
(614) 469-5582

US Department of Labor—OSHA
Federal Office Building, Room 734
234 North Summit Street
Toledo, OH 43604
(419) 259-7542

US Department of Labor—OSHA
420 West Main Place, Suite 300
Oklahoma City, OK 73102
(405) 231-5351

US Department of Labor—OSHA
1220 S.W. Third Avenue, Room 640
Portland, OR 97204
(503) 326-2251

US Department of Labor—OSHA
850 N. 5th Street
Allentown, PA 18102
(610) 776-0592

US Department of Labor—OSHA
3939 West Ridge Road

Suite B-12
Erie, PA 16506-1857
(814) 833-5758

US Department of Labor—OSHA
Progress Plaza
49 N. Progress Avenue
Harrisburg, PA 17109
(717) 782-3902

US Department of Labor—OSHA
U.S. Custom House, Room 242
Second and Chestnut Street
Philadelphia, PA 19106
(215) 597-4955

US Department of Labor—OSHA
Federal Building, Room 1428
1000 Liberty Avenue
Pittsburgh, PA 15222
(412) 395-4903

US Department of Labor—OSHA
Penn Place, Room 2005
20 North Pennsylvania Avenue
Wilkes-Barre, PA 18701-3590
(717) 826-6538

US Department of Labor—OSHA
BBV Plaza Building
1510 F. D. Roosevelt Avenue
Guaynabo, PR 00968
(787) 277-1560

US Department of Labor—OSHA
380 Westminster Street, Room 543
Providence, RI 02903
(401) 528-4669

US Department of Labor—OSHA
1835 Assembly Street, Room 1468
Columbia, SC 29201-2453
(803) 765-5904

US Department of Labor—OSHA
2002 Richard Jones Road
Suite C-205
Nashville, TN 37215-2809
(615) 781-5423

US Department of Labor—OSHA
903 San Jacinto Blvd., Suite 319
Austin, TX 78701
(512) 916-5783

US Department of Labor—OSHA
Wilson Plaza
606 N. Carancachua, Suite 700
Corpus Christi, TX 78476
(512) 888-3420

US Department of Labor—OSHA
8344 East R.L. Thornton Freeway
Suite 420
Dallas, TX 75228
(214) 320-2400

US Department of Labor—OSHA
Commons Building C
4171 North Mesa St., Room C-119
El Paso, TX 79902
(915) 534-7004

US Department of Labor—OSHA
North Star II Building
Suite 302
8713 Airport Freeway
Fort Worth, TX 76180-7604
(817) 428-2470

US Department of Labor—OSHA
17625 El Camino Real, Suite 400
Houston, TX 77058
(281) 286-0583

US Department of Labor—OSHA
350 N. Sam Houston Pkwy., Suite 120
Houston, TX 77060
(281) 591-2438

US Department of Labor—OSHA
Federal Building, Room 804
1205 Texas Avenue
Lubbock, TX 79401
(806) 472-7681

US Department of Labor—OSHA
1783 South 300 West
Salt Lake City, UT 84115-1802
(801) 487-0680

US Department of Labor—OSHA
Federal Office Building, Room 835
200 Granby Mall
Norfolk, VA 23510
(757) 441-3820

US Department of Labor—OSHA
505 106th Avenue, N.E.
Suite 302
Bellevue, WA 98004
(206) 553-7520

US Department of Labor—OSHA
405 Capitol Street, Suite 407
Charleston, WV 25301
(304) 347-5937

US Department of Labor—OSHA
2618 North Ballard Road
Appleton, WI 54911-8664
(920) 734-4521

US Department of Labor—OSHA
Federal Office Building
U.S. Courthouse
500 South Barstow St., Room B-9
Eau Claire, WI 54701
(715) 832-9019

US Department of Labor—OSHA
4802 East Broadway
Madison, WI 53716
(608) 264-5388

US Department of Labor—OSHA
Henry S. Reuss Bldg.
Suite 1180
310 West Wisconsin Ave.
Milwaukee, WI 53203
(414) 297-3315

Index

ABOUT THE AUTHOR

Rick Kaletsky is owner of *Rick Kaletsky, Safety Consultant*, which specializes in hazard recognition, site inspections, and citation resolution. Prior to founding his own firm, he worked for the USDOL/OSHA for twenty years, where he served as an occupational safety and health specialist and assistant area director in OSHA's Connecticut offices. A sought-after speaker on occupational safety and health, he is a faculty member of the Council on Education in Management, the National Business Institute, and the Institute of Business Law. He is an editorial advisory panel member for the Bureau of Business Practice and an advisory panel member for Business & Legal Reports. Mr. Kaletsky has also served as chapter vice president of the American Society of Safety Engineers and program committee chairman of the Connecticut Safety Society. He is the sole lecturer in the 1997 videotape "Employee Safety: Protecting Yourself and Your Fellow Workers" (produced/sponsored by Wausau Insurance and the Metal Treating Institute). He resides in Bethany, Connecticut.